Flexible and Wearable Sensors: Design, Fabrication Methods, and Applications

Flexible and Wearable Sensors: Design, Fabrication Methods, and Applications

Guest Editors

Yichun Ding
Tingting Zhao
Ziran Wang

MDPI

Basel • Beijing • Wuhan • Barcelona • Belgrade • Novi Sad • Cluj • Manchester

Guest Editors

Yichun Ding
School of Physics and Materials Science
Nanchang University
Nanchang
China

Tingting Zhao
School of Mechatronic Engineering and Automation
Shanghai University
Shanghai
China

Ziran Wang
School of Mechanical Engineering
Shandong University
Jinan
China

Editorial Office
MDPI AG
Grosspeteranlage 5
4052 Basel, Switzerland

This is a reprint of the Special Issue, published open access by the journal *Sensors* (ISSN 1424-8220), freely accessible at: https://www.mdpi.com/journal/sensors/special_issues/9MQ3C2F3X0.

For citation purposes, cite each article independently as indicated on the article page online and as indicated below:

Lastname, A.A.; Lastname, B.B. Article Title. *Journal Name* **Year**, *Volume Number*, Page Range.

ISBN 978-3-7258-7705-8 (Hbk)
ISBN 978-3-7258-7706-5 (PDF)
https://doi.org/10.3390/books978-3-7258-7706-5

Contents

About the Editors

Yichun Ding

Yichun Ding is a full professor at the School of Physics and Materials Science, Nanchang University. His research interests lie in the development of high-performance fibers, bio-interface materials, and flexible electronic devices. He has published more than 100 peer-reviewed papers and has garnered more than 8000 citations, with an H-index of 43. He also serves as an associate editor for *Engineering Reports* and a guest editor for *Sensors*.

Tingting Zhao

Tingting Zhao is an associate professor at Shanghai University. She is primarily engaged in research on flexible sensor devices and systems for robotic tactile sensing. She has led and participated in over 10 projects, including the General Program and Youth Program of the National Natural Science Foundation of China, and sub-projects of the National Key Research and Development Program. She has published more than 30 high-impact papers and been granted four invention patents. She was selected for the National Postdoctoral Program for Innovative Talents in 2018.

Ziran Wang

Ziran Wang received his bachelor's degree from University of Jinan, Jinan, China, and his Ph.D. from Harbin Institute of Technology, Harbin, China, all in mechanical engineering. He previously worked as a Postdoctoral Researcher with The Hong Kong Polytechnic University. He is currently a Full Professor with the School of Mechanical Engineering, Shandong University, Jinan, China. His research interests include the design and application of micro/nanosensors and health-monitoring devices.

sensors

MDPI

Editorial

Flexible and Wearable Sensors—Design, Fabrication Methods, and Applications

Yichun Ding [1,*], Tingting Zhao [2] and Ziran Wang [3]

[1] School of Physics and Materials Science, Nanchang University, Nanchang 330031, China
[2] Key Laboratory of Advanced Display and System Applications, Ministry of Education, School of Mechatronic Engineering and Automation, Shanghai University, Shanghai 200444, China; ttzhao@shu.edu.cn
[3] Key Laboratory of High Efficiency and Clean Mechanical Manufacture, Ministry of Education, School of Mechanical Engineering, Shandong University, Jinan 250000, China; wangziran@sdu.edu.cn
* Correspondence: yichun.ding@ncu.edu.cn

1. Introduction

The rapid advancement of the Internet of Things (IoT), artificial intelligence (AI), and personalized healthcare has catalyzed an unprecedented demand for flexible and wearable sensor technologies [1,2]. Unlike conventional rigid sensors, flexible and wearable sensors offer unique advantages, including lightweight construction, excellent mechanical compliance, high stretchability, and the ability to conform to irregular surfaces such as human skin [3,4]. These distinctive characteristics have positioned flexible sensors as key enablers for a wide range of emerging applications, including electronic skin, human–machine interfaces, continuous health monitoring, sports performance tracking, and soft robotics [5–8]. The convergence of innovative materials science, advanced fabrication techniques, and intelligent system integration has accelerated the development of next-generation wearable sensors that are not only high-performing but also comfortable, durable, and scalable for real-world deployment [9–12]. This Special Issue, "Flexible and Wearable Sensors: Design, Fabrication Methods, and Applications", presents a collection of cutting-edge research and comprehensive review articles that address critical challenges and highlight recent breakthroughs in the field, spanning from fundamental material design to practical application demonstrations.

2. An Overview of Published Articles

A total of eighteen manuscripts were submitted for consideration for this Special Issue, all of which underwent a rigorous review process. Ultimately, eleven papers were accepted for publication and inclusion, comprising seven research articles and four reviews, each offering unique insights into different aspects of flexible and wearable sensor technology.

As summarized in Table 1, the contributions cover a wide spectrum of flexible and wearable sensor technologies, encompassing diverse material systems, fabrication methods, and application domains. Contributions 1, 2, 3, 5, 6, 7, 9, and 11 focus on the development of high-performance flexible sensors utilizing various transduction mechanisms, including piezoresistive strain sensors based on silver nanowire–polyurethane composites, iontronic pressure sensors with engineered fiber microstructures, Kirigami-structured piezoelectric sensors for heart sound monitoring, capacitive electrodes for non-contact ECG, and triboelectric nanogenerators (TENGs) for mechanical energy harvesting and self-powered sensing. In detail, contributions 4 and 8 explore the integration of wearable sensors into novel application areas such as musical expression and provide comprehensive reviews of

https://doi.org/10.3390/s26072196

design and fabrication methodologies; contributions 6 and 7 emphasize advanced computational approaches, including knowledge graph-based sensor design and time-reversal algorithms for ultrasound focusing; contribution 9 specifically reviews electrospun thermoplastic polyurethane (TPU) strain sensors and their applications; and contribution 10 provides a focused review on optical fiber sensors for rehabilitation monitoring. Below are specific highlights for each contribution.

Table 1. Analysis of the published contributions in the Special Issue.

# of Contribution	Research Area (Type of Sensor)	Focus	Type of Article
1	Piezoresistive strain sensor	Materials (conductive silver nanowires, electrospun nanofibers as substrate); human motion monitoring	Research
2	Iontronic pressure sensor	Materials (iontronic fiber materials); human motion monitoring	Research
3	Piezoelectric pressure sensor	Structure design (enabling stretchability by Kirigami design); finite element simulations; heart sound monitoring	Research
4	Piezoresistive strain sensor, IMU motion sensor	Musical expression (translates hand motions and gestures into musical outputs)	Research
5	Capacitive electrode for ECG acquisition	Materials (multilayer Kapton-insulated Cu foil structure on a flexible polyurethane structural support); ECG signal acquisition	Research
6	Piezoresistive strain sensor	Data-driven (machine learning) materials exploration for strain sensor; performance prediction; human motion monitoring	Research
7	Flexible ultrasound transducer	Structure design and device fabrication (piezoelectric material and island-bridge device structure); time-reversal algorithm; transcranial focused ultrasound simulation	Research
8	Versatile sensors; integrated sensors	Fabrication methods; human body phantoms for evaluating wearable device performance	Review

https://doi.org/10.3390/s26072196

Table 1. *Cont.*

# of Contribution	Research Area (Type of Sensor)	Focus	Type of Article
9	Piezoresistive strain sensor	Conductive materials; electrospun; polyurethane; motion detection; healthcare monitoring; human–machine interaction	Review
10	Optical fiber sensors	Working principle of wearable optical fiber sensor; Fiber Bragg Grating sensors, self-luminescent stretchable optical fiber sensors, Fabry–Perot sensors, and polymer optical fiber sensors	Review
11	Triboelectric nanogenerators (TENGs)	Working principles and operational modes of TENGs; applications (mechanical energy harvesting, self-powered sensors for human motion detection, healthcare monitoring, and human–machine interaction)	Review

Shan et al. (contribution 1) present a systematic study on the preparation and applications of silver nanowire and TPU nanofibers for flexible strain sensors. By synthesizing ultra-high-aspect-ratio silver nanowires (aspect ratio > 1500) and combining them with electrospun TPU nanofiber membranes in a sandwich structure, the authors achieved a flexible sensor with exceptional properties: low initial resistance (26.7 Ω), high sensitivity (gauge factor > 19.21), rapid response time (112 ms), and remarkable cycling stability over 3000 stretching cycles. The sensor demonstrated a stable electrical signal output for applications in motion monitoring, human–computer interaction, and healthcare.

Liu et al. (contribution 2) conduct a comparative investigation into the impact of microstructure on the sensing performance of fiber-based iontronic pressure sensors. Using four distinct fiber materials (woven fabric, cotton nonwoven, viscose spunlace, and electrospun TPU nanofiber mats), the authors systematically evaluated how fiber diameter, porosity, and thickness influence ionic liquid absorption and consequent sensing performance. The optimized sensor, fabricated with a 33 μm-thick TPU/IL nanofiber mat, achieved remarkable sensitivities of 3.10 kPa^{-1}, 1.85 kPa^{-1}, and 1.02 kPa^{-1} across pressure ranges of 0–200 kPa, 200–400 kPa, and 400–700 kPa, respectively, with an ultrafast response time of 2.71 ms. This work provides valuable guidance for selecting and utilizing fiber materials to develop high-sensitivity iontronic pressure sensors for wearable electronics.

Zhang et al. (contribution 3) introduce a stretchable piezoelectric sensor with a Kirigami-inspired design for heart sound monitoring. The five-layer sandwich structure (PI–electrode–PVDF–electrode–PI) incorporates periodic Kirigami cuts that significantly enhance stretchability while maintaining piezoelectric conversion efficiency. Finite element simulations revealed that the Kirigami structure's stretchability is more sensitive to hinge length and thickness than to hinge width. The sensor exhibited excellent stability, with a maximum amplitude variation of approximately 11% under 0–30% strain and less than 9% output fluctuation during 1200 s of continuous excitation. Seven-day monitoring confirmed reliable detection of the first (S1) and second (S2) heart sounds, demonstrating its potential for wearable long-term cardiovascular monitoring.

https://doi.org/10.3390/s26072196

Wexler et al. (contribution 4) explore the innovative concept of wearable technologies as a medium for musical expression. The authors developed a wearable device equipped with 32 touch-sensitive pressure sensors, a nine-axis inertial measurement unit motion sensor, and LED and vibrational haptic feedback components, which can be worn as a wrist strap, hand strap, or surface drum pad. The device translates hand motions and gestures into musical outputs, offering an intuitive and accessible interface for music creation. Through iterative prototyping and user evaluation, the study demonstrates how wearable technology can democratize music learning and expression, bridging the accessibility gap in musical education and performance.

Iliev et al. (contribution 5) present the design and assessment of flexible capacitive electrodes for reusable ECG monitoring, with particular attention to the effects of sweat and adapted front-end configuration. Using a simple, cost-effective multilayer structure consisting of a copper conductor laminated with a polyimide dielectric layer on a polyurethane support, the authors evaluated electrode performance under varying textile moisture levels and exposure to common disinfectants. Electrochemical impedance spectroscopy confirmed the stable capacitive behavior and chemical stability of the polyimide–copper interface. A custom front-end readout circuit incorporating digitally adjustable gain and baseline correction enabled robust through-textile ECG signal acquisition, demonstrating the feasibility of this low-cost, reusable electrode platform for wearable biosignal monitoring.

Ke et al. (contribution 6) propose a data-driven approach to strain sensor design using a knowledge graph framework. By constructing a knowledge graph that captures relationships among materials, structures, fabrication methods, and performance metrics from the published literature, the authors employed graph representation learning to predict sensor properties and discover new design combinations. The framework achieved a prediction precision up to 0.81 and successfully identified a TPU/graphene/carbon black combination with a high correlation (0.93). The designed sensor demonstrated a wide strain range (300%) and closely matched the predicted performance, showcasing the potential of AI-assisted approaches to accelerate sensor development and reduce trial-and-error experimentation.

Jia et al. (contribution 7) report a flexible ultrasound transducer with a controllable acoustic field enabled by a unique time-reversal algorithm. The device incorporates a low-impedance piezoelectric material and an island-bridge structure, with each pixel achieving a consistent sound pressure of approximately 36 kPa. To address the challenge of accurate focusing when the device is bent, the authors developed a Virtual Point Source Time-Reversal Algorithm (VPSTRA) that compensates for time differences in ultrasound transmission through the skull. The simulation results demonstrated successful focal formation at both single and multiple virtual source points, offering a promising approach for non-invasive brain disease treatment and deep-tissue sensing.

Ali et al. (contribution 8) provide a thorough overview of wearable and flexible sensor devices, focusing on recent advances in designs, fabrication methods, and applications. The review covers state-of-the-art wireless communication technologies for wearable devices, smart wearable sensors for physiological monitoring, and emerging additive manufacturing techniques including direct ink writing (DIW) and 3D to 6D printing. The authors also discuss the importance of human body phantoms for evaluating wearable device performance and address challenges related to biocompatibility, power supply, and specific absorption rate (SAR) considerations for next-generation healthcare applications.

Zhou et al. (contribution 9) present a focused review on flexible strain sensors based on TPU nanofibers fabricated using electrospinning. The authors systematically summarize recent advances in incorporating various conductive fillers—including carbonaceous materials (carbon black, carbon nanotubes, and graphene), MXene, metallic nanomaterials

https://doi.org/10.3390/s26072196

(silver nanowires, liquid metals), and conductive polymers (PPy, PEDOT:PSS, and PANI)—into electrospun TPU nanofibers. The review discusses the working principles of strain sensors and electrospinning technology, highlights strategies for optimizing sensor performance, and explores applications in motion detection, health monitoring, human–computer interaction, and electronic skin.

Li et al. (contribution 10) review wearable optical fiber sensors for rehabilitation monitoring. The authors introduce the principles and technologies of various fiber sensors, including Fiber Bragg Grating (FBG) sensors, self-luminescent stretchable optical fiber sensors, Fabry–Perot sensors, and polymer optical fiber sensors. The review discusses specific applications in detecting finger movements, physiological parameters (blood oxygen saturation, pulse waveform, and skin temperature), and tactile responses during rehabilitation. The authors also address current challenges and future research directions, including improving sensitivity and flexibility, reducing costs, and enhancing system integration for clinical applications.

Punnarao and Yoon (contribution 11) present a review of flexible triboelectric mechanical energy harvesters for wearable and self-powered sensing applications. The review explains the working principles and four operational modes of TENGs (contact–separation, single-electrode, lateral-sliding, and freestanding modes), with detailed simulation results and working mechanisms. The authors systematically analyze fabrication techniques for triboelectric flexible films, including hydrothermal synthesis, electrospinning, 3D printing, and spray coating. The review covers applications in mechanical energy harvesting, self-powered sensors for human motion detection, healthcare monitoring, and human–machine interaction, while also addressing the current limitations and future directions for the field.

3. Conclusions

The collection of articles in this Special Issue reflects the remarkable progress and interdisciplinary nature of flexible and wearable sensor research. From the development of high-performance materials, such as silver nanowires and nanofibers, to innovative structural designs, including Kirigami patterns and sandwich architectures, these contributions demonstrate the diverse strategies being pursued to enhance sensor performance, flexibility, and wearability. Advances in fabrication methods, ranging from electrospinning and direct ink writing to 3D and 4D printing, are enabling the scalable production of sensors with precisely controlled microstructures and multifunctional capabilities. The integration of these sensors into practical applications, including cardiac monitoring, rehabilitation, musical expression, and human–machine interaction, illustrates their transformative potential across healthcare, entertainment, and beyond. Moreover, the emergence of data-driven approaches, such as knowledge graph-based design and machine learning-assisted signal processing, promises to accelerate innovation and enable the rational design of next-generation wearable sensors. While challenges remain, including long-term stability, biocompatibility, power supply, and cost-effective manufacturing, the collective efforts of researchers worldwide are steadily addressing these obstacles. We anticipate that continued advancements in materials science, manufacturing technologies, and system integration will further expand the capabilities and adoption of flexible and wearable sensors, ultimately improving quality of life through personalized, continuous, and unobtrusive monitoring and interaction.

Author Contributions: Conceptualization, Y.D., T.Z. and Z.W.; writing—original draft preparation, Y.D.; writing—review and editing, Y.D., T.Z. and Z.W.; project administration, Y.D.; funding acquisition, Y.D. All authors have read and agreed to the published version of the manuscript.

Funding: This research was funded by the National Science Foundation of China (62461042, 52305606, and 52575662).

 https://doi.org/10.3390/s26072196

Acknowledgments: We would like to express our sincere gratitude to all the authors and scholars who contributed to this Special Issue. The success of this collection, reflected in this Editorial, attests to the dedication, expertise, and hard work of each author in advancing scientific objectives. The valuable insights, innovative methodologies, and groundbreaking findings presented here enrich the field of wearable sensors across a range of applications. We also wish to thank the editorial and technical teams for their steadfast support throughout the development of this Special Issue, which ensured a smooth and efficient publication process. Finally, we extend our appreciation to the readers and the broader scientific community for their interest and engagement in this Special Issue, as it is their enthusiasm that drives the advancement of knowledge and fosters progress in wearable sensors for healthcare applications.

Conflicts of Interest: The authors declare no conflicts of interest.

List of Contributions:

1. Shan, J.; Qian, J.; Lin, L.; Wei, M.; Xia, J.; Fu, L. Preparation and Applications of Silver Nanowire-Polyurethane Flexible Sensor. *Sensors* **2025**, *25*, 5191.
2. Liu, C.; Xu, J.; Yang, S.; Xu, Y.; Wang, J.; Liu, X.; Wang, L.; Ding, Y. Impact of Microstructure on Sensing Performance of Fiber-Based Iontronic Pressure Sensors: A Comparative Study. *Sensors* **2025**, *25*, 6711.
3. Zhang, X.; Ye, X.; Lei, X.; Hu, H.; Liu, H.; Jin, S.; Ye, G.; Zhao, T. Fabrication of Stretchable Piezoelectric Sensor with a Kirigami Design for Heart Sound Monitoring. *Sensors* **2025**, *25*, 7253.
4. Wexler, D.; Yip, J.; Lee, K.-P.; Li, X.; Wong, Y.-H. A Touch on Musical Innovation: Exploring Wearables and Their Impact on New Interfaces for Musical Expression. *Sensors* **2024**, *24*, 250.
5. Iliev, I.; Nikolov, G.T.; Tomchev, N.; Stefanov, B.I.; Tzaneva, B. Design and Assessment of Flexible Capacitive Electrodes for Reusable ECG Monitoring: Effects of Sweat and Adapted Front-End Configuration. *Sensors* **2025**, *25*, 5856.
6. Ke, J.; Liu, F.; Xu, G.; Liu, M. Data-Driven Strain Sensor Design Based on a Knowledge Graph Framework. *Sensors* **2024**, *24*, 5484.
7. Jia, L.; Yan, Y.; Xu, J.; Gao, Y. A Unique Time-Reversal Algorithm-Enabled Flexible Ultrasound Transducer with a Controllable Acoustic Field. *Sensors* **2024**, *24*, 5635.
8. Ali, S.M.; Noghanian, S.; Khan, Z.U.; Alzahrani, S.; Alharbi, S.; Alhartomi, M.; Alsulami, R. Wearable and Flexible Sensor Devices: Recent Advances in Designs, Fabrication Methods, and Applications. *Sensors* **2025**, *25*, 1377.
9. Zhou, Z.; Tang, W.; Xu, T.; Zhao, W.; Zhang, J.; Bai, C. Flexible Strain Sensors Based on Thermoplastic Polyurethane Fabricated by Electrospinning: A Review. *Sensors* **2024**, *24*, 4793.
10. Li, X.; Li, Y.; Wei, H.; Wang, C.; Liu, B. A Review of Wearable Optical Fiber Sensors for Rehabilitation Monitoring. *Sensors* **2024**, *24*, 3602.
11. Punnarao, M.; Yoon, H.-J. Flexible Triboelectric Mechanical Energy Harvesters for Wearable and Self-Powered Sensing Applications: A Review. *Sensors* **2026**, *26*, 1166.

References

1. Kim, J.; Campbell, A.S.; Esteban-Fernandez de Ávila, B.; Wang, J. Wearable Biosensors for Healthcare Monitoring. *Nat. Biotechnol.* **2019**, *37*, 389–406. [CrossRef] [PubMed]
2. Xiao, X.; Yin, J.; Xu, J.; Tat, T.; Chen, J. Advances in Machine Learning for Wearable Sensors. *ACS Nano* **2024**, *18*, 22734–22751. [CrossRef] [PubMed]
3. Kim, D.-H.; Lu, N.; Ma, R.; Kim, Y.-S.; Kim, R.-H.; Wang, S.; Wu, J.; Won, S.M.; Tao, H.; Islam, A.; et al. Epidermal Electronics. *Science* **2011**, *333*, 838–843. [CrossRef] [PubMed]
4. Lee, J.-H.; Cho, K.; Kim, J.-K. Age of Flexible Electronics: Emerging Trends in Soft Multifunctional Sensors. *Adv. Mater.* **2024**, *36*, 2310505. [CrossRef] [PubMed]
5. Fu, X.; Cheng, W.; Wan, G.; Yang, Z.; Tee, B.C.K. Toward an AI Era: Advances in Electronic Skins. *Chem. Rev.* **2024**, *124*, 9899–9948. [CrossRef] [PubMed]
6. Yang, X.; Chen, W.; Fan, Q.; Chen, J.; Chen, Y.; Lai, F.; Liu, H. Electronic Skin for Health Monitoring Systems: Properties, Functions, and Applications. *Adv. Mater.* **2024**, *36*, 2402542. [CrossRef] [PubMed]
7. De Fazio, R.; Mastronardi, V.M.; De Vittorio, M.; Visconti, P. Wearable Sensors and Smart Devices to Monitor Rehabilitation Parameters and Sports Performance: An Overview. *Sensors* **2023**, *23*, 1856. [CrossRef] [PubMed]

 https://doi.org/10.3390/s26072196

8. Qu, J.; Cui, G.; Li, Z.; Fang, S.; Zhang, X.; Liu, A.; Han, M.; Liu, H.; Wang, X.; Wang, X. Advanced Flexible Sensing Technologies for Soft Robots. *Adv. Funct. Mater.* **2024**, *34*, 2401311. [CrossRef]
9. Ding, Y.; Jiang, J.; Wu, Y.; Zhang, Y.; Zhou, J.; Zhang, Y.; Huang, Q.; Zheng, Z. Porous Conductive Textiles for Wearable Electronics. *Chem. Rev.* **2024**, *124*, 1535–1648. [CrossRef] [PubMed]
10. Luo, Y.; Abidian, M.R.; Ahn, J.-H.; Akinwande, D.; Andrews, A.M.; Antonietti, M.; Bao, Z.; Berggren, M.; Berkey, C.A.; Bettinger, C.J.; et al. Technology Roadmap for Flexible Sensors. *ACS Nano* **2023**, *17*, 5211–5295. [CrossRef] [PubMed]
11. Vaghasiya, J.V.; Mayorga-Martinez, C.C.; Pumera, M. Wearable Sensors for Telehealth Based on Emerging Materials and Nanoarchitectonics. *npj Flex. Electron.* **2023**, *7*, 26. [CrossRef] [PubMed]
12. Ates, H.C.; Nguyen, P.Q.; Gonzalez-Macia, L.; Morales-Narváez, E.; Güder, F.; Collins, J.J.; Dincer, C. End-to-End Design of Wearable Sensors. *Nat. Rev. Mater.* **2022**, *7*, 887–907. [CrossRef] [PubMed]

https://doi.org/10.3390/s26072196

sensors

MDPI

Article

Preparation and Applications of Silver Nanowire-Polyurethane Flexible Sensor

Jiangyin Shan, Jianhua Qian *, Ling Lin, Mengrong Wei, Jingyue Xia and Lin Fu

Textile Science and Engineering, Zhejiang Sci-Tech University, Hangzhou 310020, China;
2022327100122@mails.zstu.edu.cn (J.S.); 2022327100130@mails.zstu.edu.cn (L.L.);
weimengrong2021@163.com (M.W.); 18154307082@163.com (J.X.); 202230202095@mails.zstu.edu.cn (L.F.)
* Correspondence: qianjianhua@zstu.edu.cn

Abstract

To expand the application of silver nanowires (AgNWs) in the field of flexible sensors, this study developed a stretchable flexible sensor based on thermoplastic polyurethane (TPU). Initially, the TPU nanofiber membrane was prepared by electrospinning. Subsequently, high-aspect-ratio AgNWs were synthesized via a one-step polyol reduction method. The AgNWs with the optimal aspect ratio were selected for the conductive layer and spray-coated onto the surface of the TPU nanofiber membrane. Another layer of TPU nanofiber membrane was then laminated on top, resulting in a flexible thin-film sensor with a "sandwich" structure. Through morphological, chemical structure, and crystallinity analyses, the primary factors influencing AgNWs' growth were investigated. Performance tests revealed that the prepared AgNWs had an average length of approximately 130 μm, a diameter of about 80 nm, and an average aspect ratio exceeding 1500, with the highest being 1921. The obtained sensor exhibited a low initial resistance (26.7 Ω), high strain range (sensing, ε = 0–150%), high sensitivity (GF, over 19.21), fast response and recovery time (112 ms), and excellent conductivity (428 S/cm). Additionally, the sensor maintained stable resistance after 3000 stretching cycles at a strain range of 0–10%. The sensor could output stable and recognizable electrical signals, demonstrating significant potential for applications in motion monitoring, human–computer interaction, and healthcare fields.

Keywords: silver nanowires; flexible sensors; high sensitivity

1. Introduction

With the current deep integration and widespread application of the Internet of Things (IoT), artificial intelligence (AI), and wearable and implantable technologies [1–6], the development of flexible sensors has garnered significant attention. Compared to conventional rigid sensors, flexible sensors, as a novel category of electronic devices, exhibit distinctive characteristics such as high resilience, excellent tensile properties, superior ductility, and light weight [7,8]. These features provide them with unique advantages in various cutting-edge fields. Yao et al. [9] developed a flexible sensor capable of conforming to human skin, and Lu et al. [10] fabricated a flexible temperature sensor using printing technology [11–14]. These studies undeniably demonstrated the immense developmental potential of flexible sensors in various domains, including smart wearable devices, electronic skin, healthcare, smart homes, and robot tactile sensing [11–14].

AgNWs [15,16], as an emerging nanomaterial, exhibit outstanding conductivity, high specific surface area, stable chemical properties, and excellent flexibility, making them an ideal choice for flexible conductive materials. YANG et al. [17] combined AgNWs with polydimethylsiloxane (PDMS) to fabricate sensors with a sandwich structure, demonstrating high stability in electrical properties. Li et al. [18] prepared a composite flexible film by combining AgNWs with hydroxypropyl cellulose (HPC), showing that the flexible sensor fabricated with AgNWs exhibits high current sensitivity and excellent cycling stability. However, AgNW-based flexible thin-film sensors have faced certain challenges in terms of integration and compatibility, necessitating further research and optimization. Therefore, it is important to select a substrate with appropriate elasticity and flexibility for flexible sensors [19,20].

In recent years, researchers have prepared various flexible substrates through multiple methods, such as electrospinning and piezoelectrics [21,22]. WANG et al. [23] fabricated a substrate based on P(VDF-TrFE) electrospun membranes, and CAI et al. [24] employed hydrogel films as the substrate for flexible sensors. However, with the progression of the era, there has been increasing demand for stability, conductivity, and strain range in flexible sensors.

Therefore, a flexible sensor with a "sandwich" structure was fabricated using a thermoplastic polyurethane (TPU) electrospun film as a substrate to enhance the adhesion of AgNWs. The morphology and crystal structure of AgNWs synthesized via the one-pot polyol method were analyzed through SEM, TEM, and XRD tests. Ultimately, AgNWs with an average aspect ratio of 1500 and a maximum aspect ratio of 1921 were selected as the conductive layer, which is much higher than most of the reported AgNWs (mostly <1000). The sensor was subjected to a series of rigorous evaluations, including tensile testing, cyclic stability testing, and other assessments. Furthermore, the electrical signal exhibited fluctuations due to its high sensitivity (GF max. 19.21), excellent conductivity (428 S/cm), and fast response time (112 ms). These properties underscore the potential applications of flexible sensors in various fields, including human–computer interaction, motion monitoring, and healthcare. Furthermore, the present study expanded the 3×3 array sensors, providing a novel perspective and direction.

2. Experimental Section

2.1. Materials and Chemicals

Polyvinylpyrrolidone (PVP, $M_w = 1.3 \times 10^6$, 5.4×10^4), anhydrous copper chloride ($CuCl_2$, 98%), and polyurethane (TPU) were supplied by Shanghai Aladdin Biochemical Technology Co., Ltd. in Shanghai, China. Ethylene glycol (EG), ethanol (C_2H_5OH), and N,N-dimethylacetamide (DMAC) were provided by Hangzhou Gaojing Fine Chemicals Co., Ltd. in Hangzhou, China. Silver nitrate ($AgNO_3$) was obtained from Guangzhou Guanghua Technology Co., Ltd. In Hangzhou, China. All chemicals were used without further purification.

2.2. Preparation of TPU Membranes and Nanofiber Membranes

First, 9 g of TPU and 41 g of dimethylacetamide (DMAC) were weighed into a conical flask to prepare an 18 wt% solution. The solution was stirred in a lab water bath at 70 °C for 9 h until it became transparent and homogeneous. Then, the solution was left to stand in a lab oven at 70 °C overnight. By setting the table-top film applicator to 70 μm, a TPU film was successfully fabricated.

Next, 1.8 g of TPU and 8.2 g of DMAC were weighed into a sample bottle to prepare an 18 wt% solution. Then, 5 mL of the aforementioned spinning solution was loaded into a syringe and fixed onto the electrospinning apparatus. The spinning voltage was set to 14 kV, the distance between the flat needle tip and the collecting roller was maintained at 15 cm, and the rotation speed of the collecting roller was set to 250 rpm. The injection pump flow rate was adjusted to 0.8 mL/h, and the spinning temperature was maintained at 25 ± 2 °C. After 6 h of electrospinning, a TPU nanofiber membrane was successfully fabricated. The detailed preparation process of the TPU nanofiber membrane is illustrated in Figure 1b.

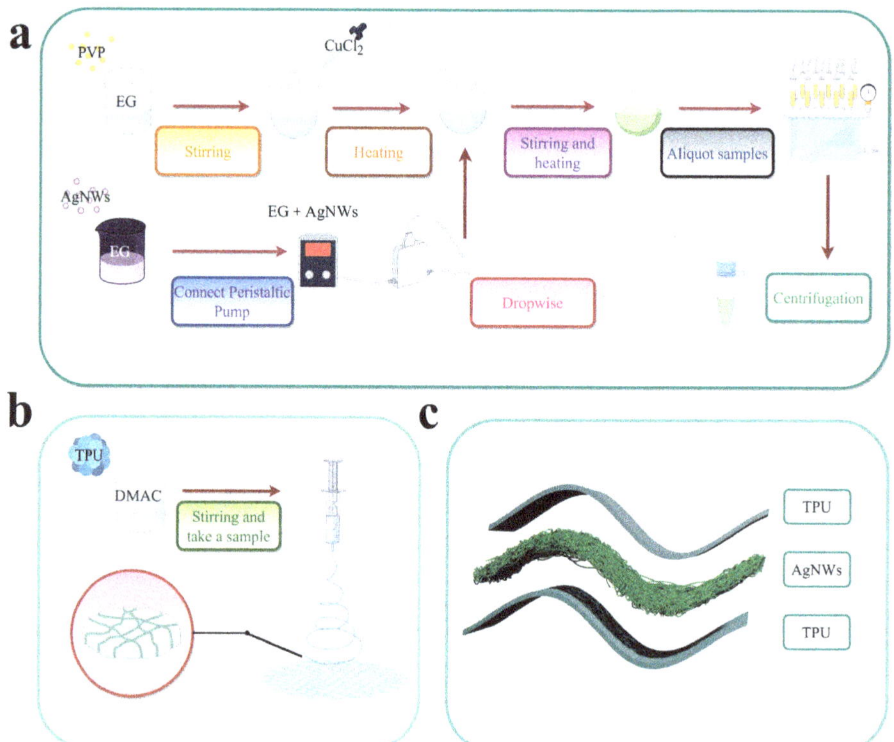

Figure 1. The production process of AgNWs (**a**). The production process of TPU-based nanofiber membranes (**b**). Architecture of the AgNW-TPU flexible film sensors (**c**) (by Figdraw, accessed on 10 June 2025).

2.3. The Preparation of AgNWs

Here, 0.69 g of PVP (Mw: 5.8 × 104) and 1.17 g of PVP (Mw: 1.3 × 106) were separately weighed and dissolved in 50 mL of EG. The PVP powder was dissolved using a magnetic stirrer (From Shanghai Aladdin Biochemical Technology Co., Ltd. in Shanghai, China) until the solution became homogeneous and stable. The prepared solution was slowly poured into the three-necked flask, which had been rinsed with EG. Subsequently, 1 mL of 3 mmol/L $CuCl_2$ (dissolved in EG) was added dropwise into the three-necked flask. A magnetic stir bar (From Hangzhou Gaojing Fine Chemicals Co., Ltd. in Hangzhou, China) was placed inside, and the flask was sealed with a glass stopper (From Hangzhou Gaojing Fine Chemicals Co., Ltd. in Hangzhou, China). The flask was then fixed in a lab oil bath at 150 °C and stirred continuously at 200 rpm for 1 h (Figure 1a).

In a completely light-free environment, 0.5 g of $AgNO_3$ was weighed, and 25 mL of EG was measured as the solvent. The solution was stirred using a magnetic stirrer until it became homogeneous. The reaction solution of AgNWs and EG was dripped into a three-necked flask at 75 rpm using a peristaltic pump while maintaining a constant temperature of 150 °C. AgNWs were prepared with reaction times of 6 h, 7 h, and 8 h, and labeled, respectively, as A6, A7, and A8. Of these, $A8_1$ and $A8_2$ represent, respectively, the underdeveloped and successfully grown AgNWs at 8 h (Figure 1a).

The three-necked flask was opened and allowed to cool to room temperature in the fume hood. The solution was then aliquoted into centrifuge tubes. To each centrifuge tube, 1 mL of the AgNW mixed solution and 2 mL of ethanol were added, followed by centrifugation in a centrifugal machine at 8000 rpm for 20 min. The supernatant was discarded. Subsequently, 6 mL of anhydrous ethanol was added, and the centrifugation process was repeated four times to obtain purified AgNWs. The purified AgNWs were dispersed in anhydrous ethanol to prevent aggregation.

2.4. The Fabrication of Flexible Sensors

Two pieces of TPU film, each measuring 3.5 cm × 3.5 cm, were cut. One piece of TPU film was fixed in a Petri dish, and the ethanol-dispersed AgNWs solution was uniformly coated onto the surface of the TPU film using the drop-casting method. After the ethanol evaporated, the AgNWs formed a network structure, resulting in a transparent film. A copper sheet (thickness: 0.02 mm; length: 2 cm; width: 0.5 cm; from Hangzhou Gaojing Fine Chemicals Co., Ltd. in Hangzhou, China) was placed on the TPU-AgNWs film and secured with conductive silver paste (From Hangzhou Gaojing Fine Chemicals Co., Ltd. in Hangzhou, China). Subsequently, the other piece of TPU film was placed on top, forming a "sandwich" structure (Figure 1c) for the flexible thin-film sensor.

3. Results and Discussion

3.1. Analysis of TPU and AgNW SEM Image

In this study, two preparation methods (casting method and electrostatic spinning method) were utilized to produce TPU films, and the following is an analysis of the TPU films produced by the two different preparation methods.

SEM images of the TPU material are shown in Figure 2. As shown in Figure 2b, the TPU film fibers prepared by the electrostatic spinning method exhibit uniform thickness and a smooth surface. The micro-topography of the film shows a fibrous three-dimensional network structure, and the fiber arrangement exhibits anisotropy and the formation of multiple pores of varying sizes between the fibers, thus reducing the incidence of agglomeration phenomena. The homogeneity and stability of the TPU film prepared by the electrostatic spinning method are excellent. Figure 2c demonstrates that there is a certain scale of agglomerate structures on the surface of TPU film prepared by the casting method, and this structure is evidence that polymer crystallization partially exists on the surface of the film; it can be observed that the TPU particles appear as spheroids of varying sizes. Additionally, the agglomeration phenomenon is more serious, resulting in a rough and uneven surface on the film. The presence of pores with suitable apertures has been shown to enhance the conductivity, permeability, and impermeability of the film as well as its strength. However, heterogeneity of the thickness has been demonstrated to affect the performance of the flexible sensor. Consequently, the TPU nanofiber membrane functions better as a flexible substrate.

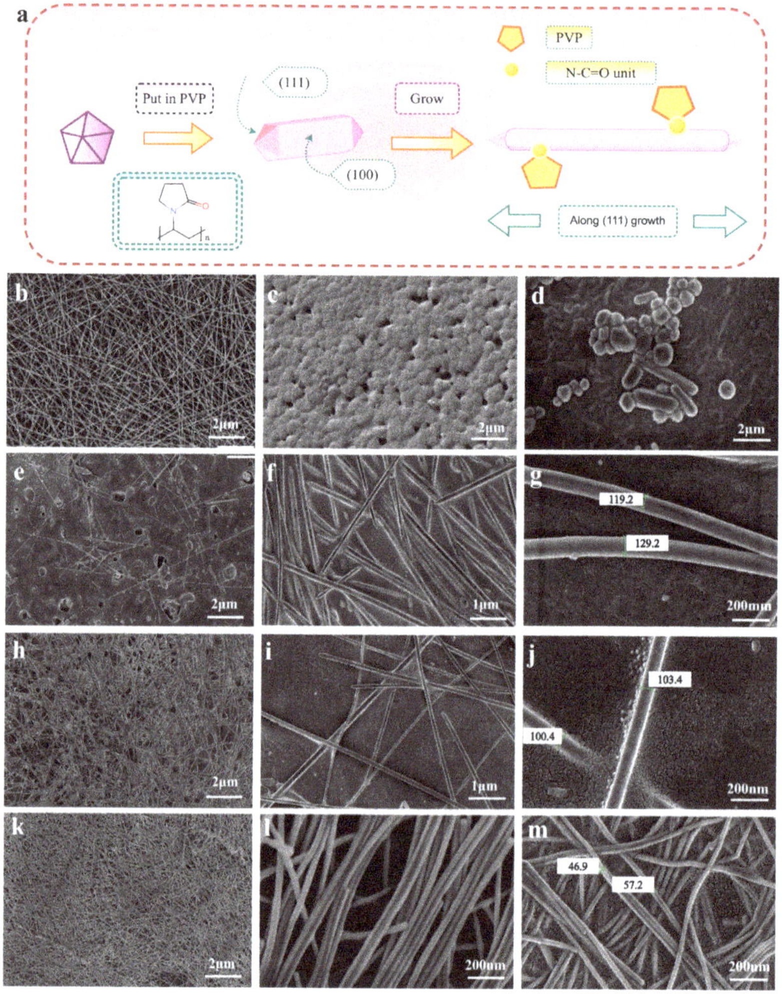

Figure 2. Schematic illustration of the growth of AgNWs (**a**) (by Figdraw, accessed on 10 June 2025); SEM images of (**b**) TPU-based nanofiber; (**c**) TPU made by the casting method; A6 (**e**–**g**), A7 (**h**–**j**), $A8_1$ (**d**), and $A8_2$ (**k**–**m**).

As the reaction time increases, the diameter of AgNWs decreases from approximately 120 nanometers at A6 to approximately 80 nanometers at $A8_2$. Following the analysis of five replicate samples, a statistically significant finding emerged: the diameter of $A8_2$ (~80 nm) was notably smaller than that of A7 (~100 nm) and A6 (120 nm). A visual examination of the SEM images further corroborated these findings, clearly demonstrating that the samples possessed varying concentrations. This observation indicated that $A8_2$ exhibited superior suitability for utilization as a raw material for the conductive layer when compared to A6, which exhibited severe agglomeration, and A7, which exhibited sparse AgNWs. This phenomenon can be attributed to the combined effects of surface ligand-mediated growth and the Ostwald ripening process. Polyvinylpyrrolidone (PVP) plays a crucial role as a capping agent. PVP demonstrates a marked preference for adsorption onto the (100) crystal faces of Ag [25–27]. During the initial reaction stage (A6), the PVP concentration relative to Ag^+ is inadequate to fully passivate the (100) faces due to a short reaction time of only 6 h. Consequently, silver atoms are deposited uncontrollably on these lateral surfaces, resulting in an increased initial diameter. As the reaction progresses to A7 (7 h) and $A8_2$ (8 h), PVP gradually saturates the (100) plane. This saturation forms a steric barrier layer that effectively inhibits lateral growth. Concurrently, the (111) crystal plane, which exhibits a weak interaction with PVP, persists in facilitating axial growth. The Gibbs–Thomson effect [28] has also been observed to result in elevated surface energy in small Ag nanoparticles or irregular protrusions on the AgNW surface (more prevalent at A6 and A7). These high-energy species undergo dissolution into Ag^+ ions, subsequently redepositing on the low-energy (111) axial surfaces of larger AgNWs in subsequent stages. This Ostwald ripening-like process serves to further refine the diameter, resulting in the more uniform and thinner AgNWs observed in $A8_2$. The enhanced crystallinity of AgNWs with prolonged reaction times is attributable to several processes. In the initial stages (A6), the rapid reduction of $AgNO_3$ by ethylene glycol results in a high nucleation rate, leading to the formation of numerous defects, including vacancies, dislocations, and twin boundaries. As the reaction time increased from A6 to $A8_2$, the system was maintained at a stable temperature of 150 °C. This temperature is sufficient to provide activation energy for atomic diffusion. Silver atoms have been observed to migrate to defect sites, where they have been shown to fill vacancies, annihilate dislocations, and reorganize diatoms and twin boundaries into more coherent interfaces.

3.2. *Analysis of TEM Images of AgNWs*

Figure 3 shows the pattern of the two-dimensional dot composition of AgNWs. From the features of this pattern, it can be clearly seen that the AgNWs have a typical face-centered cubic structure (FCC). Figure 3a,d show that the prepared AgNWs have a thinner diameter, longer length, smoother surface, and uniform thickness, which indicates that the AgNWs prepared under the experimental optimal process are excellent. There is no agglomeration phenomenon in the reaction-prepared AgNWs, and the back layer of AgNWs can be observed by transmittance observation, which proves that the reaction-prepared AgNWs are more transparent and have better optical properties, and these good transmittance and optical properties indicate high purity to a certain extent. Figure 3c,f show the electron diffraction patterns of the AgNW samples in the composition of the two-dimensional array of dots; it can be seen that there are bright diffraction spots in the patterns, and these diffraction spots are distributed with a certain regularity, which proves that the silver nanowires have good crystallinity.

3.3. *XRD Analysis of AgNWs*

The XRD images of AgNWs are shown in Figure 3. The XRD patterns of A6 (Figure 3g), A7 (Figure 3h), and $A8_2$ (Figure 3i), respectively, were tested in the range of 20°~90°, and the resulting patterns were compared with the standard XRD pattern of silver (No. #87-0597); it can be observed that the positions of the diffraction peaks in the patterns of A6, A7, and $A8_2$ are consistent with those of the standard card, but the intensities of the diffraction peaks are different. As shown in Figure 3g–i, five diffraction peaks appeared at $2\theta = 38.28°$, 44.56°, 64.27°, 77.64°, and 81.65°; no other impurity peaks appeared, indicating that the prepared AgNWs were relatively purer. These five diffraction peaks corresponded to the FCC silver structure (111), (200), (220), (311), and (222) on the five crystal faces, which proves that the synthesized AgNWs have a face-centered cubic structure and eminent crystallinity. Furthermore, XRD results (Figure 3i) demonstrate that the (111) diffraction peak exhibits the greatest intensity at $A8_2$ (intensity $\approx$ 4314), which is about 2.34 times that of A6 (1842), exhibiting a (111)/(200) intensity ratio of approximately 4.24, suggesting preferential growth along the (111) direction. The extended reaction time facilitates the continuous supply of Ag^+ ions, allowing for ordered deposition along the low-energy (111) plane [29]. A thorough examination of the XRD spectra of the three samples reveals that only the intensity increased. The positions of the diffraction peaks are determined by the lattice spacing, which is a characteristic of the crystal structure of AgNWs. It is evident that, in the absence of alterations to the crystal structure, the lattice spacing will remain constant. The enhancement in peak intensity with reaction time is predominantly attributable to the elevated crystallinity. As the reaction progresses, the number of ordered crystals increases, and defects are gradually repaired. This results in an increased number of crystal planes capable of coherent diffracting X-rays, leading to an increase in peak intensity [30]. Furthermore, the enhanced orientation growth along the (111) direction in $A8_2$ suggests the presence of a greater number of (111) planes conducive to crystallization with a diffraction orientation, thereby resulting in higher (111) peak intensity compared to A6 and A7 [31].

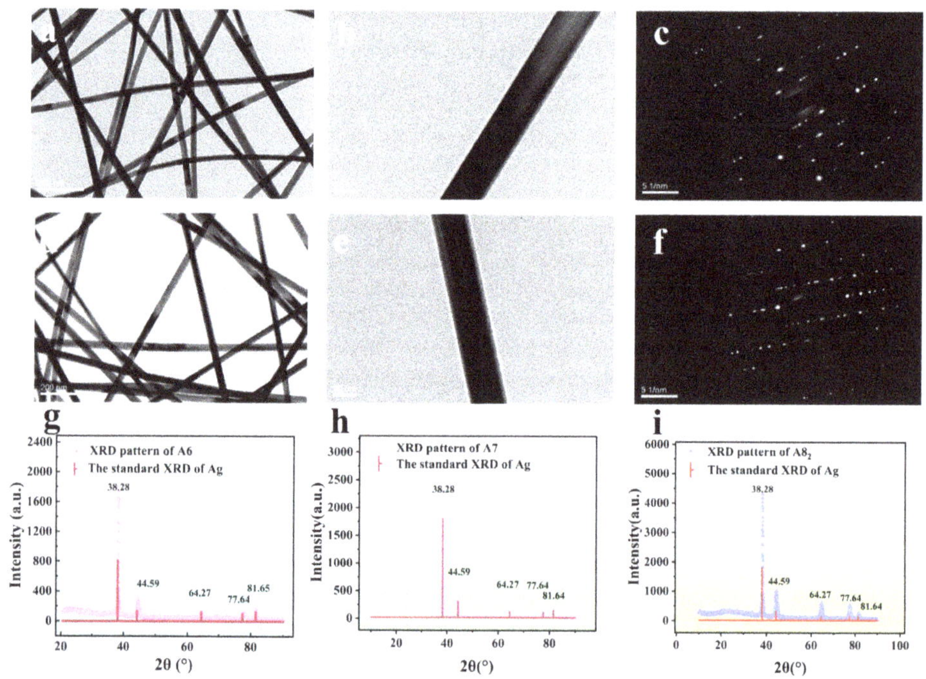

Figure 3. The surface state of AgNWs and the electron diffraction patterns composed of two-dimensional lattices (**a**,**b**,**d**,**e**), as photographed by a transmission electron microscope (**c**,**f**). XRD of A6 (**g**), A7 (**h**), and $A8_2$ (**i**).

3.4. Visualization of the Performance of AgNW-TPU Flexible Sensors

The preparation of multiple sets of samples is imperative, with each set exhibiting a uniform coating of 3 mL of AgNWs at varying concentrations. Given the tendency of alcohol to evaporate, the thickness of the final system can be regulated by adjusting the concentration and dosage of the alcohol component. When the AgNW concentration is below 18 wt%, the resistivity undergoes a precipitous decline with increasing concentration. The phenomenon can be attributed to the fact that, at low concentrations, AgNWs are sparsely dispersed, resulting in insufficient formation of conductive pathways. However, when the concentration exceeds 18 wt%, the decrease in resistivity slows down and stabilizes. At this stage, AgNWs will have formed a continuous, dense conductive network, and further increasing the concentration has a limited effect on the optimization of the pathways (Figure 4a). When the dosage is controlled between 1 and 2 mL, the conductive network is insufficiently developed, and the sensor cannot be reused after multiple cycles of stretching, resulting in a short lifespan. As the dosage increases, the sensor resistance undergoes a slight increase; however, it demonstrates commendable durability, with a recorded number of over 3000 cycles. In order to achieve the lowest possible cost while ensuring optimal performance, this experiment selected 18 wt% AgNWs at a dosage of 3 mL as the conductive layer (Figure 4b). Following the implementation of the SEM measurement of the thickness of the AgNWS conductive layer, it was determined that the optimal thickness falls between 2.3 and 3.5 μm.

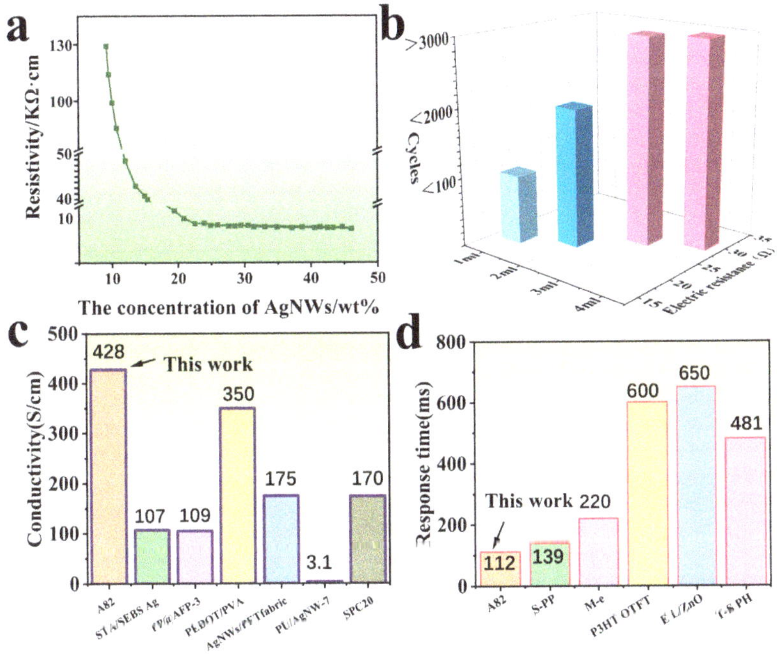

Figure 4. AgNWs' concentration versus resistivity curves (**a**), resistance and number of cycles plotted for sensors at different doses (**b**), comparison of conductivity with this work (STA/SEBS Ag [32], PP@AFP-3 [33], PEDOT/PVA [34], AgNW/PET fabric [35], PU/AgNW-7 [36], and SPC20 [37]) (**c**), comparison of response times with this work (S-PP [38], M-e [39], P3HT OTFT [40], E L/ZnO [41], and T-S PH [42]) (**d**).

Conductivity and response time are critical performance indicators for sensors. A comparative analysis of the conductivity and response time of diverse flexible conductive materials reveals that the $A8_2$ sample in this study exhibited a conductivity of 428 S/cm [32–37] and a minimum response time of 112 ms [38–42], thereby demonstrating a substantial superiority over other composite materials and fabrics depicted in the figure. Consequently, the conductivity and response time of AgNW-TPU sensors further substantiate their potential value in future markets.

3.5. Performance Analysis of Flexible Sensors

As shown in Figure 5f, the stress–strain curve presents a linear relationship within 0–100% strain. This modulus reflects the sensor's stiffness in the elastic phase, enabling it to resist excessive deformation under external forces while maintaining recoverability—upon removal of the force, the sensor regains its original form without permanent deformation, which underpins its fast response/recovery time (112 ms) and stable cyclic performance (3000 cycles). When strain increases to 100–400%, the curve deviates from linearity, indicating the onset of plastic deformation characterized by irreversible structural damage, such as the partial fracture of AgNW networks and permanent slippage of TPU molecular chains. At approximately 500% strain, the stress plummets to zero, signifying complete fracture of the sensor material.

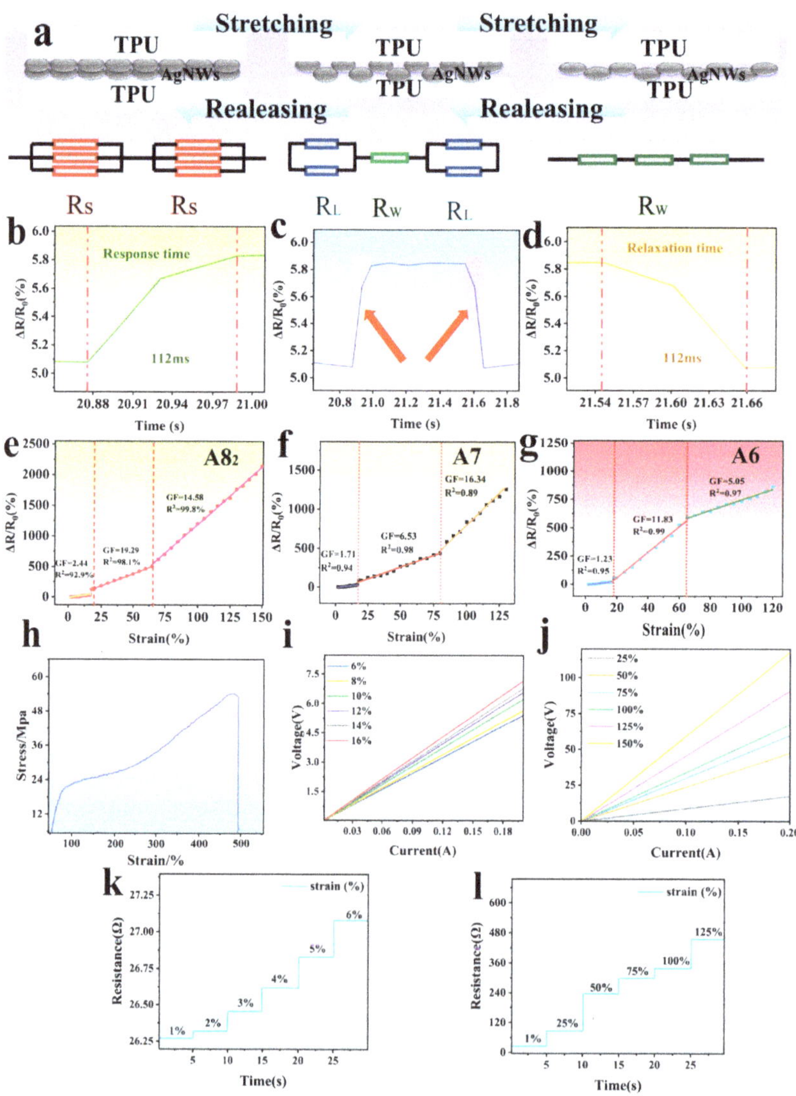

Figure 5. Schematic diagram of microscopic changes in sensor tensile resistance (**a**). Sensor response time curve (**b**–**d**). Tensile–resistance change rate curve of AgNW-TPU sensor response: $A8_2$ (**e**), A7 (**f**), and A6 (**g**). Stress–strain curve of the sensor (**h**). The V-I curves under 0%~16% (**i**) and 25%~150% tensile strain (**j**). Dynamic tensile curve of the sensor from 0% to 5% (**k**). Dynamic tensile curve from 0% to 125% (**l**).

AgNW-TPU sensors' resistance response to tensile strain (0% to 150%) is due to the evolution of their series–parallel structure. The correlation between this structure and substrate deformation directly affects the sensors' sensitivity.

In the 0% to 17% strain range, the TPU substrate barely deforms, and the AgNWs stay densely overlapping (Figure 5a). This creates a dominant multi-path parallel structure with many cross-contact points. At this point, the contact resistance is low, and the multi-path redundancy stabilizes the overall resistance at 26.7 Ω. Slight stretching slightly increases the spacing between some contact points, but the parallel dominant effect causes $\Delta R/R_0$ to grow slowly: GF = 2.44. Additionally, the uniform distribution of contact points ensures a high linear fitting degree (R^2 = 92.9%).

As strain increases from 17% to 65%, the oriented stretching of TPU molecular chains triggers network restructuring. Most parallel contact points slip or break away, and the remaining contact points have smaller contact areas. This leads to increased contact resistance. The unbroken AgNWs form a "chain-like series structure" along the stretching direction (Figure 5a). The intrinsic resistance of a single nanowire and the contact resistance are connected in series and superimposed, making the series resistance the dominant factor and the parallel resistance the secondary factor. This significantly increases the GF to 19.29. As the strain continues to increase, microdefects stabilize, and the linear relationship between the changes in resistance and strain strengthens. The fitting degree increases to 98.1%.

In the high strain range of 65% to 150%, TPU deformation is maximal, and AgNWs undergo local fractures due to excessive stretching. The original long series chains disintegrate into short segments, forming new series structures. A small portion of the short nanowires re-contact in the wrinkled regions, forming temporary parallel branches that partially compensate for the increase in resistance. The combined effect of these two factors reduces GF to 14.58, and the regularity of the structural reorganization achieves a linear fitting degree of 99.8%. This demonstrates the adaptability of the network structure.

Additionally, tensile–sensitivity curves of sensors prepared from batches A6 and A7 were examined. In Figure 5e–g, high-aspect-ratio AgNWs ($A8_2$) exhibited a substantially higher GF (19.29) compared to those of A7 (GF ≈ 16.34) and A6 (GF ≈ 11.84). This elevated GF can be attributed to the presence of denser network connections and enhanced tensile tolerance. Additionally, it was observed that the conductive network of $A8_2$ exhibited resilience, maintaining its integrity even under strain levels of up to 150%.

Response time is a critical metric of the dynamic performance of flexible sensors and is fundamentally related to the rate at which electronic transmission paths are reconfigured within the conductive network. The AgNW-TPU sensor demonstrates a response time and recovery time of 112 ms at 1% strain (Figure 5b–d). According to the RC delay theory [43], the signal transmission time constant is reduced by a low-resistance network (26.7 Ω). Furthermore, the three-dimensional porous structure of the electrospun TPU membrane provides anchoring for AgNWs. During minor deformations, the contact points of the nanowires undergo elastic displacement without breaking, enabling rapid restoration of the conductive pathway without requiring atomic diffusion or interface reconstruction.

The disparities in V-I characteristics under varying static strains (Figure 5i,j) are attributable to variations in the topological structure of the conductive network. At low strains (6–18%), the AgNW network displays a dense parallel configuration, with electron conduction primarily via direct tunneling and metal contact, resulting in a linear and stable V-I curve. At a current of 0.2 A, the voltage increase is only 6.32 V, as the redundant parallel paths average out local changes. At elevated strain levels, the network predominantly functions in a series configuration, necessitating the overcoming of elevated tunneling barriers by electrons. Furthermore, periodic slippage at contact points leads to fluctuations in tunneling resistance, resulting in significant nonlinearity in the V-I curve. The enhanced electric field effect in the high-voltage region further amplifies instability.

As demonstrated in Figure 5k,l, the resistance change stability tests validate the strain dependence of the network structure. In the range of small strains between 1% and 6%, there is an observed increase in resistance from 26.2732 Ω to 27.0810 Ω. This fluctuation is recorded to be less than 0.5% within a span of five seconds. This phenomenon can be attributed to the "elastic compression" contact between AgNWs, where alterations in area are found to be reversible, and the percolation threshold remains stable [44]. At strains ranging from 25% to 150%, the resistance exhibited exponential growth with irregular fluctuations, attributed to partial plastic deformation or fracture of the nanowires. Fractured segments reconnect via substrate wrinkles, and oxidation at contact points and atomic migration cause resistance drift [45]. However, at 150% high strain, the resistance remains macroscopically stable thanks to the elastic recovery force of TPU enabling broken nanowires to form "bridge-like connections" to maintain connectivity [46].

In summary, the sensor's electrical response is the result of "material–structure–performance" synergy. With small strains, the redundant conduction of the parallel network provides stable linear behavior, suitable for high-precision micro-deformation detection. At large strains, the tunneling mechanism of the series network leads to a nonlinear response, and the wide measurement range characteristics can meet the requirements for monitoring violent movements, providing performance boundaries and a theoretical basis for scenario-based applications.

3.6. Cyclic Stability of the Sensor

The cyclic stability of the resistor is a pivotal indicator of the sensor's performance. It exerts a direct influence on the measurement accuracy and reliability of the sensor in addition to determining its service life and application range. This section presents the findings of a study examining the dynamic resistance cycling performance of AgNW-TPU flexible sensors. To this end, the resistance changes of AgNW-TPU sensors when they are stretched at different rates under the same strain were detected. The fiber membrane was stretched at different rates (0.5, 1.0, 1.5, and 2.0 mm/s) at 50% strain, and the cyclic resistance curves were measured. As demonstrated in Figure 6a, the cyclic resistance curves at constant rates exhibited high consistency, with resistance measurements at varying rates displaying a modest increase, though this is not statistically significant. This finding suggests that the relative resistance change in AgNW-TPUs is largely unaffected by the stretching rate, thereby demonstrating the resistance response's independence from the stretching rate. The sensing stability is crucial for the repeatability of the sensor in practical applications. During deformation processes such as stretching and bending, the resistance change directly affects the sensor output signal. Excessive or unstable resistance changes can lead to signal drift and hysteresis, consequently reducing the sensor's repeatability. The investigation into the cyclic testing of the sensor at 0–10% tensile strain sensing, as illustrated in Figure 6b–d, is a crucial aspect of this study. The outcomes demonstrate that the sensor undergoes a resistance change within the range of 0% to 2%, following 3000 tensile cycles within the 0% to 10% strain range. The sensor's satisfactory cyclic stability indicates its capacity to preserve consistent resistance fluctuations during repetitive tensile and recoil procedures, ensuring precise signal surveillance and delivering dependable performance across diverse application domains.

3.7. Monitoring of Human Health

AgNW-TPU flexible sensors demonstrate considerable potential for integration into textiles, with the embedding of sensors into cuffs or collars enabling the real-time and continuous monitoring of vital signs for health assessment. As demonstrated in Figure 7d,e, the sensor output exhibited a distinct periodic waveform with minimal noise interference, effectively differentiating between a higher pulse rate of 113 beats per minute (bpm) following exercise and 65 bpm at rest; the sensor demonstrated an error margin of less than 2% when compared to commercial pulse testers, as determined by a group of volunteers in a series of trials, thereby, substantiating the precision of its measurements. The functionality of the sensor was further extended to encompass cervical spine monitoring through its attachment to the neck, enabling the detection of the degree of flexion. As demonstrated in Figure 7b, the $(R - R_0)/R_0$ ratio exhibited a substantially divergent response at 30° and 90° of neck flexion, while the signal pattern remained consistent across all angles. This outcome validates the sensor's angle-independent stability. Of particular note is the observation that the 90° bend produced proportionally higher signal amplitudes, a finding that lends itself to the ability to perform quantifiable posture tracking. This feature has the potential to be incorporated into a health monitoring system, whereby a preset threshold is exceeded, triggering a posture adjustment alarm. The efficacy of the spinal motion monitoring function was evaluated through the attachment of the sensor to the lumbar region. Participants were requested to adopt an upright sitting posture, followed by flexion of the spine, and then a return to an upright sitting posture. Figure 7c demonstrates a discernible peak in bending time, which is in contrast to the signal consistent with the baseline in the upright sitting position. This disparity can be attributed to the fact that different poses result in different levels of stretching and bending of the sensor, leading to reorganization of the network inside the flexible sensor and the subsequent output of different electrical signals. These findings confirm the sensor's ability to detect biomechanical anomalies and translate them into viable feedback information with great potential for habit correction and preventive healthcare applications.

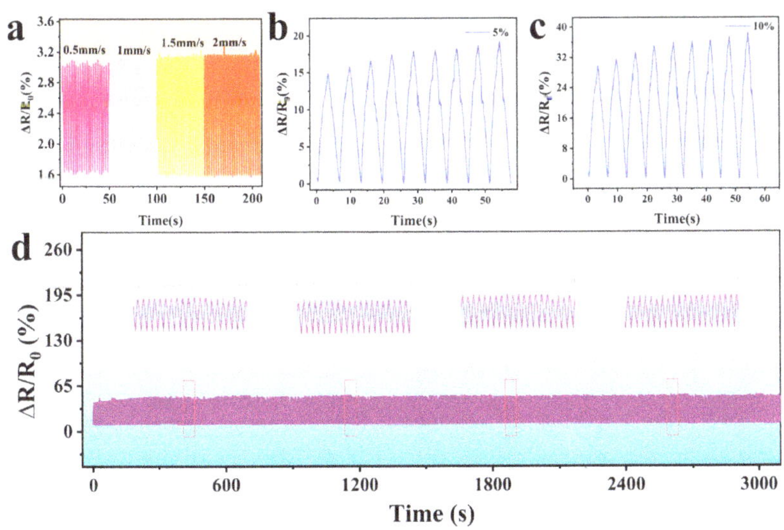

Figure 6. Cyclic resistance curves at the same strain (50%), and different speeds (0.5, 1.0, 1.5, and 2.0 mm/s) (**a**). Long-term strain sensing stability under 0–10% strain (**b**–**d**).

3.8. Application of AgNW-TPU Sensor Arrays

In the domain of human–machine interaction, the objective is to facilitate the more natural and intelligent deployment of robots on a wider scale. To this end, our research was expanded to encompass the development of a 3 × 3 array sensor. As illustrated in

Figure 1, the array sensor maintains a "sandwich" configuration, with the salient difference being that the macro structure of the inner layer network consists of 3 × 3 connection strips, denoted with A, B, and C and 1, 2, and 3, respectively, in both the horizontal and vertical directions (Figure 8). When the sensor is subjected to an external force, the horizontal 1, 2, and 3 and vertical A, B, and C points experience pressure; however, the pressure at the intersection points is the greatest. Consequently, the increased density of the array corresponds to a greater number of stress concentration points and more substantial feedback information from the robotic arm. This sensor is capable of determining an object's weight and contact shape by assessing its pressure distribution characteristics. When the weight of the dumbbell is 10 g, the $\Delta R/R_0$ concentrated at point (1, A) is 7.67%, while stress at other points exists but is not zero. The signal difference between the intersection points and non-intersection points exceeds 5.1 times. When a doll with a wider base is used, the pressure at points (2, B), (2, C), (3, B), and (3, C) is the highest, with a matching degree of 85% with the doll's base contact contour (Figure 8e). Therefore, when this sensor is attached to a glove to simulate a robot's grasping motion, it can be observed that, compared to a single sensor that can only feedback total stress, its force control accuracy is better, providing more information and further proving its feasibility in human–machine interactions (Figure 8g).

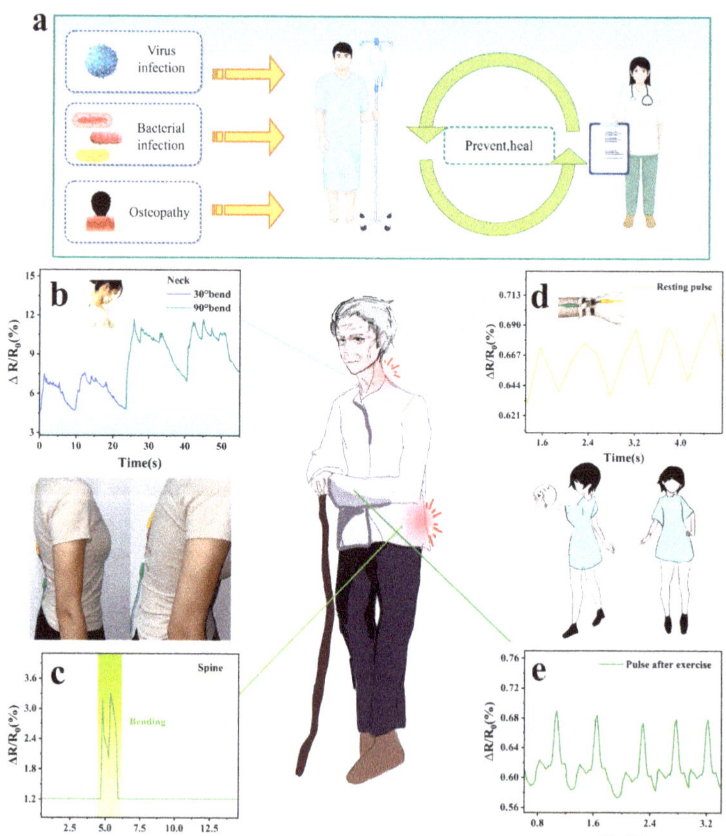

Figure 7. Flexible sensors can play a role in disease prevention (**a**) (by Figdraw, accessed on 10 June 2025). Electrical signals generated when the human neck bends (**b**). The electrical signals outputted due to the curvature of the spine (**c**). Flexible smart wristband for real-time monitoring of human pulse signal (**d**,**e**).

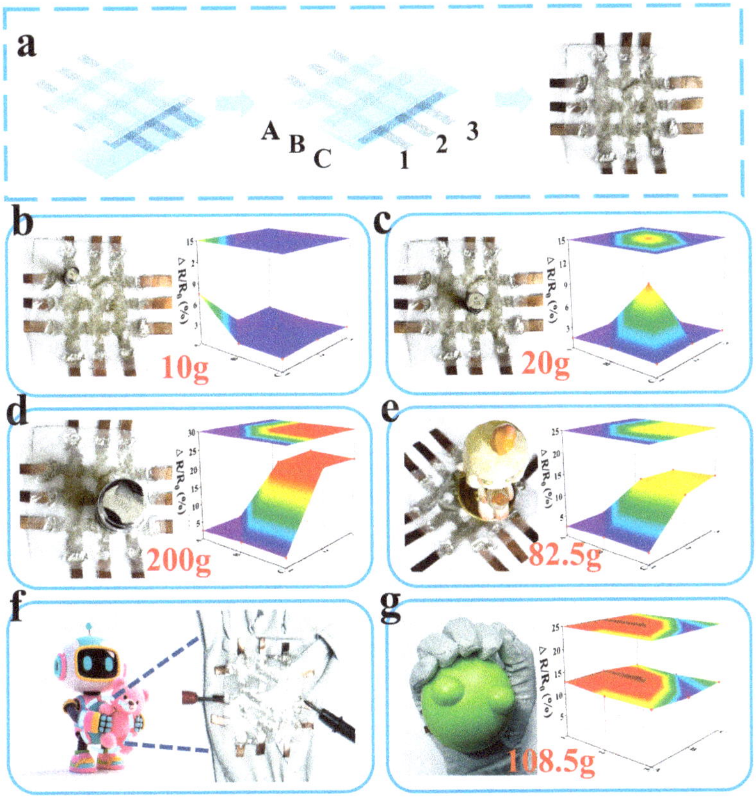

Figure 8. Schematic and physical drawings of the array sensors (**a**); electrical signals reflected on the array by different weights: 10 g (**b**), 20 g (**c**), 200 g (**d**); electrical signals generated by pressure from the dolls (**e**), sewing the sensors to the gloves to simulate movement (**f**); electrical signals generated by grasping the objects (**g**).

3.9. Field of Human–Machine Interaction

AgNW-TPU flexible sensors have great advantages in human–computer interaction, where the flexible thin-film sensors can be attached to the robotic arm for swinging, and it can be seen that the bending and recovery phases show two peaks (Figure 9d). This may be due to the reduction in the conductive pathways within the sensor during bending, resulting in an increase in the relative resistance of the sensor. As the elbow joint of the robot arm recovers, the resistance of the sensor also recovers. Flexible sensors were attached to the human knuckle for 45° and 90° bending for mechanical finger simulation, which produced two electrical signals of different heights, reflecting good stability and fast response (Figure 9b). The flexible sensors were also attached to human wrist joints for both grasping and swinging actions, which also demonstrated their detectability and sensitivity (Figure 9a). To ensure the rigor of the experiment, we attached the sensor to the arm of the mechanical puppet and let the mechanical puppet perform such simple actions as rock, paper, and scissors using the manipulation command; it can be seen from Figure 9c that the sensor can still output stable and recognizable electrical signals to the machine. A team of volunteers was recruited to interpret the observed phenomena by solely monitoring the electrical signals. Subsequent to the repetition of the experimental procedure, the results indicated that the sensor demonstrated an electrical signal error of less than 1%.

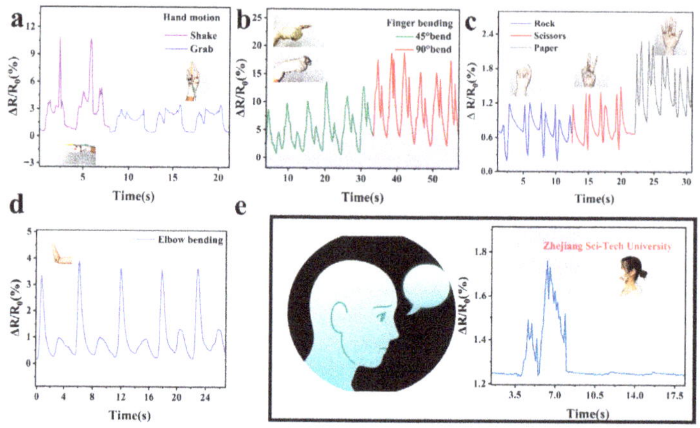

Figure 9. Electrical signals from simulated mechanical grasping and swinging motions (**a**). Finger bends at 45° and 90° (**b**). Robot plays rock, paper, and scissors (**c**). Elbow bending (**d**). Recognizing language through the movement of the masseter muscle, which can be used in healthcare (**e**).

Therefore, flexible AgNW-TPU sensors have a great advantage in human–computer interaction. In addition, the flexible sensor can output electrical signals through the tiny movement of the masseter muscle when it is attached to the face (Figure 9e). For example, when a volunteer says "Zhejiang Sci-tech University", the electrical signal quickly outputs a series of folded lines, indicating that the flexible sensor can also be used for applications involving aphasia patients or crypto-language, among others.

4. Conclusions

In this study, the sensors were constructed with thermoplastic polyurethane (TPU) as the substrate and AgNW as the conductive layer, arranged in a "sandwich" structure. TPU films were fabricated using the electrospinning method. AgNWs with an aspect ratio of up to 1921 were prepared via the alcohol reduction method. An AgNW-TPU flexible thin-film sensor was subsequently developed using the drop-casting technique. The sensor exhibited a low initial resistance (26.7 Ω), a high strain range (0–150%), high sensitivity (GF up to 19.29), a fast response and recovery time (112 ms), excellent cycling durability (more than 3000 cycles), and excellent conductivity (428 S/cm). This sensor can be integrated with textiles for monitoring human health and motion as well as applied in human–computer interaction and encrypted communication scenarios. This work provides new directions for the application of AgNW-TPU flexible thin-film sensors.

Author Contributions: Conceptualization, J.Q. and J.S.; methodology, J.Q.; software, L.L. and M.W.; validation, J.X., M.W. and L.L.; formal analysis, J.S.; investigation, J.S.; resources, J.Q. and J.S.; data curation, L.F. and J.S.; writing—original draft preparation, J.S., J.X., M.W. and L.L.; writing—review and editing, J.S.; visualization, L.F.; supervision, J.Q. and L.F.; project administration, J.S.; funding acquisition, J.Q. All authors have read and agreed to the published version of the manuscript.

Funding: This research was funded by Zhejiang University Student Science and Technology Innovation Activity Plan, grant number 2024R406A008 And The APC was funded by Zhejiang Sci-Tech University.

Data Availability Statement: The data belongs to an ongoing research project. Due to data storage issues, the data presented in this article is not readily available at this time.

Conflicts of Interest: The authors declare no conflicts of interest.

References

1. Han, S.T.; Peng, H.; Sun, Q.; Venkatesh, S.; Chung, K.S.; Lau, S.C.; Zhou, Y.; Roy, V.A.L. An overview of the development of flexible sensors. *Adv. Mater.* **2017**, *29*, 1700375.
2. Li, D.; Fan, Y.; Han, G.; Guo, Z. Superomniphobic Silk Fibroin/Ag Nanowires Membrane for Flexible and Transparent Electronic Sensor. *ACS Appl. Mater.* **2020**, *12*, 10039–10049. [CrossRef]
3. Zheng, Q.; Lee, J.; Shen, X.; Chen, X.; Kim. Graphene-based wearable piezoresistive physical sensors. *Mater. Today* **2020**, *36*, 158–179. [CrossRef]
4. Huang, C.Y.; Chiu, C.W. Facile fabrication of a stretchable and flexible nanofiber carbon film-sensing electrode by electrospinning and its application in smart clothing for ECG and EMG monitoring. *ACS Appl. Electron. Mater.* **2021**, *3*, 676–686. [CrossRef]
5. Wang, Y.; Hao, J.; Huang, Z.; Zheng, G.; Dai, K.; Liu, C.; Shen, C. Flexible electrically resistive-type strain sensors based on reduced graphene oxide-decorated electrospun polymer fibrous mats for human motion monitoring. *Carbon* **2018**, *126*, 360–371. [CrossRef]
6. Teel, B.C.-K.; Chortos, A.; Berndt, A.; Nguyen, A.K.; Tom, A.; McGuire, A.; Lin, Z.C.; Tien, K.; Bae, W.-G.; Wang, H. A skin-inspired organic digital mechanoreceptor. *Science* **2015**, *350*, 313–316. [CrossRef]
7. Di, C.A.; Zhang, F.; Zhu, D. Multi-functional integration of organic field-effect transistors (OFETs): Advances and perspectives. *Adv. Mater.* **2013**, *25*, 313–330. [CrossRef]
8. Schwartz, G.; Tee, B.C.K.; Mei, J.; Appleton, A.L.; Kim, D.H.; Wang, H.; Bao, Z. Flexible polymer transistors with high pressure sensitivity for application in electronic skin and health monitoring. *Nat. Commun.* **2013**, *4*, 1859. [CrossRef]
9. Yao, Y.; Dai, H.; Ji, M.; Han, Y.; Jiang, B.; Cheng, C.; Song, X.; Song, Y.; Wu, G. Flexible Strain Sensor Based on Ag-NWs/MXene/SEBS with High Sensitivity and Wide Strain Range. *Electron. Mater. Lett.* **2024**, *20*, 684–693. [CrossRef]
10. Sun, Y.; Mayers, B.; Herricks, T.; Xia, Y. Polyol synthesis of uniform silver nanowires: A plausible growth mechanism and the supporting evidence. *Nano Lett.* **2003**, *3*, 955–960. [CrossRef]
11. Shi, L.; Zhang, N.; Li, L.; Ding, S.; Li, W.; Li, K. Printed flexible temperature sensors: Principles, materials, processes, and applications. *Appl. Mater. Today* **2025**, *43*, 102665. [CrossRef]
12. Yang, Y.; Shen, H.; Yang, Z.; Yang, J.; Wang, Z.; Gao, K. Highly flexible and sensitive wearable strain and pressure sensor based on porous graphene paper for human motion. *J. Mater. Sci. Mater. Electron.* **2022**, *33*, 17637–17648. [CrossRef]
13. Zhu, M.; Yu, H.; Tang, F.; Li, Y.; Liu, Y.; Yao, J. Robust Natural Biomaterial Based Flexible Artificial Skin Sensor with High Transparency and Multiple Signals Capture. *Chem. Eng. J.* **2020**, *394*, 124855. [CrossRef]
14. Tian, H.; Shu, Y.; Cui, Y.L.; Mi, W.T.; Yang, Y.; Xie, D. TL Ren Scalable fabrication of high-performance and flexible graphene strain sensors. *Nanoscale* **2014**, *6*, 699–705. [CrossRef]
15. Wang, H.; Cheng, C.; Zhang, L.; Liu, H.; Zhao, Y.; Guo, Y.; Hu, W.; Yu, G.; Liu, Y. Inkjet printing short-channel polymer transistors with high-performance and ultrahigh photoresponsivity. *Adv. Mater.* **2014**, *26*, 4683–4689. [CrossRef]
16. Yan, B.; Guo, Y.; Cheng, S.; Dou, L.; Yang, Y.; Guo, X.; Chen, Y.; Cai, W.; Zhang, Y.; Liu, Z. Ag nanowires boost graphene aerogel as anode for charge-transfer in nonclassical electroactive microbial fuel cells. *Fuel* **2025**, *390*, 134650. [CrossRef]
17. Yang, Z.; Guo, Y.; Guo, W.; Zhao, M.; Wang, H.; Wei, B., Miao, Y.; Guo, K. MXene/AgNWs/MXene Sandwich-Structured Transparent Electrode for High-Performance Flexible OLED. *Small* **2025**, *21*, 2409621. [CrossRef] [PubMed]
18. Wu, M.; Li, Y.; An, N.; Sun, J. Applied voltage and near-infrared light enable healing of superhydrophobicity loss caused by severe scratches in conductive superhydrophobic films. *Adv. Funct. Mater.* **2016**, *26*, 6777–6784. [CrossRef]
19. Guo, Y. Wearable sensors to monitor plant health. *Nat. Food* **2023**, *4*, 350. [CrossRef]
20. Yang, R.; Zhang, Q.; Du, T.; Xu, Q.; Li, C.; Wu, Z. Development of AgNWs/PDMS flexible strain sensor. *Synth. Technol. Appl.* **2021**, *36*, 17–22.
21. Liu, Z.; Qu, L.; Li, X.; Jiang, X.; Yu, L. PAN/β-CD/MnO_2 composite fibers for highly efficient adsorption of heavy metals and absorption of electromagnetic waves. *Sep. Purif. Technol.* **2025**, *363*, 132070. [CrossRef]
22. Luo, J.; Wang, Y. Effect of hierarchical layer structure design on the sensing performances of $NaNbO_3$/P(VDF-TrFE) piezoelectric composite films. *J. Alloys Compd.* **2025**, *1017*, 179071. [CrossRef]
23. Wang, J.; Zhao, C.; Cao, C.; Liu, M.; Liu, Z.; Zhou, P.; Wang, G.; Zhang, T.; Zhang, L.; Qi, Y. Boosting Sensing Performance of Flexible Piezoelectric Pressure Sensors by Sb Nanosheets and BaTiO3 Nanoparticles Co-Doping in P (VDF-TrFE) Nanofibers Mat. *Adv. Energy Mater.* **2024**, *10*, 2300718. [CrossRef]
24. Cai, J.; Huang, S.; Zhou, W.; He, X.; Liang, H.; Zhou, J.; Zhao, C.; Xu, T. A stretchable and self-healing poly(vinyl alcohol)/polyacrylamide/polyaniline multi-network hydrogel for flexible strain sensors. *Mater. Today Commun.* **2025**, *44*, 111976. [CrossRef]
25. Zhao, S.; Wang, Q.; Qu, Z. Controlled Synthesis of Uniform Silver Nanowires via a Simple Polyol Process. In *IOP Conference Series: Materials Science and Engineering*; IOP Publishing: Bristol, UK, 2018; Volume 301.
26. Sun, Y.; Xia, Y. Shape-Controlled Synthesis of Gold and Silver Nanoparticles. *Science* **2002**, *298*, 2176–2179. [CrossRef] [PubMed]

27. Nikoobakht, B.; El-Sayed, M.A. Preparation and growth mechanism of gold nanorods (NRs) using seed-mediated growth method. *Chem. Mater.* **2003**, *15*, 1957–1962. [CrossRef]
28. Lifshitz, I.M.; Slyozov, V.V. Kinetics of precipitation from supersaturated solid solutions. *Phys. Chem. Solids* **1961**, *19*, 35–50. [CrossRef]
29. Wiley, B.; Sun, Y.; Mayers, B.; Xia, W. Cover Picture: Shape-Controlled Synthesis of Metal Nanostructures: The Case of Silver (Chem. Eur. J. 2/2005). *Chem. Eur. J.* **2004**, *11*, 431. [CrossRef]
30. Murph, S.E.H.; Murphy, C.J.; Leach, A.; Gall, K. A Possible Oriented Attachment Growth Mechanism for Silver Nanowire Formation. *Cryst. Growth Des.* **2015**, *15*, 1968–1974. [CrossRef]
31. Amjadi, M.; Pichitpajongkit, A.; Lee, S.; Ryu, S.; Park, I. Highly Stretchable and Sensitive Strain Sensor Based on Silver Nanowire-Elastomer Nanocomposite. *ACS Nano* **2014**, *8*, 5154–5163. [CrossRef]
32. Wu, L.; Wang, L.; Guo, Z.; Luo, J.; Xue, H.; Gao, J. Durable and Multifunctional Superhydrophobic Coatings with Excellent Joule Heating and Electromagnetic Interference Shielding Performance for Flexible Sensing Electronics. *ACS Appl. Mater. Interfaces* **2019**, *11*, 34338–34347. [CrossRef] [PubMed]
33. Luo, J.; Huo, L.; Wang, L.; Huang, X.; Li, J.; Guo, Z.; Gao, Q.; Hu, M.; Xue, H. Superhydrophobic and multi-responsive fabric composite with excellent electro-photo-thermal effect and electromagnetic interference shielding performance. *Chem. Eng. J.* **2020**, *391*, 123537. [CrossRef]
34. Peng, Y.; Yan, B.; Li, Y.; Lan, J.; Shi, L.; Ran, R. Antifreeze and moisturizing high conductivity PEDOT/PVA hydrogels for wearable motion sensor. *J. Mater. Sci.* **2020**, *55*, 1280–1291. [CrossRef]
35. Hong, X.; Zhao, W.; Yu, R.; Wang, Q.; Zeng, F.; Tao, Y.; Jin, Z.; Zhu, C. Multifunctional silver nanowire coated fabric capable of electrothermal, resistance temperature-sensitivity, electromagnetic interference shielding, and strain sensing. *J. Ind. Text.* **2022**, *51* (Suppl. 4), 6153S–6172S. [CrossRef]
36. Zhu, G.J.; Ren, P.G.; Guo, H.; Jin, Y.L.; Yan, D.X. Highly Sensitive and Stretchable Polyurethane Fiber Strain Sensors with Embedded Silver Nanowires. *ACS Appl. Mater. Interfaces* **2019**, *11*, 23649–23658. [CrossRef]
37. Ouyang, Z.; Li, S.; Liu, J.; Yu, H.-Y.; Peng, L.; Zheng, S.; Xu, D.; Tam, K.C. Bottom-up reconstruction of smart textiles with hierarchical structures to assemble versatile wearable devices for multiple signals monitoring. *Nano Energy* **2022**, *104*, 107963. [CrossRef]
38. Xue, C.; Sun, H.; Tang, Y.; Wei, J.; Guo, J. Self-Powered Photonic Sensor Using Cholesteric Liquid Crystal Elastomer-Based Sandwiched Architecture with Mechanochromism and Triboelectric Response. *ACS Appl. Mater. Interfaces* **2025**, *17*, 46226–46237. [CrossRef]
39. Lu, Z.; Wang, J.; Song, J.; Yang, Z.; Hammad, F.A.; Tian, Y. Microcrack-engineered flexible strain sensors enabled by carbon black/graphene/conductive carbon paste ternary nanocomposites for enhanced sensitivity. *J. Mater. Sci. Mater. Electron.* **2025**, *36*, 1315. [CrossRef]
40. Li, Q.; Li, Y.; Ding, L.; Li, X.; Ma, J.; Minami, T.; Sang, S. Structural optimization of field-effect modulated P3HT OTFT flexible pressure sensors for high sensitivity and rapid response. *Microchem. J.* **2025**, *215*, 114484. [CrossRef]
41. Zhan, Y.; Poisson, J.; Meng, X.; Wang, Z.; Chen, L.; Wu, T.H.; Zhang, K. Electrospun Lignin/ZnO Nanofibrous Membranes for Self-Powered Ultrasensitive Flexible Airflow Sensor and Wearable Device. *Adv. Mater.* **2025**, e2502211. [CrossRef] [PubMed]
42. Ke, L.; Shen, X.; Xiao, F.; Zhang, Y.; Zhang, Q.; Jiang, H.; Jin, S.; Huang, Y. High-performance dual skin temperature-sweat pH sensor based on ionic conductive hydrogel. *Sens. Actuators A Phys.* **2025**, *393*, 116729. [CrossRef]
43. Eun, R.; Byeong-Ung, H.; Doil, K.; Bo-Yeong, K.; Nae-Eung, L. Stretchable, Transparent, Ultrasensitive, and Patchable Strain Sensor for Human-Machine Interfaces Comprising a Nanohybrid of Carbon Nanotubes and Conductive Elastomers. *ACS Nano* **2015**, *9*, 6252–6261.
44. Wakasugi, K.; Maruo, Y. A Sensor Network Design Consideration Based on Percolation Theory. In *B—Abstracts of IEICE TRANSACTIONS on Communications (Japanese Edition)*; IEICE: Minato-ku, Japan, 2011; Volume J94-B, pp. 61–64.
45. Zhang, J.; Wang, M.; Yang, Z.; Zhang, X. Highly flexible and stretchable strain sensors based on conductive whisker carbon nanotube films. *Carbon* **2021**, *176*, 139–147. [CrossRef]
46. Shi, J.; Wang, S.; Wang, H.; Gu, J. Mechanically Tough and Highly Stretchable Hydrogels Based on Polyurethane for Sensitive Strain Sensor. *Polymers* **2023**, *15*, 3902. [CrossRef] [PubMed]

sensors

MDPI

Communication

Impact of Microstructure on Sensing Performance of Fiber-Based Iontronic Pressure Sensors: A Comparative Study

Cheng Liu [1,†], Jiaxin Xu [1,2,†], Shiman Yang [1], Yihan Xu [1], Jianyu Wang [1], Xiaoqing Liu [1,*], Li Wang [1] and Yichun Ding [1,3,4,*]

[1] School of Physics and Materials Science, Nanchang University, Nanchang 330031, China; chengliu02@163.com (C.L.); xujiaxin@fjirsm.ac.cn (J.X.); shimanyang1@163.com (S.Y.); 5714123050@email.ncu.edu.cn (Y.X.); jywang@ncu.edu.cn (J.W.); liwang@ncu.edu.cn (L.W.)
[2] Fujian Institute of Research on the Structure of Matter, Chinese Academy of Sciences, Fuzhou 350002, China
[3] Jiangxi Provincial Key Laboratory of Photodetectors, Nanchang University, Nanchang 330031, China
[4] Research Center for Chip Design, Nanchang University, Nanchang 330031, China
* Correspondence: liuxiaoqing@ncu.edu.cn (X.L.); yichun.ding@ncu.edu.cn (Y.D.)
† These authors contributed equally to this work.

Abstract

Highly sensitive flexible pressure sensors are crucial for wearable health monitoring and human–machine interaction. While the emerging iontronic sensors inherently offer high sensitivity, this can be further improved by engineering microstructured interfaces. In this study, we employ four different types of common fiber materials as substrates for fabricating ionic dielectric layers by a simple impregnation of ionic liquid (IL). A comparative study reveals that the porosity and microstructural architecture (e.g., fiber diameter) of the substrate material directly influences the amount of adsorbed IL and consequent sensing performance. We achieved the highest sensitivity by using a thin electrospun TPU/IL nanofiber mat (33 μm), which exhibited high sensitivities of 3.10 kPa^{-1}, 1.85 kPa^{-1}, and 1.02 kPa^{-1} in the pressure ranges of 0–200 kPa, 200–400 kPa, and 400–700 kPa, respectively. Furthermore, the sensor exhibited an excellent fast response (2.71 ms) and recovery time (8.71 ms), along with outstanding cyclic stability. This work provides valuable guidance for selecting and utilizing common fiber materials to develop high-sensitivity iontronic pressure sensors, paving the way for their practical application in next-generation wearable electronics.

Keywords: iontronic pressure sensors; ionic liquids; fiber materials; porosity; microstructure; wearable sensing

1. Introduction

With the rapid advancement of embodied artificial intelligence, robotics, and the Internet of Things, the demand for tactile sensing technology has grown dramatically. Pressure sensors, as key components for detecting external force, are widely used in numerous fields [1–6]. However, the mainstream pressure sensors based on conventional silicon-based materials and microfabrication techniques can hardly withstand large deformations and conform to curved surfaces due to their inherent rigid characteristics [7,8]. This not only causes signal inaccuracies but also greatly limits their application ranges. In contrast, flexible pressure sensors utilizing polymer substrates such as polydimethylsiloxane (PDMS) and Ecoflex [9,10] offer excellent flexibility, stretchability, and lightweight properties. These properties are essential for emerging applications in robotic tactile sensing, electronic skin, and wearable health monitoring [11–14].

Pressure sensors generally operate based on four mechanisms: piezoresistive, piezoelectric, triboelectric, and capacitive [15–18]. Among them, capacitive pressure sensors are widely used due to their advantages of low power consumption, both static and dynamic pressure monitoring capability, and simple structure [19]. Conventional capacitive pressure sensors are constructed by sandwiching a dielectric layer between two electrodes. According to the equation of capacitance, $C = \varepsilon A/d$ (where C, ε, A, and d is capacitance, dielectric constant, electrode contact area, and thickness of dielectric layer, respectively), researchers optimized the sensing performance of capacitive pressure sensors typically by manipulating the three key variables: (1) by modulating the dielectric constant of dielectric layer by the introduction of conductive fillers such as carbon nanotubes [20,21] and silver nanowires [22,23]; (2) by optimizing the compressive deformation through the use of a dielectric layer with low elastic modulus [24]; and (3) by constructing microstructured interfaces such as pyramids [25–27], porous hierarchies [28,29], and micropillar arrays [30,31] on the surface of dielectric layer. While these strategies can significantly enhance performance, the sensitivity of dielectric-based capacitive pressure sensors often remains insufficient for detecting subtle physiological signals, which is critical for health monitoring applications.

Recently, a new type of capacitive pressure sensor, termed as iontronic capacitive pressure sensor, was developed based on an ionic conduction mechanism [32,33]. Unlike conventional capacitive pressure sensors that rely on electronic conduction, iontronic pressure sensors utilize ion migration to form an electric double layer (EDL) at the electrode/electrolyte interface. This EDL mechanism generates extremely high capacitance, enabling ultra-high sensitivity. The iontronic pressure sensors are typically fabricated by replacing the conventional dielectric layer with an ionic material, such as a film incorporated with ionic liquid (IL) or a polyelectrolyte film. ILs are usually encapsulated within a polymer/gel matrix or absorbed into a porous substrate. Based on the EDL mechanism, increasing the concentration of IL can effectively enhance the capacitance of the interfacial bilayer. However, an excess of IL can lead to negative effects, including leakage [34,35] and a significant increase in the gel viscosity [36]. Similarly to dielectric-based capacitive sensors, the sensitivity of iontronic pressure sensors can also be significantly enhanced by constructing microstructures to regulate the interfacial contact area. For example, Niu et al. [37] constructed pyramidal microarrays on the surface of an ionic layer and achieved a sensitivity of up to 655.3 kPa^{-1}. Other microstructures include sandpaper textures [38–40], multistage microcolumns [41], pleats [42], and interlocks [43], all of which have been shown to greatly improve sensing performance. In particular, Guo's team [44] developed a reduced-order inverse design framework that theoretically simulates the required microstructure parameters to meet a target performance, offering a powerful tool for the on-demand customization of sensors. However, producing these microstructures typically requires complex and costly processes like photolithography, multi-step transfer printing, or high-precision 3D printing [45,46], which hinder their potential for mass production.

Fiber materials [47,48] such as nonwovens and textiles are ideal for making iontronic pressure sensors due to their innate porous structure that can load significant amounts of IL, serving as an effective ionic layer. In addition, the inherent microstructures of their constituent fibers can significantly enhance the sensitivity. Furthermore, their natural porosity enables the development of permeable electronics to enhance the comfortability. In this study, we present a comparative analysis of four types of fiber materials, including (a) woven fabric, (b) cotton nonwoven fabric, (c) viscose spunlace fabric, and (d) electrospun nanofiber mats, for fabricating iontronic pressure sensors by impregnation of IL. Their distinct structural properties (e.g., fiber diameter, arrangement, porosity, and thickness) led to significant differences in IL absorption and subsequent sensing performance. Results demonstrate that the electrospun nanofiber mat, characterized by small fiber diameter, high porosity, and low thickness, synergistically enhanced both IL loading capacity and reversible deformation. This resulted in a sensor with exceptional sensitivity, stability, and a wide detection range. The optimized sensor achieved high sensitivities of 3.10, 1.85, and 1.02 kPa^{-1} within the 0–200, 200–400, and 400–700 kPa ranges, respectively, alongside rapid response and recovery times of 2.71 ms and 8.13 ms. Owing to this comprehensive performance, the sensor was successfully deployed for wearable physiological monitoring, accurately capturing diverse physiological signals. This provides a reliable technological platform for applications in health monitoring and human–machine interaction.

2. Materials and Methods

Materials: Ionic liquid (1-ethyl-3-methylimidazolium bis(trifluoromethyl sulfonyl) imide ([EMIM][TFSI])) (99%) was obtained from Shanghai Macklin Bio-Chem Technology Co., Ltd., Shanghai, China. Anhydrous ethanol was purchased from Xilong Science Co., Ltd., Shantou, China. Cotton nonwoven fabric was acquired from Suzhou Oude Dust-free Materials Co., Ltd., Suzhou, China. Viscose spunlace fabric (100% viscose) was acquired from Jiangxi Xinwang Sanitary Products Co., Ltd., Shangrao, China. Electrospun thermoplastic polyurethane (TPU) nanofiber mats (33 μm, 43 μm, and 201 μm) were purchased from Jiangxi Xiancai Nanofiber Technology Co., Ltd., Nanchang, China. The woven fabric is sourced from discarded clothing.

Preparation of fiber-based ionic layer: The fiber substrates were first cleaned by ultrasonication in anhydrous ethanol for 1 h to remove impurities. Subsequently, they were cut into small pieces, and immersed in a mixture of [EMIM][TFSI]/ethanol (1:9) for 1 h to allow for IL absorption. Finally, the IL-infused fiber materials were dried at 70 °C to evaporate any residual ethanol.

Fabrication of pressure sensor: The pressure sensors were constructed in a sandwich architecture. PET/Au films, prepared by successive evaporation of Cr (5 nm) and Au (100 nm) layers onto a 20 μm PET film via thermal evaporation, were used as the flexible electrodes. The fiber-based ionic layer (1×1 cm^2) was sandwiched between two PET/Au electrodes. Silver wires were attached to the electrodes using silver paste to serve as external electrical connections. Finally, the entire assembly was sealed with 3M tape to ensure mechanical stability during testing. The structural diagram of the sensor is shown in Figure S1.

Characterization and measurement: The microstructure and elemental composition of the fiber-based ionic layers were characterized using a high-resolution field emission scanning electron microscope (FESEM, ZEISS/GeminiSEM 300, Carl Zeiss AG Co., Ltd., Oberkochen, Germany) equipped with an energy dispersive X-ray spectroscopy (EDS). Fourier-transform infrared (FT-IR) spectroscopy of ionic layer was performed on a Bruker spectrometer (Thermo Nicolet iS5, Bruker Tensor II, Berlin, Germany). The capacitance of the iontronic pressure sensors was measured at 5 kHz using a precision LCR meter (TH2840B, Changzhou Tonghui Electronics Co., Ltd., Changzhou, China). For pressure sensing test, mechanical compression was administered using a computer-controlled stage integrated with a force gauge (XLD-20E, Jingkong Mechanical Testing Co., Ltd., Guangzhou, China). All measurements were conducted on sensors with a size of 7 mm × 7 mm.

3. Results and Discussion

3.1. Preparation and Characterization of Fiber-Based Ionic Layers

We rationally selected four representative fiber-based substrates for preparing the ionic layer of iontronic pressure sensors: woven fabric, nonwoven fabric, spunlace nonwoven fabric, and electrospun nanofiber mat. The thicknesses of woven fabric, cotton nonwoven fabric, and viscose spunlace fabric are 410 μm, 243 μm, and 670 μm (with a semicircular surface thickness of 427 μm), respectively. The first three substrates are composed of conventional natural or synthetic fibers with diameters typically ranging from a few to tens of micrometers, but they are assembled using different techniques: weaving, hot pressing, and hydroentanglement (spunlacing), respectively. In contrast, the electrospun nanofiber mat is a specialized nonwoven material produced by the electrospinning technique, characterized by fiber diameters on the scale of hundreds of nanometers. Figure 1 presents optical and corresponding SEM images of the four selected fiber substrates. As shown in Figure 1a, the woven fabric consists of precisely interwoven twisted yarns, creating a regular macroscopic texture. The cotton nonwoven fabric is composed of randomly overlaid cotton fibers (Figure 1b). Its surface features periodic concave micro-patterns formed by the hot-pressing process, and the individual fibers appear irregular. The viscose spunlace nonwoven fabric is produced by entangling viscose rayon fibers with high-pressure water jets, resulting in a soft, strong, and absorbent material without binders. Its surface shows a regular array of microsphere bumps, with a highly hollow and porous structure (Figure 1c). The electrospun TPU nanofiber mat consists of randomly overlaid nanofibers (Figure 1d). These nanofibers are created by applying a high voltage to a polymer solution jet, which stretches and accelerates the jet to form fine fibers. This nanoscale morphology endows the mat with a high specific surface area and porosity.

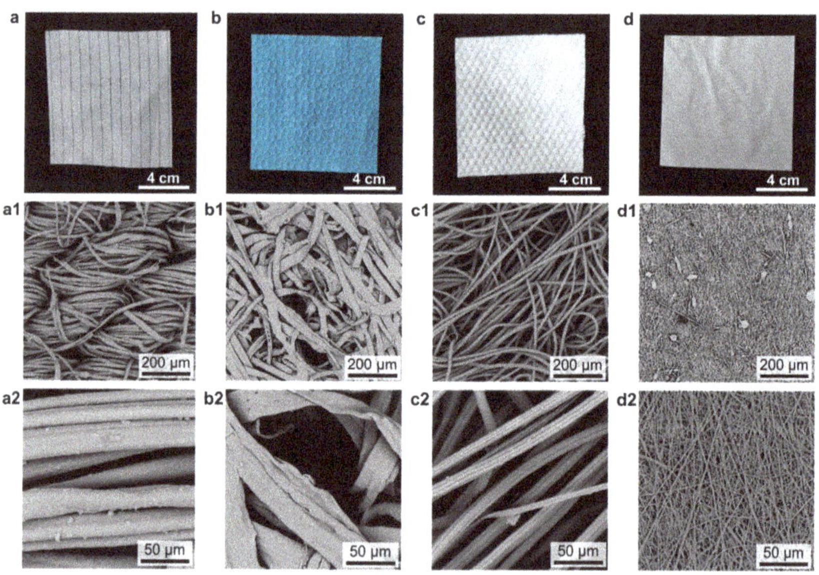

Figure 1. Optical images of the four types of fiber substrates: (**a**) Woven fabric, (**b**) Cotton nonwoven fabric, (**c**) Viscose spunlace fabric, and (**d**) Electrospun TPU nanofiber mat. (**a1**,**a2**) SEM images of woven fabric; (**b1**,**b2**) SEM images of cotton nonwoven fabric; (**c1**,**c2**) SEM images of viscose spunlace fabric; (**d1**,**d2**) SEM images of electrospun TPU nanofiber mat.

Owing to the porous structure of the fiber substrates, the ionic layer can be straightforwardly prepared via simple impregnation of IL. This process involved immersing the materials in a mixed [EMIM][TFSI]/ethanol solution followed by drying, resulting in a homogeneous IL coating. As can be seen in the SEM images (Figure 2a–d), an ultrathin IL film can be identified on the fiber surfaces. The IL absorption capacity varied significantly across the different materials (Figure S2). The woven fabric and cotton nonwoven fabric exhibited low IL adsorption (~15%), attributable to the woven fabric's dense interwoven structure and the nonwoven's hot-rolled fiber bonding. In contrast, the highly porous viscose spunlace fabric and electrospun nanofiber mat showed substantially higher IL loading. The loading on the TPU nanofiber mat increased as its thickness decreased, with a maximum loading of ~57% achieved on a 33 μm thick mat. This is primarily due to the thinner nanofiber mat providing a shorter diffusion path and higher specific surface area, which synergistically increase the IL loading capacity.

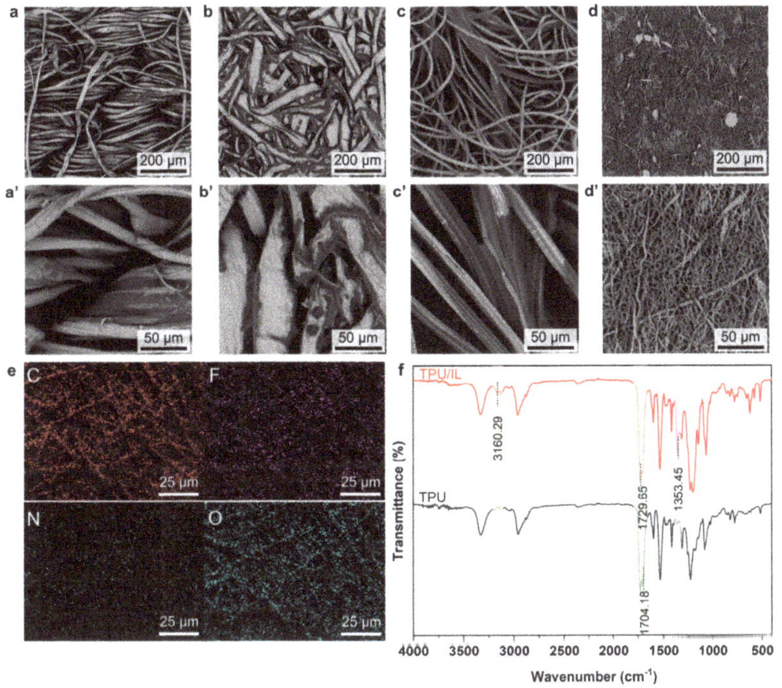

Figure 2. (**a**–**d**) SEM images of the fiber substrate absorbed with IL: (**a**) Woven fabric/IL, (**b**) Cotton nonwoven fabric/IL, (**c**) Viscose spunlace fabric/IL, and (**d**) Electrospun TPU nanofiber mat/IL. (**a′**–**d′**) Zoom-in SEM images of the fiber substrate absorbed with IL: (**a′**) Woven fabric/IL, (**b′**) Cotton nonwoven fabric/IL, (**c′**) Viscose spunlace fabric/IL, and (**d′**) Electrospun TPU nanofiber mat/IL. (**e**) SEM-EDS elemental mapping images of the TPU/IL. (**f**) FT−IR spectra of TPU and TPU/IL.

The successful coating of IL was confirmed through energy dispersive X-ray spectroscopy (EDS) and Fourier transform infrared (FT-IR) spectroscopy. For the TPU/IL nanofiber mat, EDS elemental mapping (Figure 2e) detected fluorine (F)—a key element from $[TFSI]^-$—in addition to the carbon (C), nitrogen (N), and oxygen (O) inherent to the TPU polymer. The FT-IR spectra provided further evidence: new characteristic peaks emerged at 3160 cm^{-1} and 1353 cm^{-1}, corresponding to the C-H stretch on the $[EMIM]^+$ imidazole ring and the CF_3 group in $[TFSI]^-$, respectively (Figure 2f). Crucially, the original C=O stretching vibration peaks of TPU at ~1730 cm^{-1} remained unchanged, indicating a stable, non-destructive integration of the IL with the TPU substrate [49,50]. Consistent results confirming successful IL loading were also observed for the other fiber substrates (Figures S3–S6). Importantly, despite the IL impregnation, the highly porous nature of all fiber materials was preserved, granting the composites excellent gas permeability. This was demonstrated by a water evaporation rate equivalent to that of an open container over a five-day period (Figure S7).

3.2. Performance of Fiber-Based Iontronic Pressure Sensors

The operating principle of iontronic pressure sensor differs from that of conventional capacitive sensors. While conventional sensors rely on transient electronic polarization, the iontronic pressure sensor leverages the formation of an EDL on the surface of the electrodes. The large capacitive contribution from the EDL constitutes the main source of capacitive signal. Consequently, the total capacitance of the sensor (C_{EDL}) can be expressed as [51,52]:

$$C_{EDL} = \left(\frac{1}{C_H} + \frac{1}{C_D}\right)^{-1} = \eta_A \cdot \varphi(d, \varepsilon, C, \phi, T) \cdot A = \text{UAC} \cdot A$$

where C_H and C_D represent the capacitance of the bilayer at the interface of the upper and lower electrodes, respectively; A denotes the contact area between the ionic layer and the electrode, and φ (d, ε, C, φ, T) is a composite parameter affected by the thickness of the Helmholtz layer (d), the dielectric constant (ε), the ionic species and concentration (C), the surface potential (φ), and the temperature T. When pressure is applied, the dielectric layer is compressed, resulting in an increase in its contact area with the electrode, and at the same time, the close contact decreases the effective thickness of the ionic layer, which significantly increases the interfacial capacitance and ultimately causes a change in the total capacitance. The sensor sensitivity (S) is defined as $S = (\Delta C/C_0)/\Delta P$, where ΔC is the capacitance change, C_0 is the initial capacitance, and P is the applied pressure. Based on the ion-mediated mechanism, even minimal pressure variations can trigger a substantial capacitive response, resulting in high sensitivity. Additionally, as pressure is applied, the effective permittivity of the dielectric layer also changes. Based on the Maxwell-Garnett equivalent medium theory model, we can derive the ε_{eff} of the dielectric layer:

$$\frac{\varepsilon_{eff} - \varepsilon_{fiber}}{\varepsilon_{eff} + 2\varepsilon_{fiber}} = f_{air}\frac{\varepsilon_{air} - \varepsilon_{fiber}}{\varepsilon_{air} + 2\varepsilon_{fiber}} + f_{IL}\frac{\varepsilon_{IL} - \varepsilon_{fiber}}{\varepsilon_{IL} + 2\varepsilon_{fiber}}$$

where ε_{eff}, ε_{fiber}, ε_{IL}, and ε_{air} represent the relative permittivity of the equivalent medium, fiber, interlayer (IL), and air, respectively, and f_{air} and f_{IL} denote the volume fractions of air and IL within the dielectric layer [53]. Before pressure application, air dominates, resulting in a relatively low permittivity. Under pressure, air molecules are progressively replaced by ionic liquid, causing f_{IL} to increase significantly. This leads to a sharp rise in ε_{eff}, further enhancing the sensor's sensitivity.

According to the above discussion, the significant differences in microstructure (e.g., fiber diameter, arrangement, porosity, thickness) and IL loading content of four fiber-based ionic layers would result in much difference in their sensing performance. We evaluated this by comparing the sensitivity of pressure sensors fabricated from four different fiber-based materials. As shown in the sensitivity-pressure curves (Figure 3a), the electrospun TPU nanofiber-based pressure sensor demonstrated the highest sensitivity across pressures. This superior performance is directly attributed to its highest porosity, which facilitates greater IL adsorption, and its smallest fiber diameter, which creates a pronounced surface microstructure that enables a larger change in contact area during compression. A notable observation was that the viscose spunlace fabric, despite its significantly higher IL adsorption than the woven fabric and cotton nonwoven, exhibited slightly lower sensitivity in the low-pressure regime. We attribute this to its large-sized, regular semi-circular surface structure. Under low pressure, only these surface bumps compress, resulting in limited overall deformation of the ionic layer and a negligible change in the contact area. At higher pressures, the layer compresses effectively, achieving full contact with the electrodes. This increases the number of EDLs formed, and, consequently, the sensitivity rises.

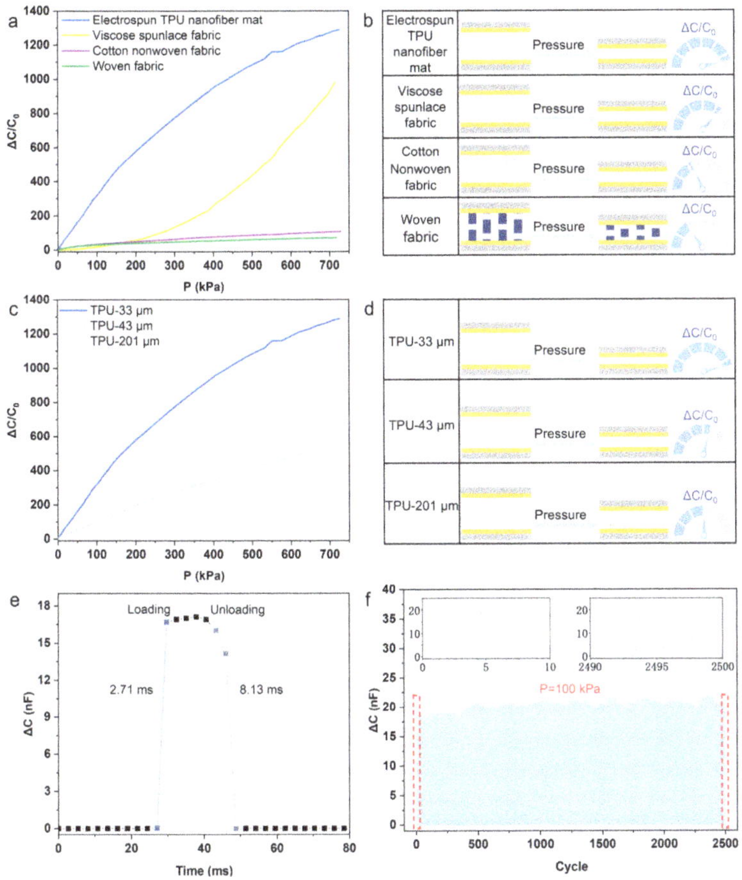

Figure 3. (**a**) Response-pressure curves of sensors with different ionic layer. (**b**) Schematics of working principle of pressure sensors with different ionic layer. (**c**) Response-pressure curves of sensors fabricated with TPU/IL with different thicknesses. (**d**) Schematics of working principle of pressure sensors fabricated with TPU/IL with different thicknesses. (**e**) Response curve of pressure to 80 kPa. (**f**) Durability test of pressure sensor under 2500 loading/unloading cycles.

The working mechanism of each sensor, correlated with material characterization results, is schematically summarized in Figure 3b. The woven fabric/IL layer exhibits low porosity and IL adsorption due to its tightly interwoven structure. Furthermore, its large-size yarns undergo minimal change in contact area under pressure, resulting in low sensitivity. In contrast, progressing from cotton nonwoven to viscose spunlace fabric and finally to TPU nanofiber mats, the fiber diameter gradually decreases while the porosity increases. This progression enhances both IL adsorption and, crucially, the magnitude of contact area change under applied pressure, leading to a corresponding increase in sensitivity.

We also investigated the influence of the thickness of the ionic layer on sensor performance. Using electrospun TPU nanofiber mat/IL of three thicknesses (33 μm, 43 μm, and 201 μm), we found that the sensor sensitivity decreases with increasing thickness (Figure 3c). The sensor fabricated with TPU nanofiber mat/IL with a thickness of 33 μm achieved sensitivities of 3.10 kPa^{-1}, 1.85 kPa^{-1}, and 1.02 kPa^{-1} in the pressure range of 0–200 kPa, 200–400 kPa, and 400–700 kPa, respectively. This relationship occurs because a thinner layer has lower flexural rigidity [54], allowing for greater compressive deformation under the same pressure. This leads to a more significant increase in the electrode contact area and, consequently, higher sensitivity (Figure 3d). To further validate the notion that thinner substrates and higher porosity can reduce flexural stiffness and enhance variations in contact area, we compared the compressive stress–strain curves of TPU fiber membranes with different thicknesses (33 μm, 43 μm, and 201 μm) (Figure S8a). The compressive modulus decreases significantly with decreasing thickness. The 33-μm film exhibits the lowest modulus, followed by the 43-μm film, while the 201-μm film is the stiffest. This implies that the low compressive modulus resulting from smaller thicknesses makes the material more prone to bending deformation under pressure. Additionally, we compared the compressive stress–strain curves of different substrates (Figure S8b). Under identical strain conditions, both the compressive modulus and stress values of the TPU fiber membrane were substantially lower than those of the textile material. Despite the textile's porous nature, its response exhibited greater rigidity compared to the TPU fiber due to the stiffness of the fiber layers and their more tightly packed arrangement. At the same time, to evaluate material repeatability and inter-device variability, we verified the reproducibility of multiple batches of sensors using TPU-33 μm as the dielectric layer (Figure S9). Overall, the sensors exhibited low repeatability error and high reliability. Additionally, we quantified the sensitivity and linearity of sensors with different dielectric layers (Tables S1 and S2). Each sample exhibited distinct performance characteristics within its corresponding pressure range, enabling a systematic evaluation of sensitivity trends and linear relationships across different materials and thicknesses.

To determine the ion transport characteristics of the electric double layer (EDL), electrochemical characterization was performed using an electrochemical workstation. Figure S10a shows the phase/Bode plots of the sensor. At low frequencies, the EDL exhibits a full response with a pronounced capacitive effect; at high frequencies, the EDL response is suppressed, with resistive components becoming prominent. Furthermore, to validate the frequency dependence of the EDL, we tested the sensor capacitance versus frequency (Figure S10b). Capacitance rapidly decays with increasing frequency, as high-frequency AC signals limit ion migration time, thereby suppressing the EDL contribution. We also examined the impact of AC amplitude/bias dependence on the EDL (Figure S10c,d). Capacitance rises to a plateau region with increasing bias voltage, facilitating accumulation of dielectric ions; further bias increases trigger Faraday reactions, disrupting this ion accumulation. For AC amplitude, capacitance tends toward stability, indicating that above a certain threshold, the AC signal no longer significantly alters the dielectric double layer structure. These trends align with typical EDL behavior, confirming that capacitance modulation originates from EDL regulation rather than geometric changes.

To summarize, optimal sensing performance was achieved by an ionic layer featuring a nanofiber-based porous architecture and reduced thickness. Therefore, a TPU/IL nanofiber mat with a thickness of 33 μm was chosen for further performance evaluation. The dynamic response speed is a vital performance criterion for pressure sensors. We tested the response capability by applying a loading/unloading force of 80 kPa to the sensor using a vibrator, and the sensor showed a response time of 2.71 ms and a recovery time of 8.13 ms (Figure 3e). Given that the response time of human skin to external stimuli is about 30–50 ms, this iontronic pressure sensor shows significant promise for applications in robotic haptics and wearable health monitoring. Figure S11 shows the detection limits of the device. Even when subjected to forces of 4.47 Pa and 3.14 Pa, the sensor maintains its ability to respond to pressure. Furthermore, long-term cyclic stability is another crucial metric. We conducted dynamic load/unload cycle tests on the sensor under varying pressures to evaluate its dynamic sensing capability. Figure S12 displays the dynamic load/unload cycle test curves for the sensor across four pressure conditions. At identical pressures, the capacitance peak variation remains stable. As pressure increases, the capacitance peak grows proportionally with pressure. The sensor's capacitance value recovers to its initial state after multiple cycles, demonstrating excellent stability and repeatability. We conducted 2500 repeated loading tests on the sensor at 100 kPa pressure. Result showed there was just a minimal signal change of 3.5% between the first and last cycles (Figure 3f), underscoring the device's reliability. Additionally, we further validated the sensor's durability under prolonged stress by subjecting it to an 80 kPa force over 10,000 cycles (Figure S13). Although the sensor capacitance exhibited a slight increase during long-term cycling, the repeatability of its dynamic response remained unaffected. Compared to recently reported capacitive pressure sensors, the device in this work demonstrates a competitive advantage, achieving a high sensitivity of 1.02 kPa^{-1} over a wide pressure range of 0–700 kPa. This performance surpasses that of most existing sensors based on the EDL sensing mechanism (Figure S14, Table S3).

3.3. Wearable Applications

Leveraging the excellent electromechanical properties, including high sensitivity, fast response, and wide detection range, the iontronic pressure sensor was applied for diverse applications, from detecting subtle physiological signals (e.g., pulse, respiration, articulation) to measuring larger bodily motions (e.g., joint flexion, compression). To demonstrate its capability in monitoring vital signs, the sensor was first attached to a subject's nasolabial fold using medical tape. As the subject breathed naturally, the weak pressure from the airflow induced periodic capacitance changes (Figure 4a). The respiratory rate, calculated from the peak-to-peak interval of the signal, was determined to be 15 breaths per minute. This ability to accurately monitor breathing patterns suggests the sensor's potential for early screening of conditions such as sleep apnea, asthma, and bronchitis.

We then applied the sensor for pulse monitoring, a critical indicator of cardiovascular health that can reveal conditions like arrhythmia and arteriosclerosis. When affixed to the wrist, the sensor recorded a heart rate of 76 beats per minute (Figure 4b), with characteristic pulse wave waveforms (P wave and T wave) clearly visible [55,56]. Furthermore, the sensor's high sensitivity enabled the detection of even weaker pulses at the fingertip, from which the corresponding characteristic waveforms were successfully extracted (Figure 4c). These results underscore the sensor's significant potential for use in disease prevention and rehabilitation assessment. It is worth noting that the pulse signal exhibits a drift, likely due to temperature effects and motion artifacts [57–59]. The observed signal drift is attributable to two primary factors: (1) ion migration from the temperature sensitivity of the ionic liquid and (2) variations in pressure and contact area caused by motion artifacts and unstable mechanical contact. Both mechanisms influence the sensor's capacitance, leading to the drift. Additionally, we have incorporated quantitative testing of sensors in wearable applications. As shown in Figure S15a, different breathing patterns yield distinct signal outputs from the sensor: deep breathing produces steady airflow, while rapid breathing accelerates air exchange frequency, reducing the interval between capacitance peaks. During breath-holding, the sensor receives no airflow-induced signals, resulting in a flatline pattern. Figure S15b,c displays pulse waveforms during different activities. Calculations indicate the test subject's pre-exercise pulse rate was 86 beats per minute. Post-exercise, accelerated blood flow due to enhanced aerobic metabolism caused a sharp pulse increase. Monitoring confirmed the subject's pulse rose to 114 beats per minute, validating the sensor's efficacy in quantitative metric testing.

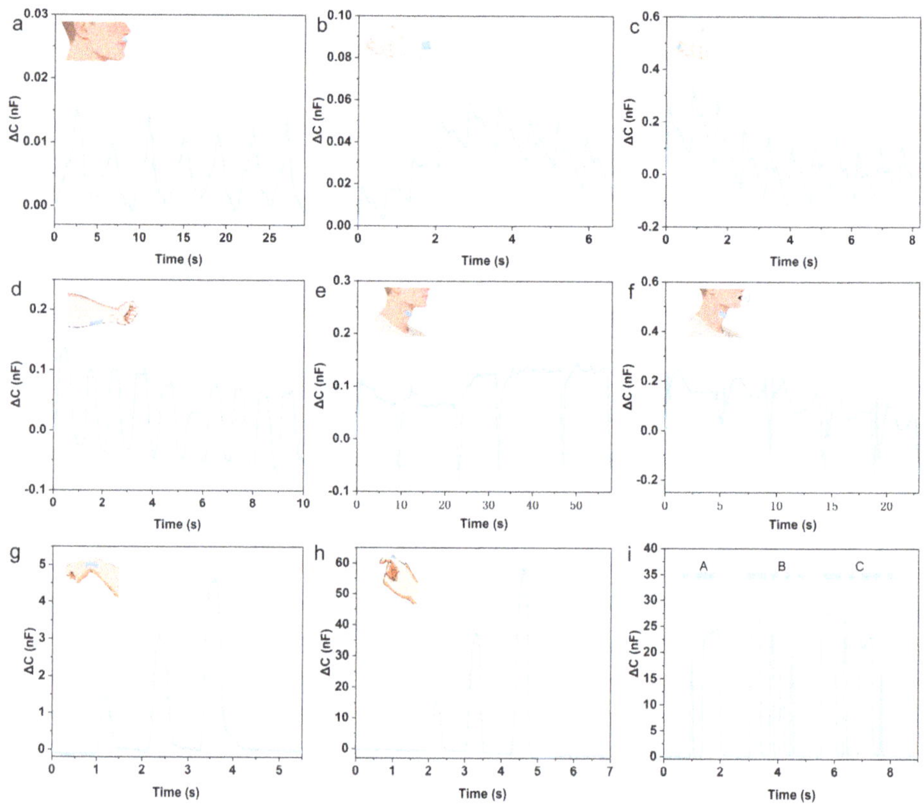

Figure 4. Wearable application of the iontronic pressure sensors. (**a**) Respiratory monitoring, (**b**) Wrist pulse wave detection, (**c**) Fingertip pulse wave detection, (**d**) contraction/relaxation muscle movement monitoring, (**e**) swallowing movement monitoring, (**f**) voice recognition, (**g**) wrist bending, (**h**) finger bending, (**i**) recognition of Morse Code output.

Then, we attached the sensor to the inner side of the forearm muscle and successfully detected the dynamic changes in capacitance during the process of clenching (muscle contraction) and relaxing (muscle relaxation), as shown in Figure 4d. When clenching, muscle contraction causes the sensor contact to weaken, resulting in a decrease in capacitance. When relaxing, the muscle restores contact and applies pressure to the sensor, causing the capacitance to increase. This functionality indicates its potential applications in the rehabilitation treatment for muscle injuries. We also recorded muscle swallowing activity (Figure 4e) and vibration phonation (Figure 4f) caused by throat movements by attaching the sensor to the throat. The resulting capacitance signals exhibited periodic changes, with phonation signals clearly distinguished between the two syllables 'a' and 'p', demonstrating the sensor's ability to detect minute physiological signals.

In addition, owing to its wide response range, the sensor also performs effectively under high-pressure conditions, such as those generated by finger bending and tactile input for safety protection (Morse code). When attached to finger and wrist joints (Figure 4g,h), the sensor measures capacitance variations correlated with joint angle and range of motion, offering valuable data for athlete training and the guidance of proper movement techniques. It is noteworthy that throat monitoring and joint flexion produce diametrically opposed capacitive signal responses. This occurs because throat movements such as swallowing and speaking cause skin stretching and increase the electrode spacing, whereas the compressive force applied to the sensor during joint flexion reduces the dielectric layer thickness. Finally, based on the principle of Morse code, we encode information by performing short and long press operations on the sensor (Figure 4i). The output capacitance signals exhibited distinct temporal characteristics and high stability, demonstrating their potential applications in information security and intelligent robot tactile interaction. Furthermore, considering the temperature sensitivity of IL, we envision the sensor's multimodal detection capabilities and aim to achieve high-resolution monitoring of the entire system through deep learning reinforcement [60–62].

4. Conclusions

In this study, we conducted a comparative study into the use of various fiber materials for fabricating ionic dielectric layers to develop highly sensitive iontronic pressure sensors. Four commonly available fiber substrates—woven fabric, cotton nonwoven fabric, viscose spunlace fabric, and electrospun TPU nanofiber mat—were selected to construct the ionic layers through a straightforward ionic liquid impregnation process. The influence of the microstructure and thickness of these ionic layers on sensor performance was systematically examined. Results indicate that the electrospun TPU nanofiber mats, characterized by high porosity and minimal thickness, serve as an ideal ionic layer. Its capability for high ionic liquid absorption and a microstructural surface composed of nanoscale fibers collectively contribute to superior sensor sensitivity. The optimized pressure sensor demonstrated a sensitivity of 3.10 kPa^{-1} in the 0–200 kPa range, with retained high sensitivities of 1.85 kPa^{-1} and 1.02 kPa^{-1} in the 200–400 kPa and 400–700 kPa intervals, respectively, underscoring both its high responsiveness and broad detection range. Furthermore, the sensor exhibited outstanding dynamic performance, with a rapid loading response time of 2.71 ms and an unloading recovery time of 8.13 ms, coupled with exceptional operational stability for up to 2500 continuous compression cycles. Leveraging these attributes, the sensor proved capable of accurately capturing subtle physiological activities such as pulse, respiration, and vocal cord vibrations, as well as larger pressure stimuli including joint movements. These results highlight its significant potential for applications in wearable health monitoring and human–machine interactions. Furthermore, to gain deeper insights into the structure-property relationship, our future work will involve systematic investigations into the optimal dielectric layer thickness. This will include fabricating substrates with identical thickness but varying material compositions, laying the groundwork for subsequent targeted designs of high-performance wearable sensors.

Supplementary Materials: The following supporting information can be downloaded at: https://www.mdpi.com/article/10.3390/s25216711/s1, Figure S1: Schematic cross-sectional illustration of the sensor structure. Figure S2: The absorbed content of IL for the different fiber substrates by simple immersing. Figure S3: FT-IR spectra of (a) woven fabric and woven fabric/IL, (b) cotton nonwoven fabric and cotton nonwoven fabric/IL, (c) viscose spunlace fabric and viscose spunlace fabric/IL. Figure S4: SEM and EDS images of the woven fabric/IL. Figure S5: SEM and EDS images of the cotton nonwoven fabric/IL. Figure S6: SEM and EDS images of the viscose spunlace fabric/IL. Figure S7: Water loss rate and water vapor transmission rate of the different fiber-based ionic dielectric layer. Figure S8: Compression stress—strain curves of (a) TPU fiber membranes of varying thicknesses and (b) different fiber materials. Figure S9: Reproducibility test of sensors based on TPU-33 μm ionic fiber membranes. Figure S10: Ion transport characteristics of the EDL. (a) Bode plot; (b) Capacitance frequency dependence; (c) AC amplitude dependence; (d) Bias voltage dependence. Figure S11: Sensing signals in response to subtle pressures as low as 3.14 Pa. Figure S12: Capacitance change in the sensor when loading and unloading periodically from 20 to 80 kPa. Figure S13: Cyclic sensing performance under the loading and unloading process under 80 kPa. Figure S14: Sensitivity performance comparison of our work with existing capacitive sensors.

Figure S15: (a) Detection of different breathing patterns; (b,c) Pulse detection before exercise (b,c) after exercise. Table S1: Sensitivity and linearity of different substrate materials at different pressure ranges. Table S2: Sensitivity and linearity of TPU/IL at different thicknesses under various pressure ranges. Table S3: Comparison of device performance of our device with literature-reported ones. References [48,63–70] are cited in the Supplementary Materials.

Author Contributions: Conceptualization, Y.D.; methodology, C.L., J.X. and S.Y.; validation, C.L. and J.X.; investigation, Y.X.; resources, Y.D.; data curation, Y.X. and S.Y.; writing—original draft preparation, C.L.; writing—review and editing, C.L. and Y.D.; supervision, J.W., X.L., L.W. and Y.D.; funding acquisition, Y.D. All authors have read and agreed to the published version of the manuscript.

Funding: This research was funded by National Science of Foundation of China (62461042), Jiangxi Provincial Natural Science Foundation (20252BAC240694), Jiangxi Provincial Key Laboratory of Photodetectors (2024SSY03041) and the start-up funding of Nanchang University (28170137).

Institutional Review Board Statement: Not applicable.

Informed Consent Statement: Informed consent was obtained from all subjects involved in the study.

Data Availability Statement: The original contributions presented in this study are included in the article/Supplementary Materials. Further inquiries can be directed to the corresponding authors.

Conflicts of Interest: The authors declare no conflicts of interest.

Abbreviations

The following abbreviations are used in this manuscript:

IL	ionic liquid
EDL	electric double layer
TPU	Thermoplastic polyurethane

References

1. Niu, H.; Li, H.; Li, Y.; Yue, W.; Gao, S.; Wei, X.; Shen, G. Cocklebur-Inspired "Branch-Seed-Spininess" 3D Hierarchical Structure Bionic Electronic Skin for Intelligent Perception. *Nano Energy* **2023**, *107*, 108144. [CrossRef]
2. Gong, Y.; Cheng, G.; Zhao, F.; Zhang, B.; Yuan, J.; Wei, D. Wide Linear Range Fabric Pressure Sensors Based on Multiscale Inorganic Structures. *IEEE Sens. J.* **2025**, *25*, 10655–10662. [CrossRef]
3. Yang, J.-S.; Chung, M.-K.; Yoo, J.-Y.; Kim, M.-U.; Kim, B.-J.; Jo, M.-S.; Kim, S.-H.; Yoon, J.-B. Interference-Free Nanogap Pressure Sensor Array with High Spatial Resolution for Wireless Human-Machine Interfaces Applications. *Nat. Commun.* **2025**, *16*, 2024. [CrossRef]
4. Wang, P.; Li, X.; Sun, G.; Wang, G.; Han, Q.; Meng, C.; Wei, Z.; Li, Y. Natural Human Skin-Inspired Wearable and Breathable Nanofiber-Based Sensors with Excellent Thermal Management Functionality. *Adv. Fiber Mater.* **2024**, *6*, 1955–1968. [CrossRef]
5. Jin, T.; Pan, Y.; Park, S.-H.K.; Fang, D.-W. Ultrathin Flexible Pressure Sensors Using Microbead Embedded Nanofibrous Membrane for Wearable Applications. *J. Alloys Compd.* **2025**, *1014*, 178609. [CrossRef]
6. Li, Y.; Zhang, Z.; Yang, F.; Zhou, H.; Zong, S.; Zhang, B.; Cao, L. High-Sensitivity and Wide-Range Biomimetic Crack Flexible Pressure Sensor. *IEEE Trans. Instrum. Meas.* **2025**, *74*, 1–8. [CrossRef]
7. Rukshana Bi, G.; Chhetri, B.; Upreti, A.; Jindal, S.K.; Patil, H.Y. Capacitance Dynamics in W-Shaped Silicon Substrate Mems Double-Touch Mode Pressure Sensor. *Meas. Sci. Technol.* **2025**, *36*, 025107. [CrossRef]
8. Rukshana Bi, G.; Motwani, M.; Jindal, S.K. Analytical Formulation and Numerical Investigation of Electromechanical Dynamics in Circular Biconvex Diaphragm Capacitive Pressure Sensor. *IEEE Sens. J.* **2025**, *25*, 18867–18875. [CrossRef]
9. Xu, J.; Ma, J.; Lian, J.; Liao, Y.; Wang, L.; Dong, L.; Sun, H. Design of Flexible Resistance Sensor Based on Mesh Convex Microstructure. *J. Phys. D-Appl. Phys.* **2022**, *55*, 315401. [CrossRef]
10. Yan, B.; Ding, H.; Zhang, Y.; Lin, P.; Wu, D.; Shi, Z.; Chen, X.; Tian, Y.; Li, X. Skin-Attachable and Flexible Mwcnt Grid/Ecoflex Strain Sensors with Fast Equilibrium of Response for Detection of Sound Vibrations and Human Motions. *J. Mater. Sci.-Mater. Electron.* **2021**, *32*, 26439–26448. [CrossRef]

11. Ying, B.; Chen, R.Z.; Zuo, R.; Li, J.; Liu, X. An Anti-Freezing, Ambient-Stable and Highly Stretchable Ionic Skin with Strong Surface Adhesion for Wearable Sensing and Soft Robotics. *Adv. Funct. Mater.* **2021**, *31*, 2104665. [CrossRef]
12. Yan, X.; Wang, S.; Nie, G.; Gao, Y.; Li, L.; Zhang, T.; Long, Y.-Z.; Han, W. The Iontronic Sensor Based on Biodegradable Polycaprolactone for Interfacial Capacitive Pressure Sensing. *J. Mater. Sci.-Mater. Electron.* **2024**, *35*, 1093. [CrossRef]
13. Li, J.; Lu, J.; Zhang, K.; Wu, J.; Zhang, C.; Cai, X.; Xu, L.; Cao, Z.; Li, Y.; Wang, X.; et al. Freeze-Drying Induced Gradient Microporous Composite Film with High Ionic Conductivity for Ultrasensitive Wearable Iontronic Pressure Sensor. *Chem. Eng. J.* **2024**, *493*, 152450. [CrossRef]
14. Tang, J.; Zhao, C.; Luo, Q.; Chang, Y.; Yang, Z.; Pan, T. Ultrahigh-Transparency and Pressure-Sensitive Iontronic Device for Tactile Intelligence. *npj Flex. Electron.* **2022**, *6*, 54. [CrossRef]
15. Zheng, Q.; Dai, X.; Wu, Y.; Liang, Q.; Wu, Y.; Yang, J.; Dong, B.; Gao, G.; Qin, Q.; Huang, L.-B. Self-Powered High-Resolution Smart Insole System for Plantar Pressure Mapping. *Bmemat* **2023**, *1*, E12008. [CrossRef]
16. Hwang, J.; Kim, Y.; Yang, H.; Oh, J.H. Fabrication of Hierarchically Porous Structured Pdms Composites and Their Application as A Flexible Capacitive Pressure Sensor. *Compos. Part B-Eng.* **2021**, *211*, 108607. [CrossRef]
17. Chen, B.; Zhang, L.; Li, H.; Lai, X.; Zeng, X. Skin-Inspired Flexible and High-Performance Mxene@Polydimethylsiloxane Piezoresistive Pressure Sensor for Human Motion Detection. *J. Colloid Interface Sci.* **2022**, *617*, 478–488. [CrossRef]
18. Seong, J.; Bak, B.-U.; Jin, J.; Kim, J. Tribo-Piezoelectric Synergistic $BaTiO_3$/Pdms Micropyramidal Structure for High-Performance Energy Harvester and High-Sensitivity Tactile Sensing. *Nano Energy* **2024**, *122*, 109264. [CrossRef]
19. Su, D.; Shen, G.; Li, J.; Qin, B.; Wang, S.; Yang, W.; He, X. Enhanced Sensitivity and Linear-Response in Iontronic Pressure Sensors for Non-Contact, High-Frequency Vibration Recognition. *J. Colloid Interface Sci.* **2024**, *659*, 1042–1051. [CrossRef]
20. Lee, J.-E.; Kim, S.-U.; Kim, J.-Y. Fabrication of A Capacitive 3D Spacer Fabric Pressure Sensor with A Dielectric Constant Change for High Sensitivity. *Sensors* **2024**, *24*, 3395. [CrossRef] [PubMed]
21. Duan, Y.; Wu, J.; He, S.; Su, B.; Li, Z.; Wang, Y. Bioinspired Spinosum Capacitive Pressure Sensor Based on Cnt/Pdms Nanocomposites for Broad Range and High Sensitivity. *Nanomaterials* **2022**, *12*, 3265. [CrossRef]
22. Chen, S.; Guo, X. Improving The Sensitivity of Elastic Capacitive Pressure Sensors Using Silver Nanowire Mesh Electrodes. *IEEE Trans. Nanotechnol.* **2015**, *14*, 619–623. [CrossRef]
23. You, B.; Han, C.J.; Kim, Y.; Ju, B.-K.; Kim, J.-W. A Wearable Piezocapacitive Pressure Sensor with a Single Layer of Silver Nanowire-Based Elastomeric Composite Electrodes. *J. Mater. Chem. A* **2016**, *4*, 10435–10443. [CrossRef]
24. Liu, S.-Y.; Lu, J.-G.; Shieh, H.-P.D. Influence of Permittivity on the Sensitivity of Porous Elastomer-Based Capacitive Pressure Sensors. *IEEE Sens. J.* **2018**, *18*, 1870–1876. [CrossRef]
25. Yang, J.C.; Kim, J.-O.; Oh, J.; Kwon, S.Y.; Sim, J.Y.; Kim, D.W.; Choi, H.B.; Park, S. Microstructured Porous Pyramid-Based Ultrahigh Sensitive Pressure Sensor Insensitive To Strain and Temperature. *ACS Appl. Mater. Interfaces* **2019**, *11*, 19472–19480. [CrossRef]
26. Zhao, K.; Han, J.; Ma, Y.; Tong, Z.; Suhr, J.; Wang, M.; Xiao, L.; Jia, S.; Chen, X. Highly Sensitive and Flexible Capacitive Pressure Sensors Based on Vertical Graphene and Micro-Pyramidal Dielectric Layer. *Nanomaterials* **2023**, *13*, 701. [CrossRef]
27. Thouti, E.; Nagaraju, A.; Chandran, A.; Prakash, P.V.B.S.S.; Shivanarayanamurthy, P.; Lal, B.; Kumar, P.; Kothari, P.; Panwar, D. Tunable Flexible Capacitive Pressure Sensors Using Arrangement of Polydimethylsiloxane Micro-Pyramids for Bio-Signal Monitoring. *Sens. Actuators A-Phys.* **2020**, *314*, 112251. [CrossRef]
28. Zhao, Z.; Guo, Q.; Sun, Y.; An, N.; Hui, P.; Yang, L.; Chen, X. Bioinspired Hierarchical Structure for an Ultrawide-Range Multifunctional Flexible Sensor Using Porous Expandable Polyethylene/Loofah-Like Polyurethane Sponge Material. *Adv. Intell. Syst.* **2023**, *5*, 2200295. [CrossRef]
29. He, S.; Wu, J.; Duan, Y.; Su, B.; Liu, S.; Teng, F.; Wang, Y. Soft-Template Synthesis of Hierarchically Porous Structured Polydimethylsiloxane Toward Flexible Capacitive Pressure Sensor. *Sci. China-Technol. Sci.* **2023**, *66*, 2696–2706. [CrossRef]
30. Luo, Y.; Shao, J.; Chen, S.; Chen, X.; Tian, H.; Li, X.; Wang, L.; Wang, D.; Lu, B. Flexible Capacitive Pressure Sensor Enhanced by Tilted Micropillar Arrays. *ACS Appl. Mater. Interfaces* **2019**, *11*, 17796–17803. [CrossRef]
31. Xu, T.; Li, K.; Zhou, Y.; Zhang, D.; Sheng, B. Flexible, Super-Stable, and Multifunctional Capacitive Pressure Sensors with Tilted Micropillar Array by Inclined Laser Ablation and Demolding. *IEEE Sens. J.* **2025**, *25*, 9367–9374. [CrossRef]
32. Li, R.; Si, Y.; Zhu, Z.; Guo, Y.; Zhang, Y.; Pan, N.; Sun, G.; Pan, T. Supercapacitive Iontronic Nanofabric Sensing. *Adv. Mater.* **2017**, *29*, 1700253. [CrossRef]
33. Gou, X.; Yang, J.; Li, P.; Zhou, Z.; Liao, C.; Zhang, C.; Dong, C.; Li, C. Biomimetic Nanofiber-Iongel Composites for Flexible Pressure Sensors with Broad Range and Ultra-High Sensitivity. *Nano Energy* **2024**, *120*, 109140. [CrossRef]
34. Fan, X.; Liu, S.; Jia, Z.; Koh, J.J.; Yeo, J.C.C.; Wang, C.-G.; Surat'man, N.E.; Loh, X.J.; Le Bideau, J.; He, C.; et al. Ionogels: Recent Advances in Design, Material Properties and Emerging Biomedical Applications. *Chem. Soc. Rev.* **2023**, *52*, 2497–2527. [CrossRef] [PubMed]
35. Nie, B.; Li, R.; Brandt, J.D.; Pan, T. Iontronic Microdroplet Array for Flexible Ultrasensitive Tactile Sensing. *Lab A Chip* **2014**, *14*, 1107–1116. [CrossRef]

36. Zhang, X. High-Sensitivity Antidrying Hydrogel Sensor with Interpenetrating Network Crosslinking Structure. *J. Mater. Sci.-Mater. Electron.* **2023**, *34*, 540. [CrossRef]
37. Niu, H.; Wei, X.; Li, H.; Yin, F.; Wang, W.; Seong, R.-S.; Shin, Y.K.; Yao, Z.; Li, Y.; Kim, E.-S.; et al. Micropyramid Array Bimodal Electronic Skin for Intelligent Material and Surface Shape Perception Based on Capacitive Sensing. *Adv. Sci.* **2024**, *11*, 2305528. [CrossRef]
38. Bai, N.; Wang, L.; Wang, Q.; Deng, J.; Wang, Y.; Lu, P.; Huang, J.; Li, G.; Zhang, Y.; Yang, J.; et al. Graded Intrafillable Architecture-Based Iontronic Pressure Sensor with Ultra-Broad-Range High Sensitivity. *Nat. Commun.* **2020**, *11*, 209. [CrossRef]
39. Zhang, M.; Gu, M.; Shao, L.; Cheng, G.; Gao, H.; Sun, B.; Li, S.; Tang, T.; Li, N.; Yi, Y.; et al. Flexible Wearable Capacitive Sensors Based on Ionic Gel with Full-Pressure Ranges. *ACS Appl. Mater. Interfaces* **2023**, *15*, 15884–15892. [CrossRef]
40. Li, P.; Xie, L.; Su, M.; Wang, P.; Yuan, W.; Dong, C.; Yang, J. Skin-Inspired Large Area Iontronic Pressure Sensor with Ultra-Broad Range and High Sensitivity. *Nano Energy* **2022**, *101*, 107571. [CrossRef]
41. Li, L.; Zhu, G.; Wang, J.; Chen, J.; Zhao, G.; Zhu, Y. A Flexible and Ultrasensitive Interfacial Iontronic Multisensory Sensor with An Array of Unique "Cup-Shaped" Microcolumns for Detecting Pressure and Temperature. *Nano Energy* **2023**, *105*, 108012. [CrossRef]
42. Yang, X.; Dai, D.; Li, J.; Luo, M.; Shu, K.; Zheng, K.; Yu, J.; Huang, Y.; Chen, X. One-Step Process of Wrinkle Microstructured Iontronic Pressure Sensor with All Fabric Wearable Electrode. *Chem. Eng. J.* **2024**, *496*, 153780. [CrossRef]
43. Wang, X.; Zi, J.; Chen, Y.; Wu, Q.; Xiang, Z.; Tu, Y.; Yang, P.; Wan, Y. Multistage Microstructured Ionic Skin for Real-Time Vital Signs Monitoring and Human-Machine Interaction. *Energy Environ. Mater.* **2024**, *7*, e12767. [CrossRef]
44. Liu, Z.; Cai, M.; Hong, S.; Shi, J.; Xie, S.; Liu, C.; Du, H.; Morin, J.D.; Li, G.; Wang, L.; et al. Data- Driven Inverse Design of Flexible Pressure Sensors. *Proc. Natl. Acad. Sci. USA* **2024**, *121*, e2320222121. [CrossRef] [PubMed]
45. Li, C.; Cheng, J.; He, Y.; He, X.; Xu, Z.; Ge, Q.; Yang, C. Polyelectrolyte Elastomer-Based Ionotronic Sensors with Multi-Mode Sensing Capabilities Via Multi-Material 3d Printing. *Nat. Commun.* **2023**, *14*, 4853. [CrossRef] [PubMed]
46. Su, Q.; Zou, Q.; Li, Y.; Chen, Y.; Teng, S.-Y.; Kelleher, J.T.; Nith, R.; Cheng, P.; Li, N.; Liu, W.; et al. A Stretchable and Strain-Unperturbed Pressure Sensor for Motion Interference-Free Tactile Monitoring on Skins. *Sci. Adv.* **2021**, *7*, eabi4563. [CrossRef]
47. Fan, L.; Yang, X.; Sun, H. Pressure Sensors Combining Porous Electrodes and Electrospun Nanofiber-Based Ionic Membranes. *ACS Appl. Nano Mater.* **2023**, *6*, 3560–3571. [CrossRef]
48. Wang, X.; Li, Y.; Wang, Y.; Huang, W.; Zhao, X.; Chen, K.; Luo, F.; Qin, Y. Fabrication Method and Various Application Scenarios of Flexible Capacitive Pressure Sensor Based on Direct Formation of Conical Structure. *Chem. Eng. J.* **2024**, *496*, 153957. [CrossRef]
49. Hwang, H.J.; Kim, J.S.; Kim, W.; Park, H.; Bhatia, D.; Jee, E.; Chung, Y.S.; Kim, D.H.; Choi, D. An Ultra-Mechanosensitive Visco-Poroelastic Polymer Ion Pump for Continuous Self-Powering Kinematic Triboelectric Nanogenerators. *Adv. Energy Mater.* **2019**, *9*, 1803786. [CrossRef]
50. Xu, Y.; Chen, L.; Chen, J.; Chang, X.; Zhu, Y. Flexible and Transparent Pressure/Temperature Sensors Based on Ionogels with Bioinspired Interlocked Microstructures. *ACS Appl. Mater. Interfaces* **2022**, *14*, 2122–2131. [CrossRef]
51. Chang, Y.; Wang, L.; Li, R.; Zhang, Z.; Wang, Q.; Yang, J.; Guo, C.F.; Pan, T. First Decade of Interfacial Iontronic Sensing: From Droplet Sensors to Artificial Skins. *Adv. Mater.* **2021**, *33*, 2003464. [CrossRef] [PubMed]
52. Ren, Z.; Liu, N.; Zhang, Q.; Yin, J.; Jia, P.; Lu, W.; Yao, Q.; Deng, M.; Gao, Y. Ionic Flexible Mechanical Sensors: Mechanisms, Structural Engineering, Applications, and Challenges. *Adv. Sens. Res.* **2023**, *2*, 2200099. [CrossRef]
53. Garnett, J. Colours in Metal Glasses, in Metallic Films, and in Metallic Solutions.—II. *Philos. Trans. R. Soc. Ser. A Contain. Pap. A Math. Phys. Character* **1906**, *205*, 237–288.
54. Ham, C.-H.; Youn, H.J.; Lee, H.L. Influence of Fiber Composition and Drying Conditions on The Bending Stiffness of Paper. *Bioresources* **2020**, *15*, 9197–9211. [CrossRef]
55. Lin, Q.; Huang, J.; Yang, J.; Huang, Y.; Zhang, Y.; Wang, Y.; Zhang, J.; Wang, Y.; Yuan, L.; Cai, M.; et al. Highly Sensitive Flexible Iontronic Pressure Sensor for Fingertip Pulse Monitoring. *Adv. Healthc. Mater.* **2020**, *9*, 2001023. [CrossRef]
56. Wei, Q.; Chen, G.; Pan, H.; Ye, Z.; Au, C.; Chen, C.; Zhao, X.; Zhou, Y.; Xiao, X.; Tai, H.; et al. Mxene-Sponge Based High-Performance Piezoresistive Sensor for Wearable Biomonitoring and Real-Time Tactile Sensing. *Small Methods* **2022**, *6*, 2101051. [CrossRef]
57. Xu, M.X.; Dou, C.; Song, T.Y.; Li, X.; Zhang, Q. A Temperature-Insensitive Silver Nanostructures@Graphene Foam for High Accuracy and Full Range Human Health Monitoring. *Rare Met.* **2024**, *43*, 5953–5963. [CrossRef]
58. You, J.; Lu, M.Y.; Dazhen, L.; Gao, M.J.; Zhang, R.Y.; Li, W.D.; Lei, F.; Ren, W.; Li, G.X.; Yang, J.L. Anti-Motion Artifacts Iontronic Sensor for Long-Term Accurate Fingertip Pulse Monitoring. *Adv. Sci.* **2025**, *12*, 2414425. [CrossRef]
59. Kato, Y.; Fukuda, K.; Someya, T.; Yokota, T. An Ultra-Flexible Temperature-Insensitive Strain Sensor. *J. Mater. Chem. C* **2023**, *11*, 14070–14078. [CrossRef]
60. Luo, H.Y.; Pang, G.Y.; Xu, K.C.; Ye, Z.Q.; Yang, H.Y.; Yang, G. A Fully Printed Flexible Sensor Sheet for Simultaneous Proximity-Pressure-Temperature Detection. *Adv. Mater. Technol.* **2024**, *6*, 2100616. [CrossRef]

61. Yang, Q.Q.; Li, B.Q.; Wang, M.K.; Pang, G.Y.; Lu, Y.Y.; Li, J.Y.; Yang, H.Y.; Lyu, H.; Xu, K.C.; Yang, G. Machine Learning-Enhanced Modular Ionic Skin for Broad-Spectrum Multimodal Discriminability in Bidirectional Human-Robot Interaction. *Adv. Mater.* **2025**, *37*, e08795. [CrossRef]
62. Kong, D.P.; Lu, Y.Y.; Zhou, S.Y.; Wang, M.K.; Pang, G.Y.; Wang, B.C.; Chen, L.P.; Huang, X.Y.; Lyu, H.; Xu, K.C.; et al. Super-Resolution Tactile Sensor Arrays with Sparse Units Enabled by Deep Learning. *Sci. Adv.* **2025**, *11*, eadv2124. [CrossRef] [PubMed]
63. Wang, B.; Wang, J.; Lou, Y.Y.; Ding, S.S.; Jin, X.; Liu, F.; Xu, Z.J.; Ma, J.Y.; Sun, Z.M.; Li, X.Y. Halloysite Nanotubes Strengthened Electrospinning Composite Nanofiber Membrane for On-Skin Flexible Pressure Sensor with High Sensitivity, Good Breathability, and Round-The-Clock Antibacterial Activity. *Appl. Clay Sci.* **2022**, *228*, 106650. [CrossRef]
64. Cui, X.H.; Chen, J.W.; Wu, W.; Liu, Y.; Li, H.D.; Xu, Z.G.; Zhu, Y.T. Flexible and Breathable All-Nanofiber Iontronic Pressure Sensors with Ultraviolet Shielding and Antibacterial Performances for Wearable Electronics. *Nano Energy* **2022**, *95*, 107022. [CrossRef]
65. AdigÜZel, S.P.; Ercan, N. Development Of PVDF-HFP Based Flexible Hybrid Iontronic Film for Wearable Capacitive Sensor And E-Skin Applications. *Adv. Mater. Technol.* **2025**, *10*, 70051. [CrossRef]
66. Zhang, X.H.; Lu, Q.X.; Zhou, L.; Zhang, W.; Zhang, X.H.; Hu, F.R. Flexible Wearable Iontronic Pressure Sensors Based on An Array of Semiellipsoids with Micropillars for Health and Motion Monitoring. *ACS Appl. Electron. Mater.* **2025**, *7*, 1820–1828. [CrossRef]
67. Zhang, Y.; Yang, J.L.; Hou, X.Y.; Li, G.; Wang, L.; Bai, N.N.; Cai, M.K.; Zhao, L.Y.; Wang, Y.; Zhang, J.M.; et al. Highly Stable Flexible Pressure Sensors with A Quasi-Homogeneous Composition and Interlinked Interfaces. *Nat. Commun.* **2022**, *13*, 1317. [CrossRef] [PubMed]
68. Huang, Y.; Zhao, L.Y.; Cai, M.K.; Zhu, J.Q.; Wang, L.; Chen, X.X.; Zeng, Y.M.; Zhang, L.Q.; Shi, J.D.; Guo, C.F. Arteriosclerosis Assessment Based on Single-Point Fingertip Pulse Monitoring Using a Wearable Iontronic Sensor. *Adv. Healthc. Mater.* **2023**, *12*, 2301838. [CrossRef]
69. Cetin, O.; Cicek, M.O.; Cugunlular, M.; Bolukbasi, T.; Khan, Y.; Unalan, H.E. Mxene-Deposited Melamine Foam-Based Iontronic Pressure Sensors for Wearable Electronics and Smart Numpads. *Small* **2024**, *20*, 2403202. [CrossRef]
70. Liu, Q.X.; Liu, Z.G.; Li, C.G.; Xie, K.W.; Zhu, P.; Shao, B.Q.; Zhang, J.M.; Yang, J.L.; Zhang, J.; Wang, Q.; et al. Highly Transparent and Flexible Iontronic Pressure Sensors Based on An Opaque to Transparent Transition. *Adv. Sci.* **2020**, *7*, 2000348. [CrossRef]

sensors

MDPI

Article

Fabrication of Stretchable Piezoelectric Sensor with a Kirigami Design for Heart Sound Monitoring

Xudong Zhang [1], Xudong Ye [1], Xi Lei [2], Hong Hu [2], Hai Liu [1], Shaobo Jin [3], Guoyong Ye [3] and Tingting Zhao [1,4,*]

1 Key Laboratory of Advanced Display and System Applications, Ministry of Education, Shanghai University, Shanghai 200072, China
2 The School of Microelectronics, Shanghai University, Shanghai 201800, China
3 Henan Provincial Key Laboratory of Intelligent Manufacturing of High-End Equipment, Zhengzhou University of Light Industry, Zhengzhou 450002, China
4 Shanghai Key Laboratory of Intelligent Manufacturing and Robotics, Shanghai University, Shanghai 200444, China
* Correspondence: ttzhao@shu.edu.cn

Abstract

Heart sounds contain critical information about valve activity and hemodynamics, serving as an essential basis for cardiovascular disease diagnosis. However, traditional heart sound sensors are either rigid or flexible but non-stretchable, limiting their ability to accommodate chest deformation and leading to signal distortion. This study proposes an easy-to-fabricate, stretchable piezoelectric heart sound sensor with a Kirigami-inspired design, a five-layer "sandwich" structure. Periodic Kirigami cuts significantly enhance stretchability while maintaining piezoelectric conversion efficiency. Finite element simulations reveal the Kirigami structure is more sensitive to hinge length and thickness than to hinge width. Electrical tests demonstrate a linear response to sound pressure, with output voltage rising from 0.11 V to 0.42 V (70–94 dB), and voltage amplitude increasing from 9 mV to 0.35 V (60–160 Hz). The sensor exhibits excellent stability, with a maximum amplitude variation of approximately 11% under 0–30% strain, a 17% voltage decrease at 11 mm bending radius, and less than 9% output fluctuation during 1200 s continuous excitation. Seven-day monitoring confirms reliable detection of the first (S1) and second (S2) heart sounds, with signals highly consistent with ECG and a commercial sensor, verifying its potential for wearable long-term monitoring and early cardiovascular disease screening.

Keywords: flexible piezoelectric sensor; PVDF; Kirigami structure; heart sound monitoring; stretchable device

1. Introduction

As direct acoustic representations of the physiological activities of the cardiovascular system, heart sound signals contain key pathological information (e.g., heart valve opening/closing and hemodynamics) and serve as an important basis for the early diagnosis of diseases such as coronary heart disease and heart failure [1–4]. Traditional stethoscopes rely on the experience of physicians, suffering from limitations including strong subjectivity and difficulty in continuous monitoring [5,6]. Existing electronic auscultation devices are mostly based on rigid sensors, which are hard to deploy portably due to their large size. Furthermore, the rigid structure has poor adherence to the skin and cannot adapt to skin deformation during human movement, failing to meet the requirements of wearable long-term continuous monitoring [7–9]. The development of flexible sensors provides a solution to this problem—their flexibility enhances adherence to the skin, enabling adaptation to long-term monitoring scenarios.

Piezoelectric polymers (e.g., Polyvinylidene Fluoride (PVDF)) have become one of the core materials for flexible sensors owing to their excellent flexibility and biocompatibility, and they exhibit unique advantages especially in cardiovascular signal monitoring [10–12]. Their thin-film form can closely adhere to the skin, realizing high-fidelity capture of weak physiological vibrations such as heart activity and pulses. For instance, previous studies have used sheet-like PVDF to construct a mattress-type unobstructive measurement system, which simultaneously acquires heart sounds, ballistocardiogram (BCG), and respiratory signals during infant sleep. The obtained pulse interval is highly consistent with the electrocardiogram (ECG) reference value, verifying its reliability in unobstructive vital sign monitoring [13]. Another study integrated flexible PVDF films into wearable chest patches to capture high-fidelity forcecardiography (FCG) and respiratory signals; the extracted heartbeat interval ($R^2 > 0.99$) and respiratory interval ($R^2 > 0.96$) are highly consistent with the results of clinical-grade devices, demonstrating excellent wearable monitoring potential [14]. However, pure PVDF films have limited stretchability and are prone to output signal distortion due to structural deformation. Therefore, improving the stretchability of PVDF sensors through structural design has become a key step in expanding their application in wearable physiological signal monitoring.

To address this limitation, this study designs a flexible PVDF piezoelectric sensor based on the Kirigami structure. The device adopts a five-layer "sandwich" structure of Polyimide (PI)-Electrode-PVDF-Electrode-PI, with the PVDF film as the core sensitive layer. Relying on its high sensitivity to weak mechanical vibrations, it achieves efficient transduction of heart sound signals. Meanwhile, periodic Kirigami cutting is introduced to release geometric constraints, significantly improving the sensor's stretchability and adaptability to dynamic skin deformation while ensuring mechanical stability. Benefiting from this structural optimization, the sensor can maintain stable detection of heart sound signals for an extended period, and is expected to provide a feasible path for the early screening of cardiovascular diseases and wearable physiological signal monitoring.

2. Materials and Methods

2.1. Fabrication of the Piezoelectric Sensor with a Kirigami Design

A double-sided aluminum (Al)-coated PVDF film (30 μm, Jinzhou Kexin Electronic Materials Co., Ltd., Jinzhou (Liaoning Province), China) was used as the piezoelectric sensing layer, which was cut into rectangular pieces (i). Two pieces of single-sided adhesive-backed polyimide films (20 μm, Dongguan Ronghui Insulating Materials Co., Ltd., Dongguan (Guangdong Province), China) were prepared; after peeling off the release films on their surfaces in sequence, they were attached to the upper and lower surfaces of the cut rectangular PVDF film (ii), with parts of the Al electrodes exposed. An ultraviolet nanosecond laser cutting machine (Changzhou Guangqi Laser Technology Co., Ltd., Changzhou (Jiangsu Province), China) was used to engrave the Kirigami structure on the film (iii). Finally, Cu conductive tape was used to attach wires to the exposed Al electrodes (iv), completing the fabrication of the piezoelectric sensor (4 cm × 5 cm). The detailed fabrication process of the piezoelectric sensor was shown in Figure 1.

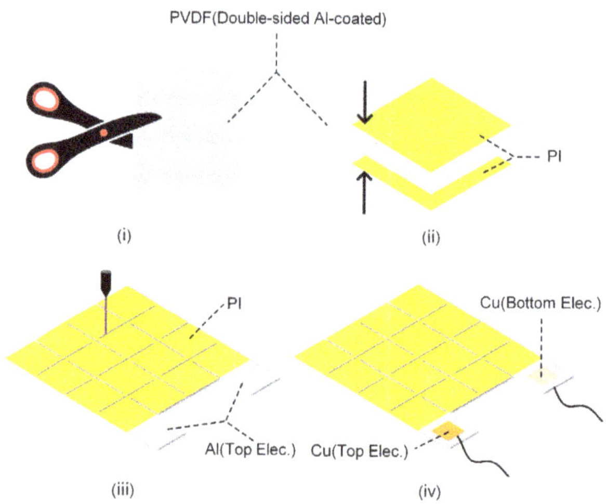

Figure 1. Simplified fabrication process of the sensor.

2.2. Finite Element Analysis

The stretchability of different structures was analyzed using the commercial Abaqus 2020. Kirigami patterns with various hinge widths, lengths, and thicknesses were first modeled in AutoCAD 2021 (R.47.0.0), and subsequently imported into Abaqus. All models were assigned the isotropic elastic material property. The materials parameters used in FEA are detailed in Section 3.2. A quasi-static analysis was performed using a dynamic implicit step, in which the kinetic energy was strictly constrained to be no more than 1% of the strain energy and the S4R elements were utilized for mesh generation [15,16]. Displacement loads were then applied to the four corners of the models. The change in stress due to the applied displacements, and changes in area due to the reaction force were numerically calculated, respectively.

2.3. Characterization and Testing

An oscilloscope (TBS2104, Tektronix Technology Co., Ltd., Shanghai, China) was used to measure the electric performance of the sensor. A speaker (3 W, Shenzhen Huayu Feifan Technology Co., Ltd., Shenzhen (Guangdong Province), China) was employed to apply a sound pressure excitation, and a sound level meter (AR844, Smart Sensor Instrument Co., Ltd., Dongguan (Guangdong Province), China) was used to record sound intensity. The piezoelectric sensor and a commercial heart sound sensor (HKY-06E, Hefei Huake Electronic Technology Institute, Hefei (Anhui Province), China) were used to simultaneously record heart sound signals. Detailed information regarding the ECG monitoring experiment is as follows: Three Ag/AgCl gel electrodes (Beijing Huarun Kangtai High-Tech Development Research Institute, Beijing, China) were placed on the left chest, right chest, and right abdomen. Electrodes were connected to a dedicated AD8232 ECG acquisition board (Sichiray Technology Co., Ltd., Wuxi (Jiangsu Province), China) via a three-lead cable. Data was transmitted via Bluetooth to a host computer for real-time display and acquisition. The output of the piezoelectric sensor was amplified and filtered by a self-made acquisition board, and then collected by the oscilloscope. The custom acquisition board operates at a sampling frequency of 1 kHz. It employs an active second-order 200 Hz hardware Butterworth low-pass filter to eliminate high-frequency noise. At the software level, an iirnotch function first designs a 50 Hz notch filter to suppress power line interference, followed by a fourth-order Butterworth filter to extract the signal. The respiratory wave is processed through a 1 Hz low-pass filter, while the heart sound signal undergoes a

20 Hz high-pass filter. The group delay of the hardware filters was measured using Texas Instruments Filter Design Tool (1.0), https://www.ti.com/tool/download/ANALOG-FILTER-DESIGNER/1.0, accessed on 17 October 2025: 1.2 ms at 20 Hz and 1.1 ms at 200 Hz, both significantly shorter than the time interval between S1 and S2 (approximately 260 ms). For software filtering, the filtfilt function was employed to perform zero-phase notch, low-pass, and high-pass filtering, compensating for phase delay. The acquisition board employs a non-inverting proportional voltage amplifier to amplify the signal with a gain of 15. An additional 1.65 V bias voltage is applied to ensure the signal falls within the ADC's sampling voltage range. The input stage utilizes a charge amplifier, featuring a CMOS precision operational amplifier with an input impedance exceeding 10 TΩ. This design effectively minimizes charge leakage and reduces signal distortion.

3. Results and Discussion

3.1. Structure and Working Mechanism of the Piezoelectric Sensor

Figure 2a illustrates the piezoelectric sensor placed on the chest for human signal monitoring. The sensor adopts a "sandwich" structure with a Kirigami design, where the top and bottom layers are protective layers made of PI films, and the middle layer is the piezoelectric sensitive layer (Figure 2b). Figure 2c shows a photograph of the sensor, where the silver-colored region corresponds to the exposed Al electrode to facilitate subsequent wire connection.

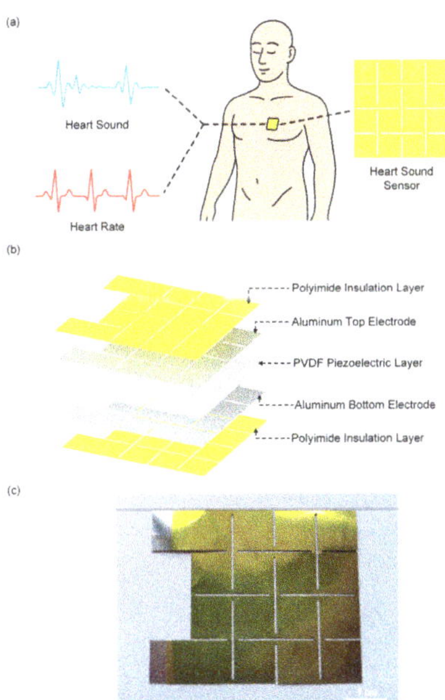

Figure 2. Structure of the piezoelectric sensor. (**a**) The Kirigami-structured sensor attached to the chest for human signal monitoring. (**b**) Exploded view of the piezoelectric sensor structure. (**c**) Photograph of the sensor.

The piezoelectric effect of PVDF originates from changes in its internal dipole moment under external force [17–19]. Initially, without external force (Figure 3(I)), the net polarization of PVDF film remains stable, free charges on the electrodes are in equilibrium, and no output signal is generated. When under pressure (Figure 3(II)), the internal dipoles shift, increasing net polarization and creating a potential difference between the electrodes, which generates a transient reverse current. Once the applied force no longer changes (Figure 3(III)), the net polarization reaches a steady state, and a new equilibrium is established with no current flow. Upon releasing the force (Figure 3(IV)), the internal dipoles gradually return to their initial state, reducing net polarization. To maintain charge balance, electrons flow from the lower to the upper electrode, producing a forward current. This completes a full pressure-electricity conversion cycle. PVDF offers unique advantages over other piezoelectric materials, such as Pb(Zr, Ti)O_3, $BaTiO_3$, and $KNbO_3$, due to its flexibility, thinness, and suitability for wearable designs. Furthermore, its wide frequency response range aligns well with the frequency range of heart sounds (20–200 Hz), making it particularly effective for heart sound detection and monitoring.

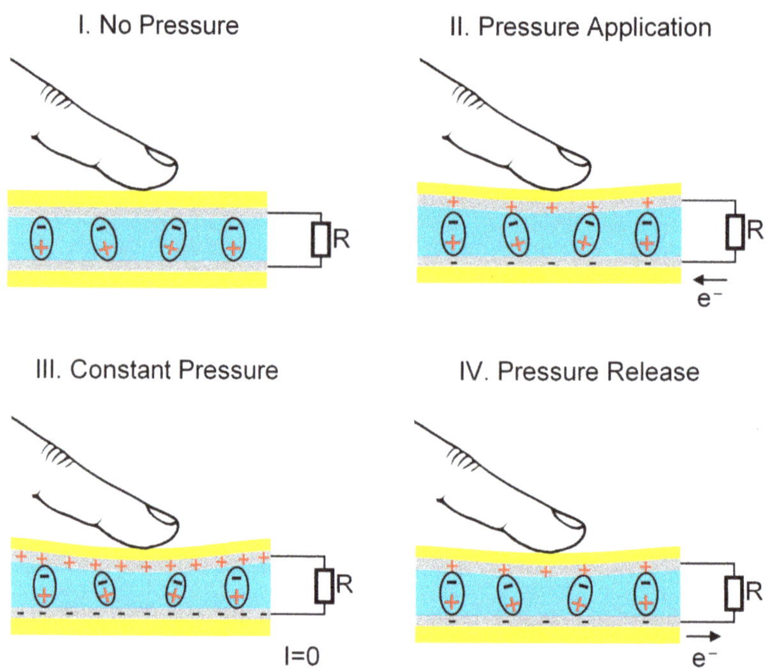

Figure 3. Working mechanism of the sensor. In the figure, the yellow regions are Polyimide (PI), the silver-gray regions are Al electrodes, and the blue regions are PVDF.

PVDF films are inherently flexible but lack stretchability. To address this limitation, a Kirigami structure was introduced into the PVDF film. By applying regular cutting patterns, the geometric constraints of the continuous PVDF film are released, endowing the overall structure with higher deformation freedom under stress and thus significantly improving stretchability. Combined with the natural flexibility of PVDF, the Kirigami design enables the device to achieve both excellent flexibility and stretchability, allowing it to conform to the complex curved surfaces (e.g., cylindrical and spherical surfaces as shown in Figure 4). This makes the device adapt well to human skin surface deformation. In addition, the Kirigami structure concentrates tensile stress in the hinge region, while the PVDF sensitive layer is mainly distributed in the sheet region. This design minimizes the impact of tensile deformation on pressure detection performance, achieving "deformation insensitivity" and ensuring stable signal output even under deformation.

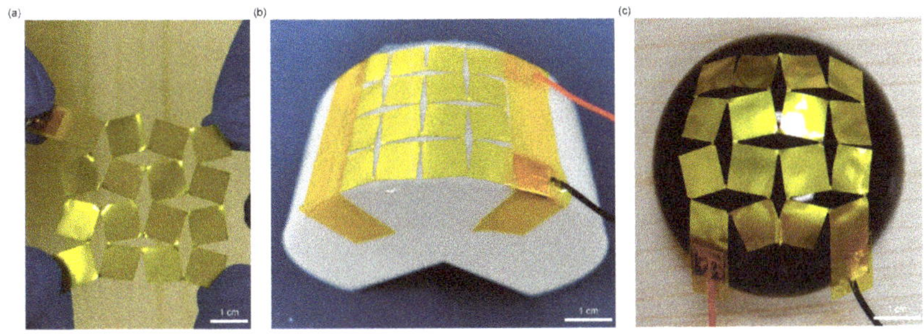

Figure 4. Deformability of the sensor: (**a**) Stretched state; (**b**) Bent state on a cylindrical surface; (**c**) Bent state on a spherical surface.

3.2. Fabrication and Mechanical Properties of the Piezoelectric Sensor with a Kirigami Structure

The Kirigami structure, a geometric design strategy derived from traditional paper-cutting art, endows materials with additional deformation freedom by introducing periodic cuts in continuous films [20–22]. This structure allows the material to undergo controlled local rotation and expansion under external force, thereby achieving macroscale stretchability [23–25]. As shown in Figure 5a, this structure adopts a biaxial design, and its basic unit is defined by geometric parameters including hinge length a, hinge width b, thickness t, and square unit length w—these parameters directly determine the deformation mode of the cuts. Benefiting from this geometric configuration, the Kirigami unit can redistribute stress during in-plane stretching, significantly improving the stretchability and conformability of the overall film. Figure 5b illustrates the stretching principle of the basic unit in the Kirigami structure. When external tensile force acts on the four corners of the structure, four adjacent square sheets rotate synergistically via the hinges, causing the originally closely arranged sheets to gradually open and form a rhombic gap in the middle. This structure transfers the overall deformation of the film to the hinge region, while the square sheet region is barely affected by stress [26–28].

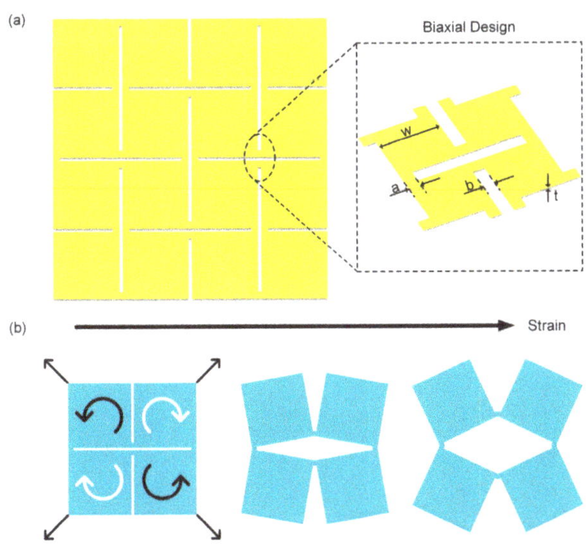

Figure 5. Geometric parameters and mechanical properties of the piezoelectric sensor. (**a**) Schematic diagram of the Kirigami structure with biaxial design on the sensor (the inset shows a magnified view of the structure). (**b**) Schematic diagram of the sensor under tensile deformation.

Figure 1 illustrates the simplified fabrication process of the piezoelectric sensor with a Kirigami design, which is divided into four key sub-steps (i)–(iv). The piezoelectric sensing layer is composed of a PVDF piezoelectric film and Al electrodes on its upper and lower surfaces. After assembling the protective layer and the piezoelectric sensing layer into an integrated structure, an ultraviolet nanosecond laser cutting machine is used to engrave crisscross "Kirigami" structure on the composite film, finally yielding the desired piezoelectric sensor (for detailed fabrication procedures, see Section 2.1).

To compare the load-bearing differences between the Kirigami structure and the planar structure, finite element simulations were conducted (Figure 6). We modeled the PI/PVDF/PI laminate as a single equivalent, homogeneous, isotropic shell that reproduces its in-plane and bending behavior within the small-strain, small-deflection regime. The equivalent material parameters, including Young's modulus, density, and Poisson's ratio of PI/PVDF/PI film, were identified by energy equivalence, as presented in Table 1 [29,30]. The thickness of the shell was 70 μm. Simulation results show that under external displacement load, the Kirigami structure effectively alleviates stress concentration and exhibits a better overall mechanical response than the traditional planar structure. For the intact planar structure, as the displacement U increases, stress concentrates rapidly in the four corner regions, exhibiting an obvious stress concentration effect. This not only limits its overall stretchability but also tends to cause premature failure of local materials. In contrast, the Kirigami structure incorporates periodic cuts, which converts external stretching into unit rotation and relative displacement, enabling stress redistribution within the film. As the displacement increases from 0 mm to 3 mm, the basic panel units gradually open up, forming uniform rhombic gaps. It can be observed that stress is mainly concentrated in the hinge regions, while the square panel regions only experience minimal stress—this effectively prevents the core sensing layer from failing, demonstrating the deformation insensitivity characteristic of the Kirigami structure. These results indicate that, based on its periodic cuts, the Kirigami structure outperforms the traditional planar structure in terms of stress concentration, making it a preferred option for stretchable structural designs involving core sensing elements.

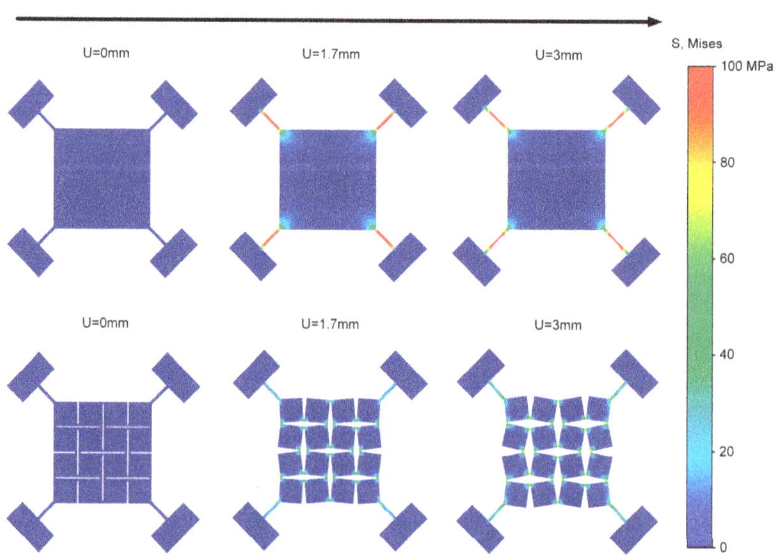

Figure 6. Tensile simulation results of the planar structure and the Kirigami structure.

Table 1. Comparison of physical parameters of PI, PVDF, and PI/PVDF/PI film.

	PI	PVDF	PI/PVDF/PI
Young's modulus (MPa)	2500	2500	2500
Poisson's ratio	0.34	0.35	0.34
Density (kg/m^3)	1.40×10^{12}	1.78×10^{12}	1.40×10^{12}
Thickness (μm)	20	30	70

To further clarify the influence of geometric parameters on structural stretchability, stretching simulations were performed by adjusting the hinge length a, hinge width b, and thickness t separately. Tensile reaction forces at each state were extracted using Abaqus tools, where the horizontal axis represents reaction forces at the four corners and the vertical axis denotes the rate of area change (calculated by dividing the area difference resulting from membrane deformation by the original area). Figure 7a plots the relationship between area change rate and tensile reaction force across varying hinge lengths a. It shows that for the same applied force, a smaller hinge length a results in a larger area change rate. Notably, under a 0.05 N tensile force, a 1.5 mm hinge length yields a 0.8% area change rate, while a 0.5 mm hinge length produces a 5.6% rate. This indicates that a smaller hinge length a correlates with superior tensile performance. Figure 7b illustrates how area change rate varies with tensile reaction force at different hinge widths b. It reveals that for the same applied force, a larger hinge width b leads to a higher area change rate. Specifically, when subjected to a 0.05 N tensile force, a 0.5 mm hinge width produces a 1.9% area change rate, whereas a 1.5 mm hinge width yields a 2.5% area change rate. This suggests that a larger hinge width b enhances tensile properties, but its impact on area change rate is relatively minor compared to that of hinge length and thickness. Figure 7c displays how the area change rate varies with tensile reaction force for different hinge thicknesses t. It indicates that for the same applied force, a thinner hinge thickness t yields a larger area change rate. In particular, under a 0.05 N tensile force, a 0.125 mm hinge thickness produces a 1.1% area change rate, while a 0.025 mm hinge thickness achieves a 5.8% area change rate. This confirms that a smaller hinge thickness t improves better tensile properties. Integrating these simulation results and analyses, we conclude that to enhance the tensile properties of Kirigami structures, it is advisable to use slender and thin hinges—specifically, a smaller hinge length a, a larger hinge width b, and a thinner hinge thickness t. Additionally, for manufacturing and design convenience, this study selected a hinge length (a) of 1 mm, a hinge width (b) of 0.5 mm, a hinge thickness (t) of 0.07 mm, and a square panel size (w) of 1 cm.

3.3. Characterization of the Electrical Properties of the Piezoelectric Sensor

To facilitate the testing of heart sound signals, a loudspeaker excitation method is adopted (The specific test apparatus is shown in Figure S1. When the piezoelectric sensor is subjected to a sound pressure of 88 dB at a frequency of 160 Hz, its output voltage is approximately 0.35 V (Figure 8a). The sensitivity curve of the sensor is plotted in Figure S2. To investigate the effect of Sound Pressure Level (SPL) on the output characteristics, we kept the frequency constant and varied the SPL; the results show that the output voltage exhibits an approximately linear increase with increasing SPL, with an R^2 value of 0.99. When the SPL increases from 70 dB to 94 dB, the output voltage rises from 0.11 V to 0.42 V (Figure 8b). Considering that cutting the sensor into a Kirigami structure might affect its piezoelectric performance, we conducted sound pressure excitation tests on both the planar-structured and Kirigami-structured sensors to verify whether the structural optimization impairs the piezoelectric conversion efficiency of PVDF. Under sound pressure excitation at 88 dB and 160 Hz, the output voltages of the two structures are 0.34 V and 0.35 V, respectively, with a relative difference of only 2.9% (Figure 8c). This result confirms that Kirigami cutting only alters the sensor's structural mechanical properties and does not cause adverse effects on its piezoelectric conversion efficiency. With the SPL fixed, we further tested the effect of excitation frequency (60 Hz to 160 Hz) on the output, obtaining a frequency response curve: as the frequency increases from 60 Hz to 160 Hz, the output voltage rises from 9 mV to 0.35 V (Figure 8d). We also characterized the Signal-to-Noise Ratio (SNR) in Figure S3, which showed that the sensor had a SNR of 10 dB at 60 Hz and 42 dB at 160 Hz, indicating good signal quality.

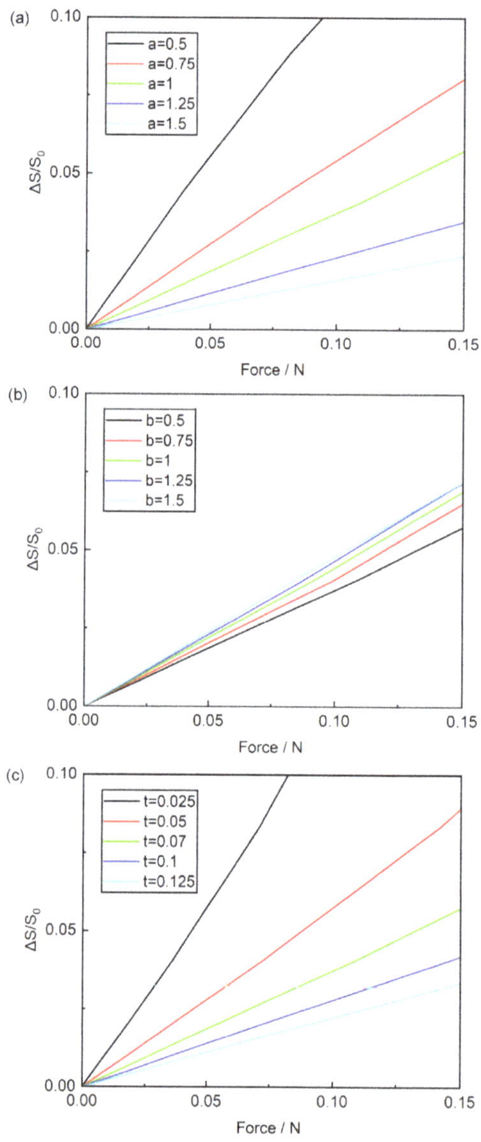

Figure 7. Effects of (**a**) hinge length a, (**b**) hinge width b, and (**c**) hinge thickness t on the area change rate ($\Delta S/S_0$).

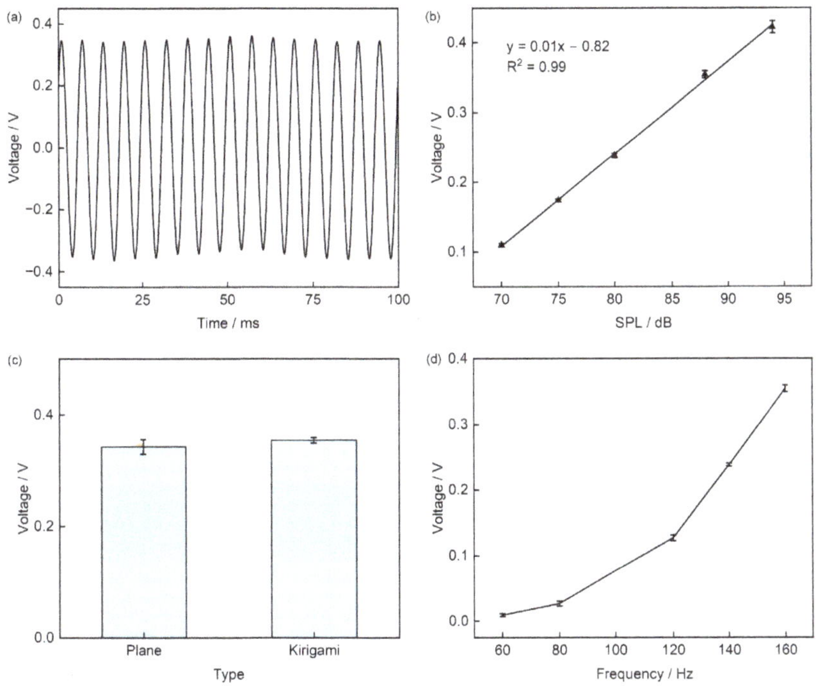

Figure 8. Electrical Performance Characterization of the Sensor. (**a**) Output voltage under sound pressure excitation at 160 Hz and 88 dB; (**b**) Effect of different Sound Pressure Levels (SPL) on output voltage; (**c**) Comparison of output voltage before and after cutting; (**d**) Frequency response of the sensor in the range of 60–160 Hz.

Notably, owing to the introduction of the Kirigami structure, the sensor exhibits the key characteristic of insensitivity of the output response to deformation—we applied four levels of tensile deformation (0%, 11%, 23%, and 30%) to four sensors, and their output voltage only showed minor fluctuations (Figure 9a and Figure S4), with a maximum variation of approximately 11% (compared to the 0% tensile state). This is because when the sensor deforms, the square panels undergo almost only in-plane rotation and displacement, without causing additional stress damage to the PVDF sensing layer. To meet the practical application requirement of heart sound signal measurement on the human skin surface, we further characterized the sensor's output performance under different bending conditions (Figure 9b). Under the same SPL and frequency, the sensor's output decreases slightly as the bending radius decreases: when the bending radius is 11 mm, the output voltage is 0.29 V, a decrease of approximately 17% compared to the non-bent state. The reason for this may be that as the bending radius decreases, some regions of the square panels become nearly perpendicular to the sound pressure plane, leading to reduced sound pressure reception efficiency; nevertheless, the sensor still maintains a relatively high output amplitude, demonstrating its stable operating capability on curved surface.

Finally, to verify the long-term reliability of the sensor, we performed a 1200 s continuous sound pressure excitation test on it (88 dB, 160 Hz). The results show that the sensor's output voltage remains essentially stable, with a fluctuation range of less than 9% (Figure 9c). Magnifying the voltage waveforms around the two time points (60 s [I] and 1000 s [II]) reveals that the output voltage decreases from the initial 0.35 V to 0.32 V. This slight decrease may stem from minor sound pressure drift caused by the long-term operation of the loudspeaker. Nonetheless, the sensor's output still remains at a relatively high voltage level (approximately 0.32 V), fully confirming its long-term operational stability.

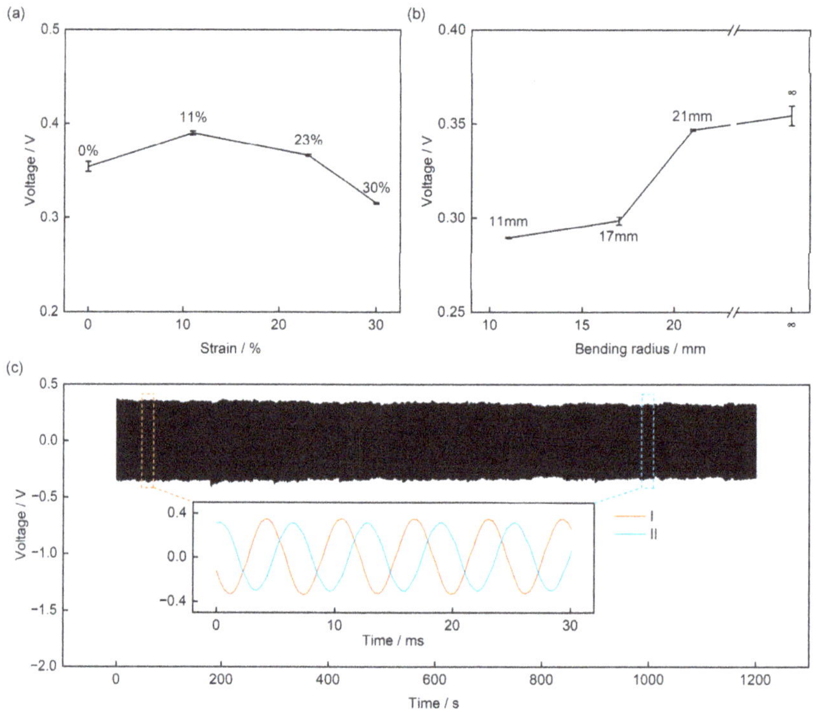

Figure 9. Electrical Performance of the Sensor under Mechanical Deformation and Long-Term Stability Tests. (**a**) Output voltage under tensile conditions. (**b**) Output voltage under bending conditions. (**c**) 1200 s cyclic stability test.

3.4. Application in Heart Sound Monitoring

Heart sound signals are acoustic manifestations of mechanical vibrations generated during cardiac contraction and relaxation, and are widely used in disease diagnosis [1–4]. In a normal cardiac cycle, the main components are the first heart sound (S1, generated by atrioventricular valves closure) and the second heart sound (S2, generated by semilunar valves closure) [4,31,32]. Figure 10a presents the original signal and its processing results recorded by the self-made piezoelectric signal acquisition board with a sampling frequency of 1000 Hz. Specifically, the original signal was processed using a 1 Hz low-pass filter and a 20–200 Hz band-pass filter to obtain the respiratory wave and heart sound signal, respectively. Figure 10b is an enlarged view of the signal in the 6–8 s time window in Figure 10a, from which the S1 and S2 components of the heart sound signal can be clearly identified. Furthermore, by expanding the signal analysis range to 5 consecutive cardiac cycles and calculating the frequencies of the two components separately, the results show that the dominant frequency of the S2 component (mean: 38.6 Hz) is slightly higher than that of S1 (mean: 30 Hz) (Figure 10c). This frequency difference aligns with the physiological mechanism of heart sound generation—semilunar valves close faster than atrioventricular valves, leading to higher-frequency mechanical vibrations, thus demonstrating the typical characteristics of normal heart sound signals.

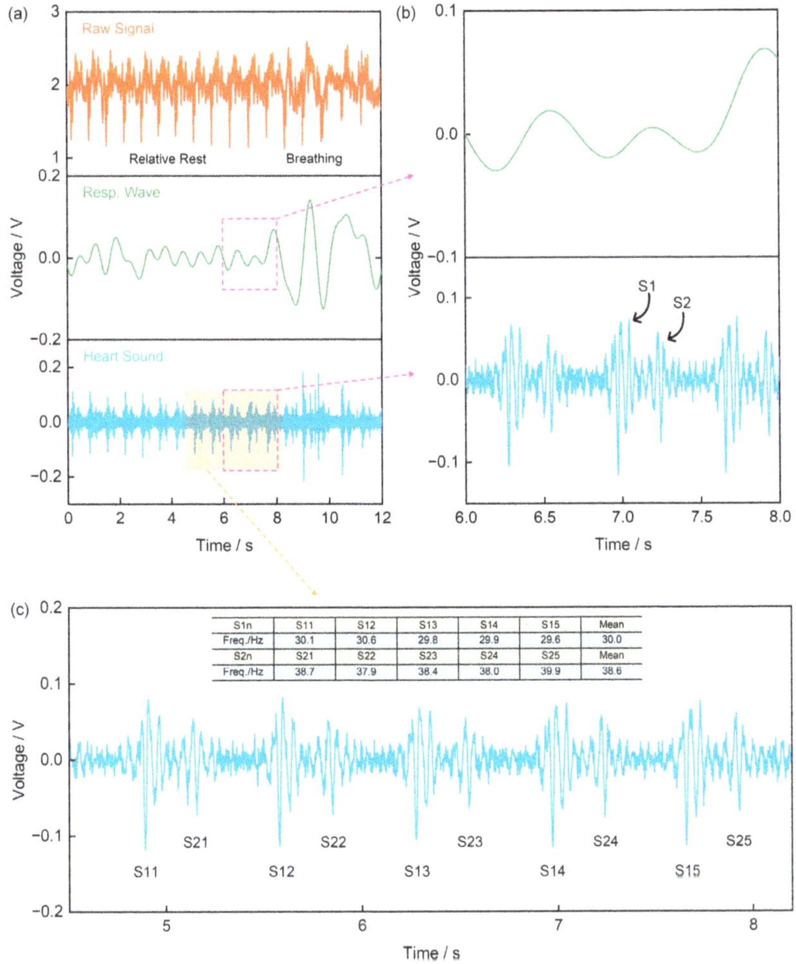

Figure 10. Application of the Piezoelectric Sensor in Heart Sound Monitoring. (**a**) The recorded original signal, along with the respiratory wave and heart sound signal obtained via a 1 Hz low-pass filter and a 20–200 Hz band-pass filter, respectively; (**b**) Enlarged view of the signal in the 6–8 s time window from (**a**), where the upper trace is the respiratory wave and the lower trace is the heart sound signal; (**c**) Average frequencies of the first heart sound (S1) and the second heart sound (S2).

To verify the reliability of the piezoelectric sensor in monitoring heart sound signals, we used an ECG sensor, our piezoelectric sensor, and a commercial heart sound sensor (HKY sensor) to record heart sounds, simultaneously (Figure 11). The results indicate a clear temporal correlation between the ECG signal and the heart sound signal. Specifically, the QRS complex in the ECG marks the start of ventricular contraction, immediately followed by S1, which is generated by the closure of the atrioventricular valves; in contrast, S2, which appears near the T wave, corresponds to the closure of the semilunar valves and the onset of ventricular diastole [4,33–35]. We compared the heart sound cycles recorded by our sensor and commercial sensor, both measuring the S1–S1 peak time interval (Table S1). Results showed that the Mean Absolute Error (MAE) between our sensor and commercial sensor was 0.015 s, with a Mean Relative Error (MRE) of 2.08%. These small errors indicate that our sensor effectively records heart sound signals. Additionally, we compared the temporal correlation between our sensor signals and ECG signals. Figure S5a shows the correspondence between the R-wave peak time of the ECG and the S1 peak time of our sensor, while Figure S5b illustrates the correspondence between the T-wave end time of the ECG and the S2 peak time of our sensor. The fitting results from both figures demonstrate that the heart sound signals measured by our sensor exhibit well temporal correspondence with the ECG signals. In summary, comparisons with ECG and commercial sensors confirm the reliability of our piezoelectric sensor in heart sound monitoring.

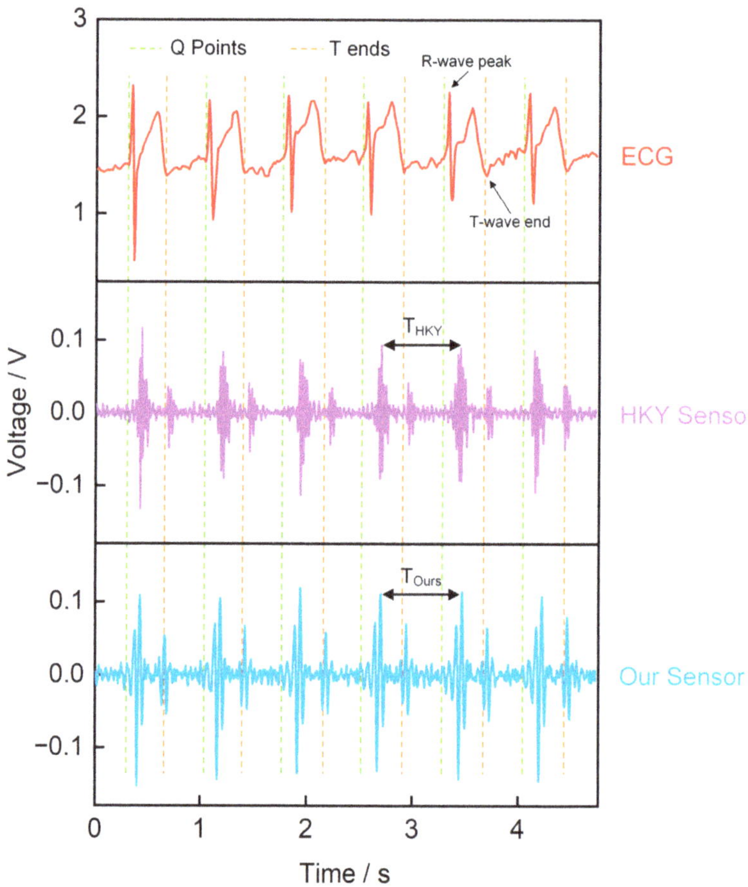

Figure 11. Simultaneously recorded waveforms. The upper trace is the ECG, the middle trace is the commercial sensor signal, and the lower trace is our piezoelectric sensor signal.

Finally, to verify the sensor's feasibility for long-term monitoring, we conducted a test in which the sensor was worn on the left precordial area over a 7-day period using 3M medical transparent adhesive tape, with measurements taken every two days (Figure 12). The acquired signals clearly show the characteristics of S1 and S2 peaks of heart sound signals. Further analysis across the four tests reveals that the sensor's output voltage reached approximately 0.13 V in the first test, slightly higher than the 0.09 V and 0.1 V observed in subsequent tests. We attribute this discrepancy to that the adhesive tape was initially stronger; after prolonged wear, its adhesiveness weakened, leading to partial detachment at the skin–device interface. This reduced contact explains the lower signal amplitudes in later measurements. Notably, while the amplitude decreased, the signal frequency content remained unchanged. Based on test results and analysis, our sensor demonstrates potential for application in long-term heart sound monitoring scenarios.

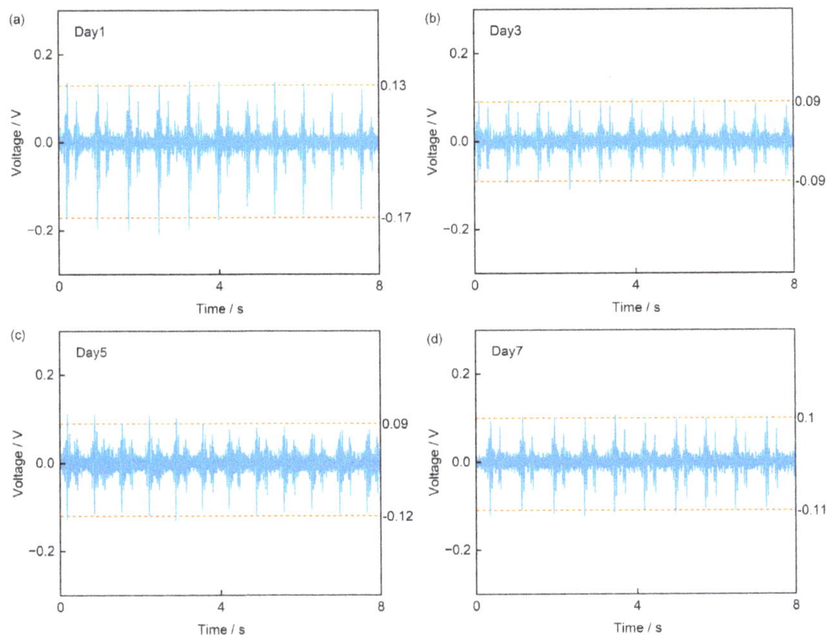

Figure 12. Comparison of heart sound signals from periodic tests over 7 days. (**a**) Day 1, (**b**) Day 3, (**c**) Day 5, (**d**) Day 7.

4. Conclusions

This study proposes and verifies a flexible PVDF piezoelectric heart sound sensor based on Kirigami structure. The device adopts a five-layer "sandwich" structure of PI-Electrode-PVDF-Electrode-PI. While maintaining flexibility, it releases geometric constraints via the Kirigami structure, achieving high adaptability to dynamic skin deformation. Finite element simulations and geometric parameter tuning further demonstrate that the Kirigami design facilitates effective stress transfer. Specifically, stretchability exhibits higher sensitivity to hinge length a and thickness t, but relatively low sensitivity to hinge width b—thereby providing guidance for subsequent designs. In terms of electrical characterization, the sensor exhibits an approximately linear output in response to sound pressure excitation. When the SPL increases from 70 to 94 dB, the output voltage rises from approximately 0.11 V to 0.42 V; when the frequency is scanned from 60 to 160 Hz, the output voltage increases from roughly 9 mV to 0.35 V. Under stretching and bending conditions, the device maintains stable output performance with "deformation insensitivity": at tensile levels of 0%, 11%, 23%, and 30%, the maximum amplitude variation is approximately 11%; When the bending radius is reduced to 11 mm, the voltage is about 0.29 V, which is a decrease of roughly 17% compared to the planar state—nevertheless, it still maintains a favorable voltage level. Long-term and cyclic tests also confirm its stability: under 1200 s of continuous excitation, the output fluctuation is within approximately 9%; in the 7-day heart sound measurements with the sensor attached to the chest, the S1 and S2 components can still be clearly identified. Additionally, the synchronized comparison with the ECG sensor and commercial heart sound sensor reveals a clear temporal correlation. Overall, the Kirigami-structured PVDF piezoelectric sensor developed in this study achieves a balance among flexibility, stretchability, and output stability. It can reliably capture heart sound signals under stretching, bending, and long-term wear conditions, providing a feasible approach for wearable long-term heart sound monitoring and early screening of cardiovascular diseases. It is worth noting that the sample size n for human subjects in this study is only 1, which may result in the limitations in terms of sample representativeness and population universality. We plan to expand the sample size and integrate clinical needs in the future to further enhance the sensor's application value.

Supplementary Materials: The following supporting information can be downloaded at: https://www.mdpi.com/article/10.3390/s25237253/s1, Figure S1: Sound Pressure Excitation Test Diagram, Figure S2: Sensor Sensitivity (mV/Pa) Curve, Figure S3: Signal-to-noise ratio (SNR) of the sensor, Figure S4: Output voltages of other three identical sensors (Figure a, Figure b, Figure c) under different tensile conditions, Figure S5: Sensor Temporal Correlation; Table S1: Average Time Difference in Heart Sound Cycles Recorded by Our Sensor and Commercial Sensor.

Author Contributions: Conceptualization, X.Z. and T.Z.; methodology, X.Z. and T.Z.; validation, X.Z., X.Y. and X.L.; investigation, X.Z.; data curation, X.Z.; writing—original draft preparation, X.Z.; writing—review and editing, T.Z.; supervision, H.H., H.L., S.J., G.Y. and T.Z.; project administration, T.Z.; funding acquisition, T.Z. All authors have read and agreed to the published version of the manuscript.

Funding: This work was supported in part by the National Key Research and Development Program under Grant 2024YFB3612102 and in part by the National Natural Science Foundation of China under Grant 52305607, Grant 51805311, and Grant 62471282.

Informed Consent Statement: Informed consent was obtained from all subjects involved in the study.

Data Availability Statement: The data presented in this study are available on request from the corresponding author.

Acknowledgments: The authors are grateful to the Key Laboratory of Advanced Display and System Application, Ministry of Education, Shanghai University.

Conflicts of Interest: The authors declare no conflicts of interest.

Abbreviations

The following abbreviations are used in this manuscript:

PVDF	Polyvinylidene Fluoride
ECG	electrocardiogram
BCG	ballistocardiogram
FCG	forcecardiography
PI	Polyimide
Al	aluminum
SPL	Sound Pressure Level
S1	the first heart sound
S2	the second heart sound
MAE	Mean Absolute Error
MRE	Mean Relative Error
SNR	Signal-to-Noise Ratio

References

1. Hui, X.; Tang, L.; Zhang, D.; Yan, S.; Li, D.; Chen, J.; Wu, F.; Wang, Z.L.; Guo, H. Acoustically Enhanced Triboelectric Stethoscope for Ultrasensitive Cardiac Sounds Sensing and Disease Diagnosis. *Adv. Mater.* **2024**, *36*, 2401508. [CrossRef]
2. Nayeem, M.O.G.; Lee, S.; Jin, H.; Matsuhisa, N.; Jinno, H.; Miyamoto, A.; Yokota, T.; Someya, T. All-Nanofiber–Based, Ultrasensitive, Gas-Permeable Mechanoacoustic Sensors for Continuous Long-Term Heart Monitoring. *Proc. Natl. Acad. Sci. USA* **2020**, *117*, 7063–7070. [CrossRef] [PubMed]

3. Li, Y.; Huang, Y.; Zhao, N. Low-Intensity Sensitive and High Stability Flexible Heart Sound Sensor Enabled by Hybrid Near-Field/Far-Field Electrospinning. *Adv. Funct. Mater.* **2023**, *33*, 2300666. [CrossRef]
4. Han, L.; Liang, W.; Xie, Q.; Zhao, J.; Dong, Y.; Wang, X.; Lin, L. Health Monitoring via Heart, Breath, and Korotkoff Sounds by Wearable Piezoelectret Patches. *Adv. Sci.* **2023**, *10*, 2301180. [CrossRef] [PubMed]
5. Arts, L.; Lim, E.H.T.; van de Ven, P.M.; Heunks, L.; Tuinman, P.R. The Diagnostic Accuracy of Lung Auscultation in Adult Patients with Acute Pulmonary Pathologies: A Meta-Analysis. *Sci. Rep.* **2020**, *10*, 7347. [CrossRef]
6. Hafke-Dys, H.; Bręborowicz, A.; Kleka, P.; Kociński, J.; Biniakowski, A. The Accuracy of Lung Auscultation in the Practice of Physicians and Medical Students. *PLoS ONE* **2019**, *14*, e0220606. [CrossRef]
7. Meng, K.; Xiao, X.; Liu, Z.; Shen, S.; Tat, T.; Wang, Z.; Lu, C.; Ding, W.; He, X.; Yang, J.; et al. Kirigami-Inspired Pressure Sensors for Wearable Dynamic Cardiovascular Monitoring. *Adv. Mater.* **2022**, *34*, 2202478. [CrossRef]
8. Wang, C.; Li, X.; Hu, H.; Zhang, L.; Huang, Z.; Lin, M.; Zhang, Z.; Yin, Z.; Huang, B.; Gong, H.; et al. Monitoring of the Central Blood Pressure Waveform via a Conformal Ultrasonic Device. *Nat. Biomed. Eng.* **2018**, *2*, 687–695. [CrossRef]
9. Wang, C.; Qi, B.; Lin, M.; Zhang, Z.; Makihata, M.; Liu, B.; Zhou, S.; Huang, Y.; Hu, H.; Gu, Y.; et al. Continuous Monitoring of Deep-Tissue Haemodynamics with Stretchable Ultrasonic Phased Arrays. *Nat. Biomed. Eng.* **2021**, *5*, 749–758. [CrossRef]
10. Parlato, S.; Esposito, D.; Centracchio, J.; Andreozzi, E.; Gragnaniello, M.; Riccio, M.; Bifulco, P. A New, Simple Wrist-Mounted PVDF Sensor for Continuous Heart Rate Monitoring. In Proceedings of the 2024 IEEE Sensors Applications Symposium (SAS), Naples, Italy, 23–25 July 2024; pp. 1–6.
11. Ansari, S.; Golbus, J.R.; Tiba, M.H.; Mccracken, B.; Wang, L.; Aaronson, K.D.; Ward, K.R.; Najarian, K.; Oldham, K.R. Detection of Low Cardiac Index Using a Polyvinylidene Fluoride-Based Wearable Ring and Convolutional Neural Networks. *IEEE Sens. J.* **2021**, *21*, 14281–14289. [CrossRef]
12. Manh, L.-N.; Li, J.; Kweon, H.; Chae, Y. Simultaneous Measurement of Two Biological Signals Using a Multi-Layered Polyvinylidene Fluoride Sensor. *Sci. Rep.* **2022**, *12*, 1507. [CrossRef]
13. Kumaki, D.; Motoshima, Y.; Higuchi, F.; Sato, K.; Sekine, T.; Tokito, S. Unobstructive Heartbeat Monitoring of Sleeping Infants and Young Children Using Sheet-Type PVDF Sensors. *Sensors* **2023**, *23*, 9252. [CrossRef]
14. Parlato, S.; Centracchio, J.; Cinotti, E.; Gargiulo, G.D.; Esposito, D.; Bifulco, P.; Andreozzi, E. A Flexible PVDF Sensor for Forcecardiography. *Sensors* **2025**, *25*, 1608. [CrossRef] [PubMed]
15. Hu, H.; Tian, H.; Gao, Y.; Wan, Z.; Wang, L.; Xu, H.; Wang, C.; Shao, J.; Zheng, Z. Revisiting the Contact Splitting Hypothesis: An Effective Route for Enhancing Adhesion on Rough Surface. *J. Mech. Phys. Solids* **2023**, *170*, 105121. [CrossRef]
16. Hu, H.; Guo, X.; Zhang, Y.; Chen, Z.; Wang, L.; Gao, Y.; Wang, Z.; Zhang, Y.; Wang, W.; Rong, M.; et al. Elasto-Plastic Design of Ultrathin Interlayer for Enhancing Strain Tolerance of Flexible Electronics. *ACS Nano* **2023**, *17*, 3921–3930. [CrossRef]
17. Hong, L.; Zhang, H.; Kraus, T.; Jiao, P. Ultra-Stretchable Kirigami Piezo-Metamaterials for Sensing Coupled Large Deformations. *Adv. Sci.* **2024**, *11*, 2303674. [CrossRef]
18. Hazarika, D.; Lu, J.; Wu, J.; Shah, M.N., Li, J., Zhang, K.; Xu, L.; Chen, C.; Cao, Z.; Jin, H.; et al. Ion Dipole Interaction and Directional Alignment Enabled High Piezoelectric Property Polyvinylidene Fluoride for Flexible Electronics. *npj Flex. Electron.* **2025**, *9*, 20. [CrossRef]
19. Huang, S.; Gao, Y.; Hu, Y.; Shen, F.; Jin, Z.; Cho, Y. Recent Development of Piezoelectric Biosensors for Physiological Signal Detection and Machine Learning Assisted Cardiovascular Disease Diagnosis. *RSC Adv.* **2023**, *13*, 29174–29194. [CrossRef]
20. Xie, Q.; Han, L.; Liu, J.; Zhang, W.; Zhao, L.; Liu, Y.; Chen, Y.; Li, Y.; Zhou, Q.; Dong, Y.; et al. Kirigami-Inspired Stretchable Piezoelectret Sensor for Analysis and Assessment of Parkinson's Tremor. *Adv. Healthc. Mater.* **2025**, *14*, 2402010. [CrossRef] [PubMed]
21. Tao, J.; Khosravi, H.; Deshpande, V.; Li, S. Engineering by Cuts: How Kirigami Principle Enables Unique Mechanical Properties and Functionalities. *Adv. Sci.* **2023**, *10*, 2204733. [CrossRef]
22. Kang, G.R.; Hwang, G.W.; Lim, D.; Jeon, S.H.; Song, M.; Hong, C.-H.; Kim, H.J.; Pang, C. Robustly Repeatable, Permeable, and Multi-Axially Stretchable, Adhesive Bioelectronics with Super-Adaptive Conductive Suction Cups for Continuously Deformable Biosurfaces. *Adv. Sci.* **2025**, *12*, 2500346. [CrossRef]
23. Mao, C.; Jin, J.; Mei, D.; Wang, Y. Development of Kirigami-Patterned Stretchable Tactile Sensor Array with Soft Hinges for Highly Sensitive Force Detection. *Adv. Sens. Res.* **2024**, *3*, 2400012. [CrossRef]
24. Deng, Y.; Xu, K.; Jiao, R.; Liu, W.; Cheung, Y.K.; Li, Y.; Wang, X.; Hou, Y.; Hong, W.; Yu, H. Rotating Square Tessellations Enabled Stretchable and Adaptive Curved Display. *npj Flex. Electron.* **2024**, *8*, 4. [CrossRef]
25. Jang, B.; Won, S.; Kim, J.; Kim, J.; Oh, M.; Lee, H.-J.; Kim, J.-H. Auxetic Meta-Display: Stretchable Display without Image Distortion. *Adv. Funct. Mater.* **2022**, *32*, 2113299. [CrossRef]
26. Cho, Y.; Shin, J.-H.; Costa, A.; Kim, T.A.; Kunin, V.; Li, J.; Lee, S.Y.; Yang, S.; Han, H.N.; Choi, I.-S.; et al. Engineering the Shape and Structure of Materials by Fractal Cut. *Proc. Natl. Acad. Sci. USA* **2014**, *111*, 17390–17395. [CrossRef]
27. Rao, Z.; Lu, Y.; Li, Z.; Sim, K.; Ma, Z.; Xiao, J.; Yu, C. Curvy, Shape-Adaptive Imagers Based on Printed Optoelectronic Pixels with a Kirigami Design. *Nat. Electron.* **2021**, *4*, 513–521. [CrossRef]

28. Jiang, S.; Liu, J.; Xiong, W.; Yang, Z.; Yin, L.; Li, K.; Huang, Y. A Snakeskin-Inspired, Soft-Hinge Kirigami Metamaterial for Self-Adaptive Conformal Electronic Armor. *Adv. Mater.* **2022**, *34*, 2204091. [CrossRef] [PubMed]
29. Nettles, A.T. *Basic Mechanics of Laminated Composite Plates*; NASA Reference Publication 1351; NASA: Washington, DC, USA, 1994. Available online: https://ntrs.nasa.gov/api/citations/19950009349/downloads/19950009349.pdf (accessed on 17 October 2025).
30. Arockia Julias, A. Rule of Mixture for Composites. 2021. Available online: https://crescent.education/wp-content/uploads/2021/03/Rule-of-Mixture-of-composites.pdf (accessed on 17 October 2025).
31. Chen, H.; Yu, S.; Liu, H.; Liu, J.; Xiao, Y.; Wu, D.; Pan, X.; Zhou, C.; Lei, Y.; Liu, S. A Two-Stage Amplified PZT Sensor for Monitoring Lung and Heart Sounds in Discharged Pneumonia Patients. *Microsyst. Nanoeng.* **2021**, *7*, 55. [CrossRef] [PubMed]
32. Tamborini, A.; Gharib, M. Listening to Heart Sounds through the Pressure Waveform. *Sci. Rep.* **2024**, *14*, 26824. [CrossRef]
33. Abdessater, E.; Balali, P.; Pawlowski, J.; Rabineau, J.; Tordeur, C.; Faoro, V.; van de Borne, P.; Hossein, A. A Novel Method for ECG-Free Heart Sound Segmentation in Patients with Severe Aortic Valve Disease. *Sensors* **2025**, *25*, 3360. [CrossRef]
34. Liu, C.; Springer, D.; Li, Q.; Moody, B.; Juan, R.A.; Chorro, F.J.; Castells, F.; Roig, J.M.; Silva, I.; Johnson, A.E.W.; et al. An Open Access Database for the Evaluation of Heart Sound Algorithms. *Physiol. Meas.* **2016**, *37*, 2181. [CrossRef] [PubMed]
35. Springer, D.B.; Tarassenko, L.; Clifford, G.D. Logistic Regression-HSMM-Based Heart Sound Segmentation. *IEEE Trans. Biomed. Eng.* **2016**, *63*, 822–832. [CrossRef] [PubMed]

sensors

MDPI

Article

A Touch on Musical Innovation: Exploring Wearables and Their Impact on New Interfaces for Musical Expression

David Wexler [1], Joanne Yip [1,2,*], Ka-Po Lee [1], Xiaolu Li [1] and Yiu-Hong Wong [1]

1 School of Fashion and Textiles, The Hong Kong Polytechnic University, Hung Hom, Kowloon, Hong Kong SAR, China; d.wexler03@gmail.com (D.W.); maple.lee@connect.polyu.hk (K.-P.L.); xiao-lu.li@connect.polyu.hk (X.L.); yiu-hong-97.wong@connect.polyu.hk (Y.-H.W.)

2 Photonics Research Institute, The Hong Kong Polytechnic University, Hung Hom, Kowloon, Hong Kong SAR, China

* Correspondence: tcjyip@polyu.edu.hk; Tel.: +852-2766-4848

Abstract: This paper explores the innovative concept of using wearable technologies as a medium for musical expression. Special emphasis is placed on a unique wearable device equipped with motion, touch, and acceleration sensors, which can be used as a wrist strap, hand strap, or surface drum pad. The aim is to create a new musical instrument that simplifies music learning and expression and makes them more intuitive. The wearable device contains 32 individual touch-sensitive pressure sensors, a nine-axis inertial-measurement-unit motion sensor, and various light-emitting diode and vibrational haptic-feedback components. The inclusion of tactile and intuitive features in the wearable device enhances the musical experience of users by enabling engaging interaction. Consequently, it is believed that this groundbreaking technology has significant potential to contribute to the field of music, providing musicians with a versatile and intuitive instrument that facilitates their creative expression.

Keywords: wearable technologies; wearable music instrument; pressure sensors; motion capture; music learning; musicians; gyroscope; accelerometer

1. Introduction

People have been making music throughout recorded history, using all kinds of instruments, from ancient flutes in 80,000 BC to Babylonian bells in 2000 BC [1]. Even today, people can still make music with nothing but their own bodies, as seen in body percussion. This idea—that anyone can make music—has driven the recent development of wearable technology designed for music creation.

Traditional ways of teaching music, which focus mostly on classical instruments, are changing. While these instruments have much to offer in terms of sound and skill, they can be difficult and expensive to learn. Their complexity can discourage those who want to learn music, especially considering issues like cost and lack of easy access to lessons. Recent data show that as many as 31% of adults and children decide not to learn to play classical instruments, a statistic that points to problems in the current approach to music education [2]. On the other hand, the burgeoning market in electronic musical instruments, which includes technologies such as finger drum pads and virtual music software, reflects the growing appetite for more accessible means of musical expression. This market was valued at US $678.9 million in 2022 and is projected to grow by another 50% within the next decade. The shift towards electronic instruments is part of an emerging trend towards simplicity and immediacy in music creation [3]. The growing convergence of wearable technology and music-making reflects a demand for simpler and more accessible means of musical expression. With the worth of wearable tech industry anticipated to reach an estimated US $931.31 billion by 2030, a new horizon of transformative possibilities lies before us [4]. Wearable technology presents an opportunity to bridge the accessibility gap in musical expression by providing a new way for anyone to make music.

Music creation through movement is not a new concept, and it has been reimagined through recent developments in wearable music technology [5]. Researchers have explored various methods for using gesture data, sensory data, and multisensory data from instrumental performers to create personalized music materials [6,7]. For example, Marlon et al. investigated physical gestures from the perspective of computer-aided composition (CAC) and proposed solutions and implementation for the integration of gesture signals as musical materials [8]. Additionally, Gunther et al. built a compositional language using the sense of touch and created a vibrotactile composition suit implementing that tactile composition system [9]. Furthermore, new musical instruments with gestural sensors have also been introduced [10], and electronic variants have transformed the musical gestalt of player–instrument and listener–music relations [11]. These advancements showcase the potential of utilizing gesture and sensory data for musical expression. Such approaches can enhance the interaction between instruments and instrument players. As a result, they have the potential to expedite the development of personalized music materials.

However, the development of wearable music instruments poses significant challenges due to various requirements, including miniaturization, seamless integration of sensors, efficient signal processing, low latency, reliable connectivity, effective power management, optimal user experience, and affordability. The implementation of this technology requires the design of a wearable device equipped with all these features, which necessitates expertise in electronics, sensor technology, signal processing techniques, and human-computer interaction. Additionally, considerations such as optimization of power consumption, intuitive interface design, and cost-effectiveness further contribute to the complexity involved in creating a successful wearable music instrument. Therefore, the development process typically involves iterative prototyping, rigorous testing, and continuous refinement to create a device that successfully addresses these multifaceted challenges.

Pioneering commercial devices such as DrumPants, Freedrum, Mi.Mu Gloves, and Remidi gloves have graced the global market since 2008 and have been primarily backed by community-funding platforms such as Kickstarter and Indiegogo. These inventions emphasize the employment of motion sensors for music creation and draw in early adopters, a group consisting of enthusiastic learners with different musical backgrounds and knowledge levels [12]. The integration of touch-sensitive pressure sensors is a secondary innovation in these devices [13] that allows users to augment their musical expression through finger and palm-pressure controls, as well as through body movement. However, these sensors have serval limitations. For instance, DrumPants digitalizes the human body by replicating the typical movements of someone drumming on a drum kit [14]. However, it experiences latency issues when large numbers of tracks and effects are added. Additionally, although Freedrum offers valuable insight into the mechanics of virtual music creation because it utilizes a gyroscope and accelerometer, it should be noted that the sensors are attached to the drumsticks rather than directly to the body of the wearer. The Mi.Mu and Remidi e-gloves manifest the essence of wearable technology for music creation by integrating flex and pressure sensors, as well as nine-axis inertial measurement units (IMUs) [15], but they are expensive.

This project aims to address the limitations of the currently available wearable music devices by introducing an innovative and more affordable miniaturized wearable music instrument that makes music learning more accessible and enhances the user experience. It can be worn as a wrist strap, hand strap, or surface drum pad, offering enhanced versatility compared to traditional glove-style devices. The primary objective, to provide musicians and enthusiasts with a unique and intuitive method of interacting with music, is achieved through the simplification of traditional musical instruments and feedback systems. This approach is believed to enhance both musical expression and creativity. The device incorporates advanced sensor technology, including 32 touch-sensitive pressure sensors and a nine-axis inertial measurement unit motion sensor into the wearable device, enabling precise motion and touch detection. Additionally, the wearable device features a light-emitting diode and vibrational haptic feedback components, providing tactile and intuitive feedback. Through the integration of these components, the wearable device offers an immersive and interactive musical experience. Furthermore, this project also seeks to explore the potential of wearable technology to drive musical innovation. By creating a symbiotic relationship between musicians and their music through this novel instrument, the project aspires to revolutionize the music industry and empower users to realize their full artistic potential.

2. Materials and Methods

This study employs a practice-based-design research methodology, with a specific focus on qualitative analysis. The research process began with an in-depth examination of related products and technologies through case studies, aiming to identify their strengths and limitations. Six case studies were included in this research: (1) KaiKU gloves, (2) Mi.Mu gloves, (3) Remidi T8 gloves, (4) DrumPants, (5) AUUG wearable device, and (6) Mogees. Following the case-study analysis, two innovative prototypes were designed and developed. The design of these prototypes, which are intended to be worn on the hands, were informed by the insights gained from the case studies. The design process was iterative, with each version of the prototypes being refined based on feedback and testing.

Subsequently, wear trials of the prototypes were conducted. These trials were designed to gather data for the evaluation and analysis of the design. A validation process was employed in which subjects were recruited to try the prototypes and evaluate their performance in playing music. Specifically, subjects were asked to play a prechosen song, and each subject's ability to accurately reproduce the melody, rhythm, and pitch was assessed. Additionally, measurements of hand size and dexterity were taken during these trials. The objective was to understand the variations in these factors and how they could influence the design of a new wearable device that could fit all hands universally and enhance interactivity. The data collected from these wear trials were then analyzed to assess the effectiveness of the prototypes and to inform further design improvements. This iterative process of design, testing, and refinement is central to the methodology of practice-based-design research employed in this study.

3. Results and Discussion

3.1. Case Study

3.1.1. KAiKU Gloves

Developers have been experimenting with the expression of music using not only software, but also hardware, for many years. One of the more recent inventions is a handheld device that translates not only motion, but also pressure from the fingers [16]. The device works impressively well. Naturally, this product was not made for the masses. Instead, like the Mi.Mu gloves described below, it was made for professionals. Its use is complex, requires specialized training, and cannot be made user-friendly, and that is why it was never widely adopted.

3.1.2. Mi.Mu Gloves

Among similar research and commercial products, the Mi.Mu gloves were among the first to use flex sensors in music production [15]. They are also equipped with pressure sensors and a shared nine-axis inertial measurement unit. The combination of these technologies in the gloves allows not only for the creation or triggering of sound, but also for the production of coherent and intelligible songs. Moreover, the gloves have a state-of-the-art design and a highly responsive electronic schematic that makes use of all five fingers and all axes of the gyroscope to detect acceleration. In a prime demonstration of the possibilities of wearable music technology, music artist Chagall showcased how using individual finger flex sensors, paired with the multidirectional controls of the Mi.Mu gloves, can create a bassline loop, percussive drumbeat sounds, and even vocal sound effects. Over a span of 3 min and 15 s in her video, Chagall provides a compelling demonstration of the inherent motion of music and illustrates how—depending on the predefined parameters of the New Interface for Musical Expression (NIME)—a beginner can swiftly transition into performing as a virtual musician [15]. While the product garnered significant attention, its primary drawback lies in its pricing. Considering that a single Mi.Mu glove costs at least US $1500, it is hardly affordable to the general public, except to dedicated electronic-music or wearable-technology enthusiasts. In an age in which we aspire to make new wearable technologies for music accessible, this pricing is a deficiency of the Mi.Mu gloves.

3.1.3. Remidi T8 Gloves

Similarly, the Remidi T8 gloves were developed by integrating eight pressure sensors and a MIDI wrist controller into ultra-light spandex gloves. Additionally, when these gloves are synchronized with haptic-feedback technology, musicians are able to play a virtual keyboard in the air by moving their fingers and hands [17]. Simultaneously, the gloves provide tactile feedback to simulate the sensation of pressing physical keys. This functionality enables the gloves to function as an independent instrument that mimics traditional musical instruments. It is believed that this wearable instrument offers a compact and immersive method for playing classical piano pieces, facilitating convenient practice sessions and live performances. However, it is important to note that the tight fit of these synthetic gloves may compromise their breathability, which could potentially result in the accumulation of perspiration within the glove and thus in malodor and discomfort.

3.1.4. DrumPants

DrumPants is a vest with integrated velocity-sensitive pressure sensors and a digital interface that is designed to replicate playing the drums, but on the body [14]. Additionally, the sensors can be affixed to users' chosen body parts with elastic straps and Velcro. Upon touch, each sensor triggers a designated MIDI output associated with a specific sound or effect. Thus, users can press on the surface of the vest to change pitches, bow virtually, and trigger various string articulations. For instance, striking one's lap simulates hitting a snare drum, and stamping one's foot recreates hitting a bass drum. Even with these simple elements alone, a user can mimic playing a drum kit, as aptly demonstrated in the Kickstarter video for DrumPants. It enables drummers to practice virtually and experiment with electronic sound effects, thus expanding their creative horizons. These features serve as a benchmark for the proposed wearable music instrument.

3.1.5. AUUG Wearable Device

The AUUG is a wearable device for the hand that combines a wristband and holder for a smartphone [18]. Musicians can strum, pluck, and bend virtual strings to produce realistic guitar sounds. The sensors of the wristband detect finger movement and pressure, which allow for intricate playing techniques. "WristStrum" is a portable and expressive alternative to traditional guitars that enables musicians to create music outside traditional settings. However, it is not an independent instrument in its own right, but is attached to traditional instruments to enrich musical expression.

3.1.6. Mogees

Mogees, an app and vibration sensor, is a splendid example of how a MIDI device in its simplest form can make an enormous impact on the community [19]. For several years, it was common for users to attach their Mogees to a large number of different materials and objects and create intelligible electronic musical sounds by transforming and translating the natural frequencies of the materials and objects. Mogees offers users the ability to generate creative, playable music, but without the regularity and organization associated with musical instruments.

3.2. *Design and Development of Prototypes*

Based on the case studies, two novel prototypes were designed and developed to create music through hand motions. The concept design was arguably influenced by, but not directly copied from, the designs of the Mi.Mu and Remidi gloves. Additionally, the placement of all sensors in the hand was completely novel, introducing a unique configuration. The details are as follows.

3.2.1. Prototype 1: Strap with Four Pressure Sensors

The aim in creating the prototype was to optimize for interactivity. Therefore, the first prototype was designed to function as both a surface controller and a piano, enabling users to interact with it by pressing on its surface. However, this implementation proved unsuccessful, leading to the development of a second prototype to address its limitations. As shown in Figure 1, the design incorporated a silicone strap that wrapped around the palm twice and split around the thumb. A silicone cover was chosen for its ability to facilitate stable signal transmission and its ease of cleaning [20]. To fulfill the criteria of minimal space utilization and high sensitivity to pressure, four individual piezo pressure sensors (FM0008-01 Long FPC-FSR) from various Alibaba suppliers (China) were used as the flexible printed circuit (FPC) pressure sensors. The sensors were positioned at specific locations: (1) on the palm, (2) on the side of the palm, and (3) on the metacarpal of the middle finger and (4) on the ring finger. Additionally, the strap was securely fastened around the index finger using a loop design.

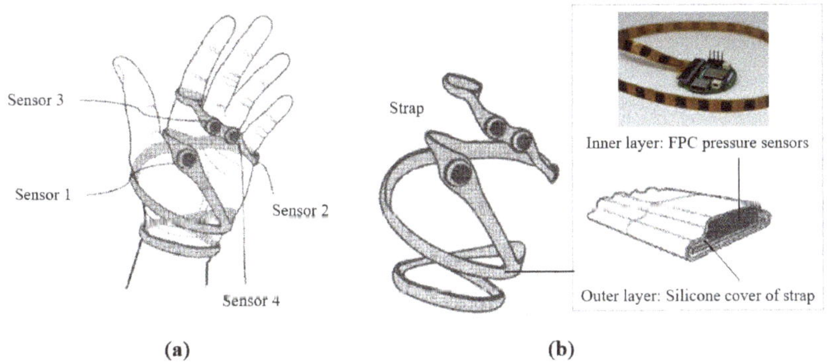

Figure 1. Design details of Prototype 1: (**a**) locations of pressure sensors and (**b**) cross-section of the strap.

Three different FM 0008-01 Long FPC-FSR pressure-sensor samples with varying resistance levels were tested at different intervals using varying amounts of force (kg). Sample 1 featured a thicker top layer of filament, while Samples 2 and 3 had similar thicknesses but were made with slightly different filament materials. As shown in Figure 2, Sample 1 displayed a greater range of resistance values, from 1 kg to 3.5 kg, indicating higher sensitivity and increased responsiveness to user touch. These results indicated that Sample 1 could be triggered at the lowest weight, which was only 1 kg.

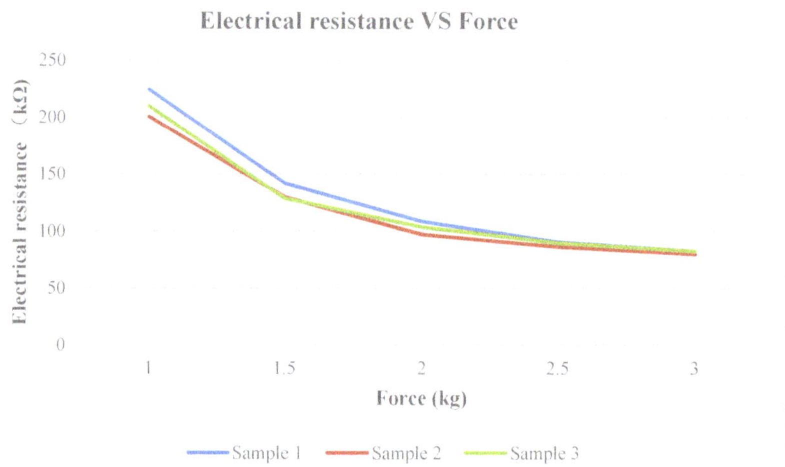

Figure 2. Comparison of electrical resistance in different pressure-sensor samples.

Despite its effectiveness in distinguishing between hand motions such as a slap, a chest-slap, and a clap, Prototype 1 encountered limitations in its functionality as a wearable hand controller and piano-like device. For example, it was necessary to affix a significant number of pressure sensors to the strap. These manufacturing challenges ultimately prompted the development of the second prototype.

3.2.2. Prototype 2: Strap with Full Length of Touch-Sensitive Pressure Sensors

The second prototype was constructed based on the same design concept as Prototype 1 and was wrapped around the hand in a similar manner. However, unlike Prototype 1, which has only four FPC pressure sensors, the second prototype contained 32 individual touch-sensitive pressure sensors along the strap (Figure 3). The incorporation of a greater number of pressure sensors expands the range of sound and light effects, as in a piano wherein each note carries a distinct sound. This modification enables Prototype 2 to function as a surface piano or drum pad while still being a wearable device that covers both the palm and knuckles of the hand for dual-control access by the fingers and the back of the hand. It is believed that this design offers enhanced versatility compared to traditional glove-style devices. Additionally, special attention was given to addressing breathability and hygiene problems, ensuring user comfort by preventing sweat accumulation and the associated clingy feeling. Furthermore, the motion processing unit (MPU) and printed circuit board assembly (PCBA) on the motherboard were used to define the origin point of the nine-axis motion controller. These components detect hand movements and translate them into musical parameters such as pitch or volume. Moreover, the inertial motion unit (IMU) measures the gyroscopic, acceleration, and magnetometer values of the user's arm gestures, enabling dynamic control over musical elements [21]. To improve the user experience and increase the degree of interactivity that is possible, Prototype 2 is also equipped with a 12-diode chip on board (CoB) LED. With this LED, the device displays different lights for each gesture when users rotate their hands between two different gestures. Meanwhile, the vibration motor provides tactile feedback by vibrating when the user squeezes a pressure sensor. These additions allowed for the exploration of motion touch and haptic feedback, utilizing lights and vibrations to enrich the process of musical creation.

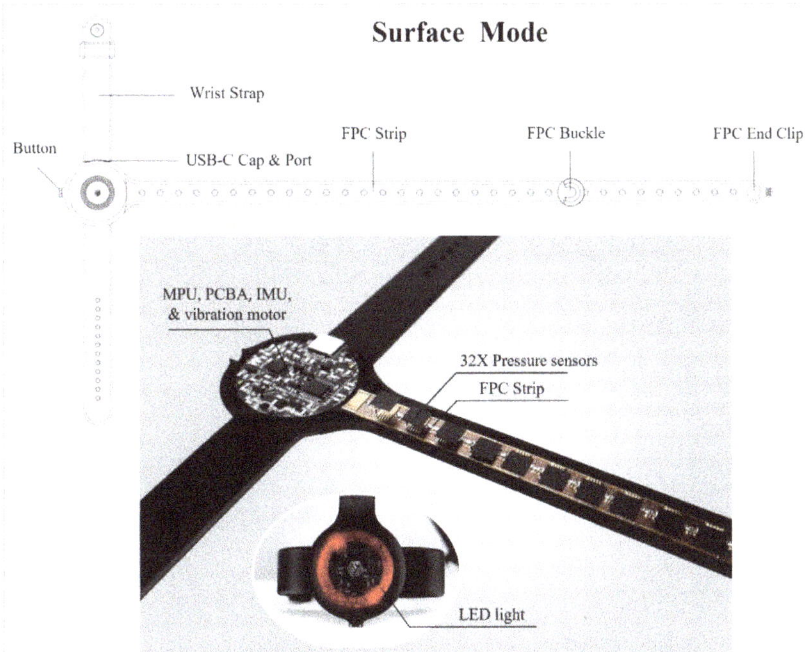

Figure 3. Design details of Prototype 2.

The integration of motion, touch, and acceleration sensors in the musical interface enriches the user experience with enhanced expression, control, and responsiveness. For example, the motion sensors enable users to physically express themselves and create music through gestures and motions, adding a tangible and embodied dimension to the experience. The touch sensors facilitate precise interaction, allowing users to shape the musical output according to their intentions. Additionally, the dynamic control provided by the acceleration sensors empowers users to infuse their music with expressive gestures, imbuing it with a sense of liveliness and responsiveness.

In the grip mode, the strap covers the palm, the side of the palm, and all the metacarpals of the hand (Figure 4a). In the wrist mode, the device can be worn as a bracelet that does not need to be removed completely when it is not in use (Figure 4b). In both cases, the device enables music creation, with the main distinction being that in wrist mode, the fingers of the same hand cannot be utilized to trigger the pressure sensors, as those sensors are positioned on the wrist.

The device employs a motion sensor to detect motion across all wearing modes. However, the range of motion varies depending on the specific mode chosen. In both the wrist and grip modes, the device is capable of capturing the full range of motion of the wrist, allowing users to leverage their natural hand and wrist movements for seamless control and interaction. Conversely, in surface mode, the motion control of the device is constrained, and the device primarily recognizes motion when the user deliberately lifts the device from the surface and adjusts it. In this mode, the focus shifts towards intentional manipulation and repositioning of the device, rather than continuous motion capture. These distinctions in motion sensing in different wearing modes empower users to adapt their interaction with the device, tailoring it to their preferred mode and intended use. This information has been incorporated into the manuscript, providing further clarity regarding the device's capabilities with regard to motion sensing.

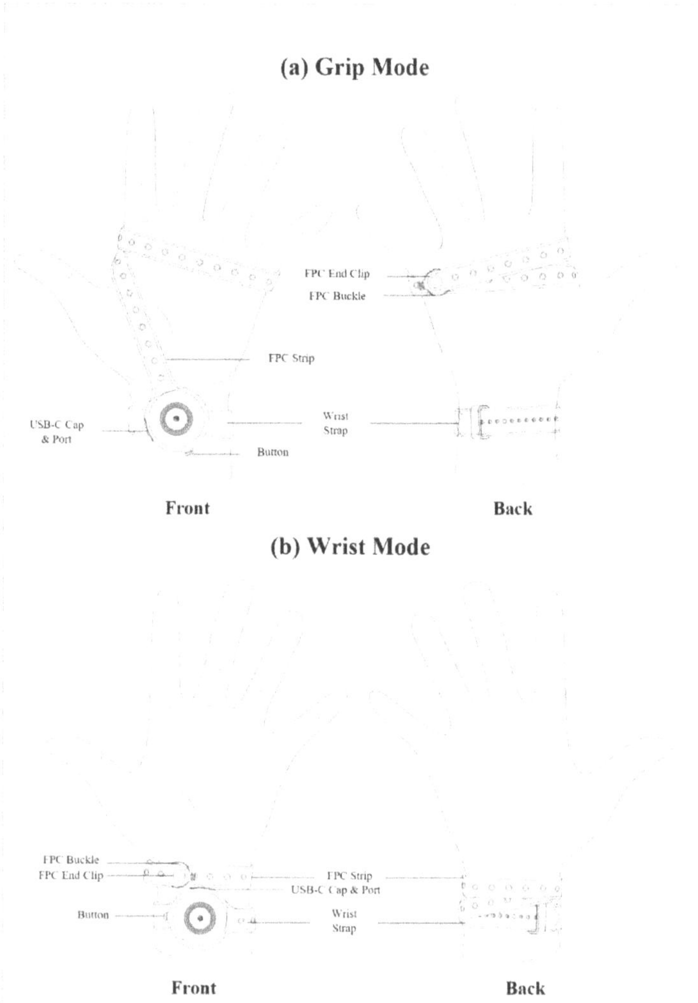

Figure 4. (**a**) Grip and (**b**) wrist modes of Prototype 2.

3.3. *Making Music through Movement*

There are infinite ways in which motion can translate into music. One method involves subdividing the angular data from the gyroscope and mapping each angular range to a specific set of musical outputs (MIDI outputs). One arrangement that employs this method maps 90° angles onto block triad chords of major and minor scales (Figure 5). In this study, this concept is referred to as Motion Static. Each 90° angle corresponds to a specific gesture, and the corresponding gestures are shown in Table 1 [22]. Moreover, within the range of each 90° gesture, the device triggers a different MIDI output. In this case, the outputs are defined as the block chords of the major and minor scales of Western music. The experiments were conducted for the left hand only.

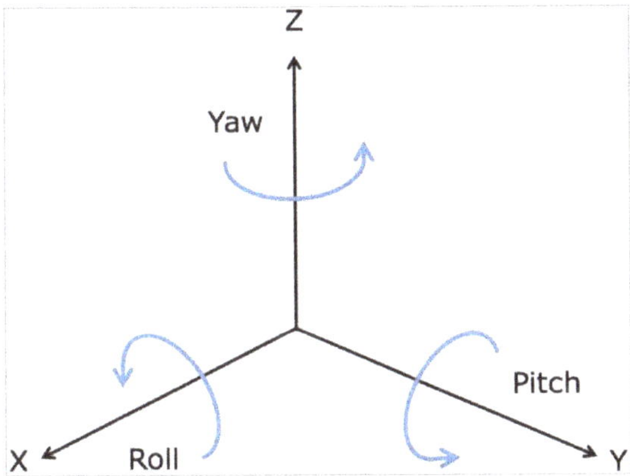

Figure 5. Making music using movement.

Table 1. Motion Static settings and MIDI Note Output.

Gesture (Point of View: Reviewer)	Orientation of X, Y, Z (°) & MIDI Note Output	Gesture (Point of View: Reviewer)	Orientation of X, Y, Z (°) & MIDI Note Output
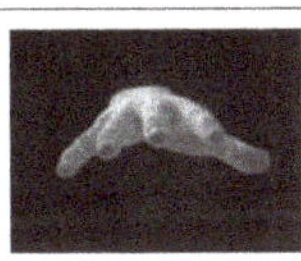 Origin	0, 0, 0 C, E, G	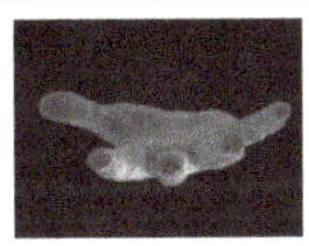 Reverse Origin	−180, 0, 0 A, C, E
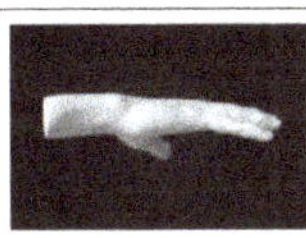 Left	0, 0, 90 Db, F, Ab	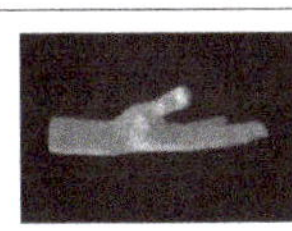 Reverse Left	−180, 0, 90 Bb, Db, F
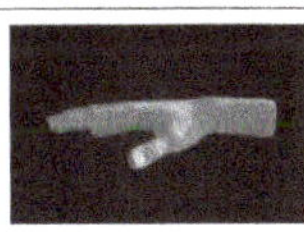 Right	0, 0, −90 A, Db, E	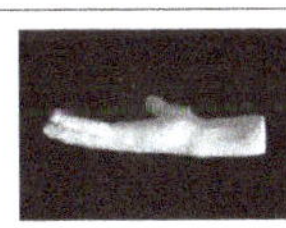 Reverse Right	−180, 0, −90 Gb, A, Db
 Up	0, 90, 0 A, Db, E	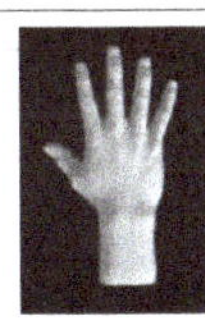 Reverse Up	−180, 90, 0 Gb, A, Db
 Down	0, −90, 0 Eb, G, Bb	 Reverse Down	−180, −90, 0 C, Eb, G

Table 1. *Cont.*

Gesture (Point of View: Reviewer)	Orientation of X, Y, Z (°) & MIDI Note Output	Gesture (Point of View: Reviewer)	Orientation of X, Y, Z (°) & MIDI Note Output
 Handshake	−90, 0, 0 G, B, D	 Reverse Handshake	90, 0, 0 E, G, B
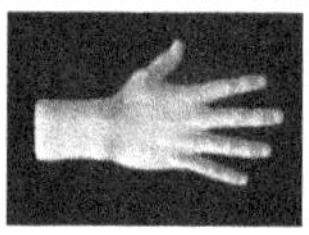 Handshake Left	−90, 0, 90 Gb, Bb, Db	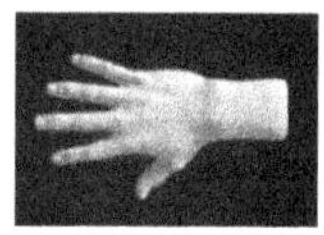 Reverse Handshake Left	90, 0, 90 Eb, Gb, Bb
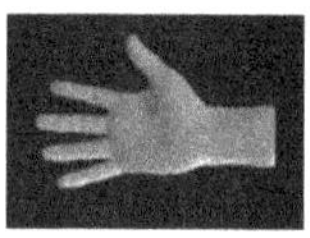 Handshake Right	−90, 0, −90 D, Gb, A	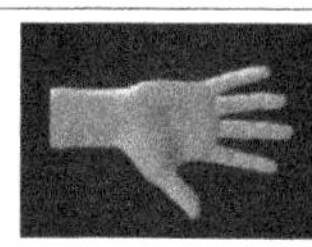 Reverse Handshake Right	90, 0, −90 B, D, Gb
 Handshake Up	−90, 90, 0 E, Ab, B	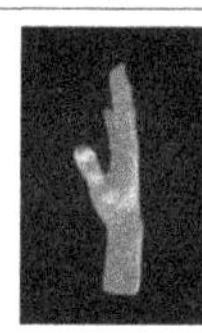 Reverse Handshake Up	90, 90, 0 Db, E, Ab
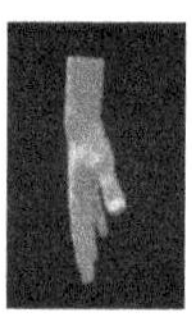 Handshake Down	−90, −90, 0 Ab, C, Eb	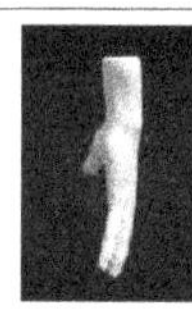 Reverse Handshake Down	90, −90, 0 F, Ab, C

In addition to the predefined values for 90° angles, an acceleration threshold in the negative Z direction is set so that the user can switch between block (all notes played together) and arpeggiated (all notes played separately in a sequence) chords. A third and final condition was implemented using pressure sensors. If the pressure threshold is exceeded, all sound outputs are muted.

3.4. Mobile Software Design

An essential component of a wearable music controller involves providing users with an interface and instructions on how to effectively utilize the controller. Consequently, it is imperative to design a tailored mobile application (app) to fulfill this purpose. The app consists of three distinct pages: a piano page for visualizing the music played, a turn-dial page for editing the types of sounds and their effects, and a home page for organization. This app is user-friendly enough so that novices can immediately start to produce music while gaining an understanding of the theoretical aspects (Figure 6).

Figure 6. Interface of the mobile software design with instructions.

The app was further enhanced after a trial run with users, who indicated that the ability to edit the actual output parameters of their devices would be ideal. Thus, a desktop application that allows users to do so was subsequently created for Windows 10/11 and Mac. Controls to map the individual gyroscopic values for pitch, roll, and yaw, along with the defining values for the minimum and maximum range positions for each movement, are shown in Figure 7. Feedback was received from over 20 users, and enhancement of the preexisting Desktop App continued with the addition of more features and functions of control. For example, a "Theremin" was implemented; this feature essentially functioned as an in-app synthesizer that allowed users to create sounds directly from the app without the need to connect it to third-party Digital Audio Workstation (DAW) software. Users tended to be between the ages of 18 and 42 and included mostly tech-savvy musicians, but the group also included complete music beginners interested in making music for the first time, especially through movement.

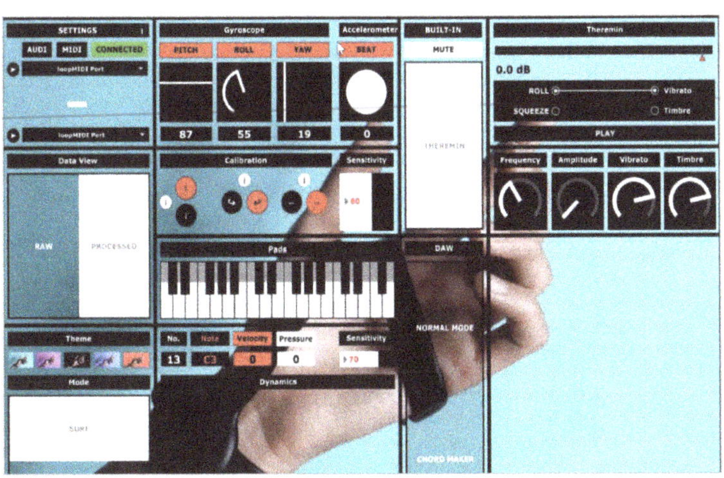

Figure 7. Modified interface.

3.5. Design Evaluation

A wear trial was implemented to evaluate the design of the prototypes. In this trial, to evaluate the prototypes, different arpeggiated chords were played according to three different gestures: handshake, origin, and origin right (Figure 8). The wear trial was conducted with over a dozen members of the public from the university who were selected through various competitions and through word of mouth, including, for example, friends of friends. Most users were interested in the product itself, which is why it was possible to conduct the trials using volunteers.

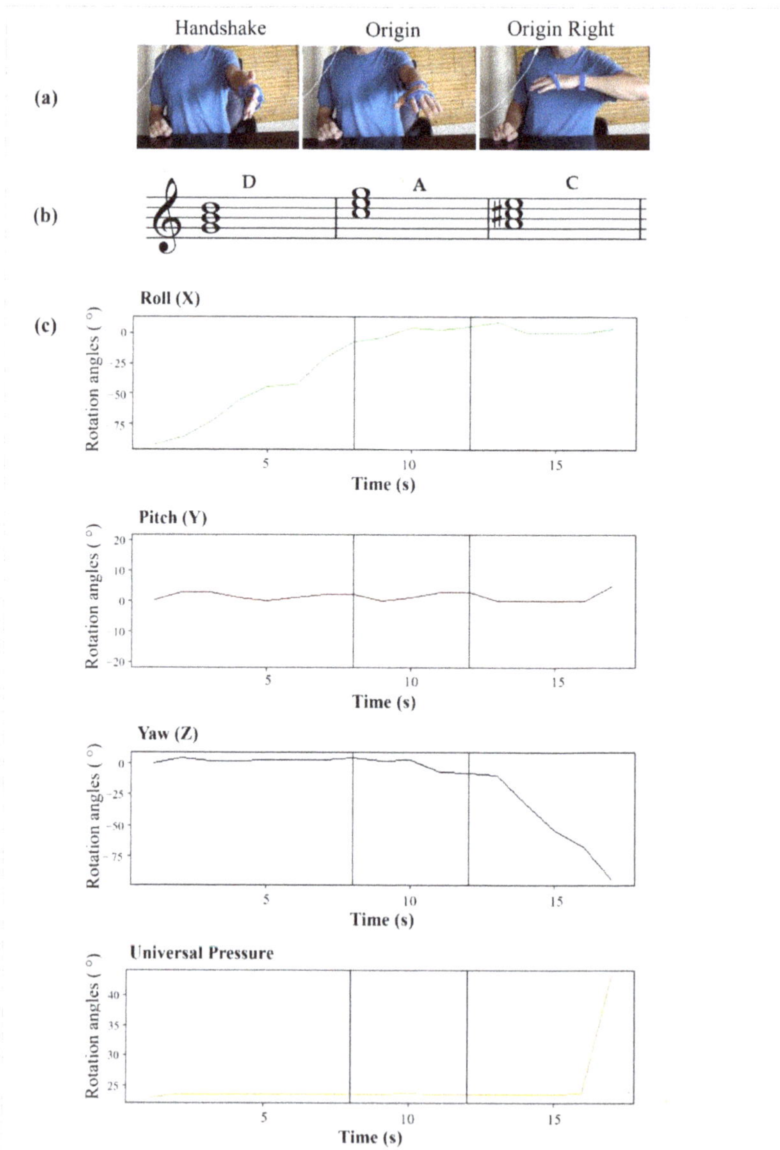

Figure 8. (**a**) Gestures mapped to (**b**) major chords and (**c**) rotation angles.

In this example, the flow of the gestures from Handshake to Origin, and then to Origin Right, was recreated to play a block chordal progression with GarageBand, a Digital Audio Workstation (DAW) developed by Apple. Throughout the performance, the changing signal data from the nine-axis IMU sensor (X, Y, and Z rotation angles) and pressure sensor are wirelessly transmitted to the computer via gestures.

However, latency was a challenge at some points, as it affected the connection between gestures and music. Additionally, the Motion Static profile did not include a large number of motions, i.e., users were meant to arrange their hands only in specific positions and not to perform any ongoing or continuous motion. As a result, practically all 90° angles output the designated musical output. Gyroscopic accuracy was more of an issue than latency, as the three-axis values showed particular sensitivity. Therefore, a threshold was implemented to limit accidental triggering of the X, Y, and Z values during slow and steady 90° gestures. This setting is also known internally as an offset calibration.

4. Challenges of Developing Wearable Music Technology

The development of wearable music technology is not predominantly hindered by the manufacturing of the hardware involved [23–25]. Many startups worldwide have developed an assortment of hardware kits that focus on electronic music controls. Although many of these startups grapple with producing commercially viable hardware, production is not the core predicament. Similarly, the creation of software that converts sensor data from the hardware into sound and music is typically uncomplicated. For instance, most software engineers have the capability to utilize basic MIDI encoding that enables a user to generate prerecorded samples or start a sine wave [26,27].

However, the real challenges in developing wearable music technology emerge when considering the creation of an intelligible musical instrument that can facilitate the production of coherent songs. These songs should uniquely amalgamate distinct musical rhythms, melodies, sounds, and effects. A wearable technology device that simply generates a sound based on a singular gesture does not serve as a universally comprehensible musical instrument. It is this criterion that differentiates a musical toy from a bona fide musical instrument.

Creating a device that transforms specific movements or gestures into unique musical sounds may initially seem exciting. However, it is crucial to consider how well users comprehend these actions and the music they generate [28]. One potential solution lies in personalization, wherein users have the ability to define their output parameters based on pre-provided input-gesture values. Without flexibility that allows users to gain personal understanding through trial-and-error mapping of gestures to musical outputs, the art of creating music through motion and touch can become simply another repetition of existing technology, rather than an intuitive, personalized learning experience. This concept of personalization has been explored in the prototypes developed in this study, with the aim of innovating in the field of wearable music technology [29].

5. Limitations of Experiments and Future Trends

5.1. Limitations of Experiments

While the findings of this study are encouraging, it is important to acknowledge several limitations that were encountered. The first limitation is the time-delay issue encountered during the study. The delay between the musician's movements and the corresponding musical output could disrupt the rhythm of the music and limit real-time music creation. Future research should aim to minimize this delay to ensure a more fluid musical experience.

The second limitation pertains to the unstable signal transmission experienced by the prototypes developed in this study. This instability, which was likely due to the wireless nature of the devices, could potentially interrupt the process of musical creation. Future research should therefore prioritize enhancing the stability and reliability of signal transmission.

The third limitation involves the suboptimal responsiveness of the pressure sensors. There were instances in which the sensors did not respond as anticipated, potentially hindering the musician's ability to accurately control the musical output. To address these issues, future iterations of the device could incorporate higher-quality sensors or refine the sensor-calibration process.

These proposed enhancements are intended to provide a smoother musical experience and superior control over the instrument. As we consider future directions, this study underscores the dynamic nature of music technology and the ways in which it can empower individuals. The prototype offers not only a means of making music, but also an immersive learning experience that can benefit both musicians and non-musicians. It challenges traditional limitations and opens up new possibilities for creative expression. The future of music technology will likely continue to push boundaries and explore innovative ways of creating music.

5.2. Future Trends

Despite the prevailing opinion among traditionalists that a musical instrument not invented prior to 1950 is considerably primitive in terms of music-making capability, modernists and their latest musical instrument inventions have evolved beyond such conservative notions. The music-technology-products industry alone has fostered more talent in the music community, even within the definitions of the traditionalists themselves, than their classical instruments ever did.

Even if traditional musical instruments are ultimately abandoned due to the time, effort, and cost required to learn how to play them, there is still a good deal of public enthusiasm and interest in making music. Most people believe that the application of this type of wearable technology is limited to music-technology enthusiasts and prospective musicians. However, many individuals, especially young people, are now seeking new means by which to share their creative and expressive talents on social-media platforms like TikTok, which are used by billions of users and influencers each month. Therefore, the question is never "Will they use it?", but rather "Can they use it?". The venture accelerators, also known as startup accelerators, are all asking the wrong questions. Non-musicians can indeed become musicians. Moreover, musicians from traditional classical or digital backgrounds can transcend their current limitations and use technology to enhance their abilities.

The prototype in this study is capable of providing such individuals with another outlet through which to do so. It offers an immersive learning experience that schools and educational institutions can use not only to develop musical skills, but also to reap the cognitive benefits of this technology. Additionally, it challenges the traditional limitations of music with its ability to simulate almost all existing means of making music using movement and touch alone. By combining some of the more complex theories and intricacies of musical creativity with the sublime simplicity of hand motion, the prototype was able to engage music enthusiasts of all backgrounds and beliefs. Aside from making music, the prototype may have other functional uses, such as empowering those who have both physical and mentally challenges, for example, individuals with ADHD. The ability to create music through movement means that those who face such limitations will be able to stimulate their creative musical abilities and express themselves if they are not able to do so vocally or physically. It is believed that the future of music technology is likely to continue pushing the boundaries imposed by traditional instruments and to explore new ways of creating music.

6. Conclusions

In conclusion, this study has effectively pioneered the application of wearable technologies as a novel medium for musical expression. The two prototypes, both equipped with pressure sensors and designed to be worn on the hand, demonstrated the potential to revolutionize music creation and experience. The first prototype served as an initial demonstration of a wearable surface controller and piano, showcasing the potential of wearable technology to create a highly interactive and intuitive musical instrument. The second prototype, with its increased number of pressure sensors and additional features such as a motion processing unit, LED lights, and a vibration motor, further enhanced the musical experience by providing a more versatile and engaging interface. These groundbreaking devices have the potential to significantly contribute to the field of music, offering musicians a new, intuitive, and versatile instrument for creative expression. Future research will involve exploring the potential of wearable technology to expand individuals' musical capabilities and transcend traditional limitations. Potential applications include leveraging wearable devices to provide immersive learning experiences in schools and educational institutions, enabling individuals with diverse backgrounds and abilities to engage in music creation. Additionally, wearable technology holds promise for empowering individuals with physical and mental challenges to stimulate their creative musical abilities and engage in self-expression. This ongoing trend in music technology aims to push the boundaries imposed by traditional instruments and explore new avenues for creating music.

Author Contributions: Conceptualization, D.W. and J.Y.; methodology, D.W. and X.L.; software, D.W.; validation, D.W., J.Y. and K.-P.L.; formal analysis, D.W., K.-P.L. and X.L.; investigation, D.W. and X.L.; resources, D.W. and J.Y.; data curation, D.W., X.L. and Y.-H.W.; writing—original draft preparation, D.W., J.Y., K.-P.L. and X.L.; writing—review and editing, D.W., J.Y., K.-P.L., X.L. and Y.-H.W.; visualization, D.W. and K.-P.L.; supervision, J.Y.; project administration, J.Y.; funding acquisition, J.Y. All authors have read and agreed to the published version of the manuscript.

Funding: This research was funded by The Hong Kong Polytechnic University Grant under the Project of Strategic Importance titled "Investigate the Mechanism Underlying the Efficacy of Conservative Management" (project code: 1-ZE2H).

Institutional Review Board Statement: The experimental procedure was approved by the Departmental Research Committee of The Hong Kong Polytechnic University, reference number HSEARS20231206004.

Informed Consent Statement: Informed consent was obtained from all subjects involved in the study. Written informed consent has been obtained from the patient(s) to publish this paper.

Data Availability Statement: The datasets that were generated and/or analyzed during the current study are available upon reasonable request from the corresponding author.

Conflicts of Interest: The authors declare no conflicts of interest.

References

1. Rossing, T.D. The Acoustics of Bells: Studying the vibrations of large and small bells helps us understand the sounds of one of the world's oldest musical instruments. *Am. Sci.* **1984**, *72*, 440–447.
2. ABRSM. Making Music. Available online: https://gb.abrsm.org/en/making-music/4-the-statistics/ (accessed on 12 October 2023).
3. Sudip, S. Electronic Musical Instruments Market Outlook (2022–2032). Available online: https://www.futuremarketinsights.com/reports/electronic-musical-instruments-market#:~:text=Electronic%20Musical%20Instruments%20Market%20Outlook,period%20(2022%20to%202032) (accessed on 18 November 2023).
4. Fortune Business Insights. Wearable Technology Market Size, Share & Covid-19 Impact Analysis, By Type (Smart Glasses, VR Headsets, Smart Jackets and Smart Rings, Smart Watches, Fit Bands), By Technology (IoT Based, AR and VR, Others), By End Use (Health and Fitness, BFSI, Gaming and Entertainment, Fashion, Travel, Education, Logistics & Warehouse), and Regional Forecast, 2023–2030. Available online: https://www.fortunebusinessinsights.com/wearable-technology-market-106000 (accessed on 18 November 2023).
5. Cavdir, D.; Wang, G. Designing felt experiences with movement-based, wearable musical instruments: From inclusive practices toward participatory design. *Wearable Technol.* **2022**, *3*, e19. [CrossRef]
6. Ometov, A.; Shubina, V.; Klus, L.; Skibińska, J.; Saafi, S.; Pascacio, P.; Flueratoru, L.; Gaibor, D.Q.; Chukhno, N.; Chukhno, O. A survey on wearable technology: History, state-of-the-art and current challenges. *Comput. Netw.* **2021**, *193*, 108074. [CrossRef]
7. Einbond, A. Musique instrumentale concrète: Timbral transcription in What the Blind See and without Words. In *The OM Composer's Book*; Bresson, J., Agon, C., Assayag, G., Eds.; Editions Delatour: Ircam-Centre Pompidou, France, 2016; pp. 155–171.
8. Schumacher, M.; Wanderley, M.M. Integrating gesture data in computer-aided composition: A framework for representation, processing and mapping. *J. New Music Res.* **2017**, *46*, 87–101. [CrossRef]

9. Gunther, E.; O'Modhrain, S. Cutaneous grooves: Composing for the sense of touch. *J. New Music Res.* **2003**, *32*, 369–381. [CrossRef]
10. Paradiso, J.A. The brain opera technology: New instruments and gestural sensors for musical interaction and performance. *J. New Music Res.* **1999**, *28*, 130–149. [CrossRef]
11. Ihde, D. A finnish turn: Digital and synthesiser musical instruments. *J. New Music Res.* **2021**, *50*, 165–174. [CrossRef]
12. Hallam, S.; Rinta, T.; Varvarigou, M.; Creech, A.; Papageorgi, I.; Gomes, T.; Lanipekun, J. The development of practising strategies in young people. *Psychol. Music* **2012**, *40*, 652–680. [CrossRef]
13. Dempsey, S.J.; Szablewski, M.; Atkinson, D. Tactile sensing in human–computer interfaces: The inclusion of pressure sensitivity as a third dimension of user input. *Sens. Actuators A Phys.* **2015**, *232*, 229–250. [CrossRef]
14. Hansen, S.H.; Jensenius, A.R. The Drum Pants. In Proceedings of the Audio Mosty Conference—A Conference on Sound in Games, Piteå, Sweden, 11–12 October 2006.
15. Mi.Mu. Mi.Mu Gloves. Available online: https://www.mimugloves.com/gloves/ (accessed on 18 November 2023).
16. Taction Enterprises Inc. KAiKU Glove. Available online: https://www.kaikumusicglove.com/product-page/copy-of-kaiku-music-glove (accessed on 20 November 2023).
17. Remidi T8 Wearable Instrument. Remidi: First Wearable Instrument to Record, Play & Perform. Available online: https://www.kickstarter.com/projects/376537646/remidi-first-wearable-instrument-to-record-play-an (accessed on 18 November 2023).
18. AUUG. New AUUG Project. Available online: https://www.auug.com/2023-project/ (accessed on 20 November 2023).
19. Mogees Ltd. Mogees-Transform anything into a musical instrument. Available online: https://www.kickstarter.com/projects/mogeesplay/mogees (accessed on 18 November 2023).
20. Lee, K.-P.; Yip, J.; Yick, K.-L.; Lu, C.; Lu, L.; Lei, Q.-W.E. A Novel Force-Sensing Smart Textile: Inserting Silicone-Embedded FBG Sensors into a Knitted Undergarment. *Sensors* **2023**, *23*, 5145. [CrossRef] [PubMed]
21. Mak, T.H.A.; Liang, R.; Chim, T.W.; Yip, J. A Neural Network Approach for Inertial Measurement Unit-Based Estimation of Three-Dimensional Spinal Curvature. *Sensors* **2023**, *23*, 6122. [CrossRef] [PubMed]
22. cgspektor. FREE Hand (3D model). Available online: https://artstn.co/m/qDn63/ (accessed on 19 December 2023).
23. Sullivan, J.; Tibbitts, A.; Gatinet, B.; Wanderley, M.M. Gestural control of augmented instrumental performance: A case study of the concert harp. In Proceedings of the 5th International Conference on Movement and Computing, Genoa, Italy, 28 June 2018.
24. Turchet, L.; Fischione, C.; Essl, G.; Keller, D.; Barthet, M. Internet of musical things: Vision and challenges. *IEEE Access* **2018**, *6*, 61994–62017. [CrossRef]
25. Seneviratne, S.; Hu, Y.; Nguyen, T.; Lan, G.; Khalifa, S.; Thilakarathna, K.; Hassan, M.; Seneviratne, A. A survey of wearable devices and challenges. *IEEE Commun. Surv. Tutor.* **2017**, *19*, 2573–2620. [CrossRef]
26. Hosken, D. *Music Technology and the Project Studio: Synthesis and Sampling*; Routledge: London, UK, 2012.
27. Ballivian, G.Z. *Creating, Capturing and Conveying Spatial Music: An Open-Source Approach*; University of California San Diego: La Jolla, CA, USA, 2023.
28. O'modhrain, S. A framework for the evaluation of digital musical instruments. *Comput. Music J.* **2011**, *35*, 28–42. [CrossRef]
29. Li, L.; Wang, G. Design and Application of Interactive Music Equipment Based on Wireless Wearable Sensors. *Sci. Program.* **2022**, *2022*, 4719884. [CrossRef]

sensors

MDPI

Article

Design and Assessment of Flexible Capacitive Electrodes for Reusable ECG Monitoring: Effects of Sweat and Adapted Front-End Configuration

Ivo Iliev [1,*], Georgi T. Nikolov [1], Nikolay Tomchev [1], Bozhidar I. Stefanov [2,*] and Boriana Tzaneva [2]

[1] Department of Electronics, Faculty of Electronic Engineering and Technology, Technical University of Sofia, Kliment Ohridski Blvd., 8, 1000 Sofia, Bulgaria; gnikolov@tu-sofia.bg (G.T.N.); nntomchev@gmail.com (N.T.)
[2] Department of Chemistry, Faculty of Electronic Engineering and Technology, Technical University of Sofia, Kliment Ohridski Blvd., 8, 1000 Sofia, Bulgaria; borianatz@tu-sofia.bg
* Correspondence: izi@tu-sofia.bg (I.I.); b.stefanov@tu-sofia.bg (B.I.S.)

Abstract

This work presents the development and characterization of a flexible capacitive electrode for non-contact ECG acquisition, fabricated using a simple and cost-effective method from readily available materials. The electrode consists of a multilayer structure with a copper conductor laminated by a polyimide (Kapton®) dielectric layer on a polyurethane support. The impedance and capacitance of the electrode were evaluated under varying textile moisture levels with artificial sweat, as well as after exposure to common disinfectants including ethyl alcohol and iodine tincture. Electrochemical impedance spectroscopy (EIS) and broadband impedance measurements (10^{-1}–10^5 Hz) confirmed stable capacitive behavior, moderate sensitivity to moisture, and chemical stability of the Kapton–copper interface under conditions simulating repeated use. A custom front-end readout circuit was implemented to demonstrate through-textile ECG signal acquisition. Simulator tests reproduced characteristic waveform patterns, and preliminary volunteer recordings confirmed the feasibility of through-textile acquisition. These results highlight the promise of the electrode as a low-cost platform for future wearable biosignal monitoring technical research.

Keywords: flexible electrode; capacitive ECG; non-contact biopotential sensing; impedance spectroscopy; Kapton dielectric; moisture effects

1. Introduction

Healthcare systems in developed countries, including Bulgaria, are increasingly challenged by the unprecedented aging of the population. According to the World Health Organization (WHO), by 2030, one in six people globally will be aged 60 years or older, and this number is projected to double to 2.1 billion by 2050. In Europe and North America, one in four individuals will fall into this age group [1]. As a result, social and healthcare expenditures are rising sharply, driven by the need to maintain health and quality of life in the aging population.

To address this issue, new technologies are being developed to shift healthcare delivery beyond traditional clinical settings—toward home- and community-based care. One promising solution is remote monitoring (telemetry) of vital signs for patients who do not require immediate hospitalization but still need regular observation to manage chronic conditions. This approach is especially valuable in rural or underserved areas where access to specialized care is limited [2].

Systematic reviews [3] have highlighted the growing adoption of wearable technologies for remote health monitoring, particularly in elderly populations. Commonly tracked parameters include heart rate, body temperature, motion, respiratory rate, blood glucose, oxygen saturation, and blood pressure. Comparative studies of commercial and prototype devices have evaluated their suitability for such applications. A related review [4] emphasized ongoing challenges in material selection, sensor design, and power supply—particularly the need for flexible, multifunctional, and biocompatible materials.

Given that cardiovascular diseases account for 32% of all global deaths [5,6], remote monitoring technologies are increasingly focused on cardiac function. Early screening and monitoring are critical, and wearable or non-contact electrocardiographic (ECG) systems are gaining traction [7,8]. When integrated with information and communication technologies (ICT) and artificial intelligence (AI), these systems can support preliminary assessment and long-term observation.

Despite these advancements, remote ECG monitoring cannot yet replace standard diagnostic tools such as clinical electrocardiographs and long-term Holter monitors. Successful use of out-of-hospital ECG systems requires that users be in sufficiently good physical and cognitive condition to operate the device, maintain regular data acquisition, and, most importantly, correctly position the electrodes. Therefore, ongoing research aims to simplify ECG self-recording, with much of the effort focused on electrode development.

Traditional Ag/AgCl gel electrodes, widely used in clinical settings, provide low-impedance skin contact but suffer from gel drying over time, leading to signal degradation. Gel and adhesive components may also cause skin irritation or allergic reactions [9–11]. To address these issues, researchers have proposed dry electrodes made from various conductive materials. These eliminate the need for gel and generally improve user comfort [12–16]. Some designs, such as the micro-spiked electrodes by Griss et al. [17] or the 3D-printed conical structures by Salvo et al. [18], aim to reduce impedance on hairy skin, though they remain susceptible to motion artifacts. Other innovations include the thin Ti/Au composite electrodes on flexible substrates by Baek et al. [19] and the hexagonal labyrinth pattern proposed by Elango et al. [20], which was found to offer superior skin contact and long-term performance. Still, even dry electrodes require direct contact between conductive material and skin.

An alternative approach utilizes galvanically isolated (non-contact) electrodes that operate primarily through capacitive coupling with the body. In this context, the capacitive ECG (cECG) architecture consists of a conductive electrode separated from the body by a dielectric layer, forming a capacitor in which the body acts as the second electrode plate. This configuration enables biopotential signal acquisition without direct electrical contact, reducing skin irritation and improving user comfort. A key advantage of the cECG approach is its ability to capture ECG signals through clothing, allowing for seamless integration of electrodes into textile-based systems and expanding the potential for wearable, unobtrusive health monitoring applications. Nevertheless, cECG has inherent limitations: it cannot measure absolute signal amplitudes reliably, very low-frequency components (below 1 Hz) are strongly attenuated, variability between measurements can be extreme, calibration across sessions is problematic, and motion artifacts are significantly more severe than in standard ECG. These limitations restrict the immediate clinical applicability of cECG and require careful consideration when interpreting experimental results.

The main technical challenge in cECG is the transmission of low-amplitude (1–5 mV), low-frequency (0.05–150 Hz) signals through a capacitance typically in the 10–100 pF range [21]. Depending on the specific purpose of the diagnostic examination, a low frequency of 0.05 Hz is determined as the limit value, and for rhythm interpretation, the accepted frequency is 0.67 Hz [22]. Such conditions demand extremely high input impedance from the signal acquisition front end. Conventional buffer stages based on operational amplifiers often include a high-value resistor (>100 GΩ) to stabilize input potential, but these resistors are costly, prone to noise, and difficult to source [23,24]. Alternative methods include using T-shaped resistor networks or bootstrap amplifier configurations [23,25], typically employing amplifiers with >1 TΩ input impedance, <2 pF input capacitance, high common-mode rejection, and minimal bias currents.

Badarov et al. [26] proposed a simplified bootstrap design using JFET op-amps that mitigates parasitic capacitance and signal drift but requires wider power supply ranges—limiting its use in compact wearable systems. While these circuit solutions make cECG feasible, practical challenges remain, particularly motion artifacts and equipment-induced noise. Power-line interference (PLI) can be minimized through shielding, guarding, and right-leg drive circuits [23,27]. Some researchers also propose adaptive filtering using electrode-tissue impedance as a reference signal [28].

Nonetheless, long-term stability remains a concern, primarily due to variations in electrode capacitance that determine the input impedance of the amplifier. These variations complicate calibration and, consequently, the accurate measurement of ECG signal amplitudes (see justification below). As noted, the small values of electrode capacitance directly affect the accuracy of recording low-frequency ECG components (below 1 Hz) and, in particular, the faithful reproduction of diagnostically important elements such as the ST segment. Motion artifacts (MAs) are another major issue in cECG recording. Various approaches for MA suppression have been reported, including the following: (1) the use of digital filtering methods implemented with DSP [29] and (2) adaptive noise filtering, where the Electrode–Tissue Impedance (ETI) is used as a reference signal [30]. In a comprehensive review on MAs in capacitive ECG monitoring systems, Khalili et al. [28] noted that most of the proposed solutions require substantial computational resources and/or additional circuitry (sensors and measurement channels, etc.), which in turn increase power consumption and often limit their applicability in wearable devices. According to the authors, a promising direction for future research lies in approaches that combine digitally controlled analog designs, which may prevent amplifier saturation and signal loss.

In addition to motion-related disturbances, a number of less obvious but equally critical factors, such as ambient humidity, electrode flexibility, and moisture content, especially in textile-based dielectrics, can strongly influence signal quality [31]. Moisture, in particular, tends to reduce impedance and enhance signal acquisition [32]. Several studies recommend maintaining optimal humidity conditions [9,31,33,34]. Cui et al. [35] showed that high humidity combined with controlled pressure improves signal quality and impedance consistency. It is reasonable to assume that sweat could have a similar effect; however, the influence of its chemical composition on signal integrity and electrode longevity during continuous or repeated use requires further study.

This work presents the evaluation of a flexible capacitive ECG electrode, comprising a copper conductor insulated by a polyimide (Kapton®, produced by DuPont, Wilmington, DE, USA) dielectric layer on a polyurethane support, with a focus on its stability and performance under varying moisture exposure. The feasibility of through-textile ECG signal acquisition is examined here as a technical demonstration by testing the electrode's impedance characteristics and electrochemical behavior in conditions simulating real-world use, including exposure to artificial sweat and contact with common disinfectants such as ethanol and iodine tincture. Particular attention is given to the effects of repeated wetting and cleaning, assessing the electrode's durability and functional reliability. In parallel, a dedicated hardware solution is implemented to provide real-time digital adjustment of amplifier gain asymmetry, enabling improved common-mode interference suppression and enhancing signal quality during non-contact ECG measurements through clothing.

2. Materials and Methods

2.1. Flexible Capacitive Electrode Implementation and Testing Protocol

The capacitive electrode evaluated in this study consists of a flexible multilayer structure, schematically illustrated in Figure 1.

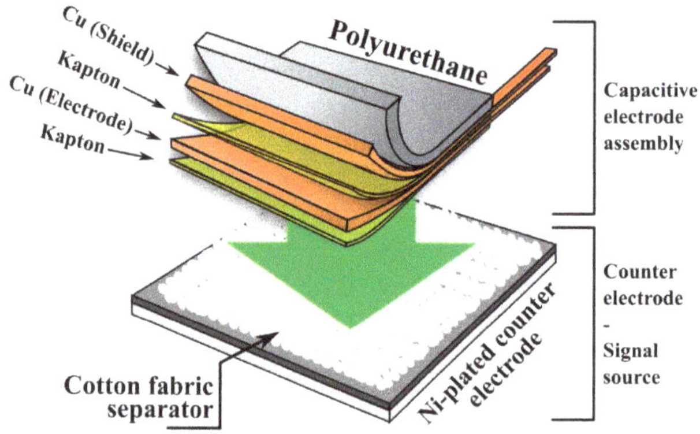

Figure 1. Schematic representation of the capacitive multilayer electrode structure and the test configuration, including the cotton textile separator and the Ni-plated counter-electrode, used as a signal source.

The capacitive electrode was formed from copper (Cu) foil (50 × 50 × 0.05 mm), laminated onto a Kapton insulating sheet, which also acts as the dielectric barrier in the capacitive interface. The copper layer serves as the conductive plate of the capacitor, while the human body acts as the opposing plate during ECG signal acquisition. A secondary Cu/Kapton bilayer is included for electrical shielding. The complete multilayer structure was laminated onto a polyurethane-based flexible substrate (50 × 50 × 0.7 mm) to provide mechanical support and ensure even pressure distribution during testing. Based on the bill of materials, the raw material cost of one such electrode (50 × 50 mm) is approximately EUR 0.37, with unit costs of ~EUR 23.5/m^2 for copper tape, ~EUR 9.9/m^2 for Kapton tape, and ~EUR 78.2/m^2 for polyurethane sheets. For comparison, single-use Ag/AgCl electrodes typically cost in the range of EUR 0.50–1.02 per unit, depending on the fabrication method, while carbon-based printed electrodes are reported at EUR 0.56 per unit, and laser-induced graphene (LIG) electrodes can reach as low as EUR 0.27 per unit [36]. This demonstrates the cost-efficiency of the proposed reusable electrode design.

A textile separator was used in all test setups to simulate a clothing interface, mimicking conditions for through-garment ECG acquisition. The separator was made of knitted cotton fabric (jersey-type, 50 × 50 × 0.7 mm), chosen for its hygroscopic properties. The counter-electrode was fabricated using a nickel-plated FR-4 substrate to ensure chemical resistance in the artificial sweat environment. During testing, the capacitive electrode was pressed into contact with the textile layer and positioned against the counter-electrode under a constant pressure of approximately 425 g-force (1.67 kPa). This configuration was used across all electrochemical and frequency-dependent impedance measurements, as well as in ECG signal acquisition tests using a simulator.

To assess changes in the electrical properties of the electrode under moisture exposure, a model solution of artificial sweat was used. The formulation contained 20 g L^{-1} NaCl, 17.5 g L^{-1} NH_4Cl, 5 g L^{-1} acetic acid, and 15 g L^{-1} D,L-lactic acid, following ISO 3160-2:2015 recommendations for testing the corrosion resistance of metals in contact with skin [37]. Tests using the "dry textile" separator were conducted under ambient atmospheric conditions, at a relative humidity of 55 ± 5% and room temperature

(25 ± 3 °C). Under these conditions, the naturally adsorbed moisture in the textile (moisture regain of cotton of 6.61%) is neglected and is therefore referred to in this work as "dry textile". Artificial sweat was applied to the textile separator using a fine mist sprayer in volumes ranging from 0.2 mL to 0.8 mL, evenly distributed over the entire 25 cm^2 surface area. Accordingly, the volume of sweat per unit area varied from 8 $\mu L\ cm^{-2}$ to 32 $\mu L\ cm^{-2}$, corresponding to a moisture content of approximately 26% to 58% when normalized to the weight of the textile separator (573 mg). The upper value of 32 $\mu L\ cm^{-2}$ corresponds to the maximum liquid retention capacity of the textile separator (before dripping) and to the intense sweating rate in the chest area of an adult individual after physical exertion [38].

To simulate cyclic reuse conditions and evaluate disinfectant compatibility, additional tests were conducted on electrodes subjected to standard cleaning agents. These included rubbing alcohol (aqueous ethyl alcohol, 98% v/v) and tincture of iodine (5 wt. % I_2, 2 wt. % KI in 50 v/v aqueous ethyl alcohol), which were applied to the Kapton surface for a duration of 30 min, followed by rinsing with distilled water and thorough drying. The high concentrations of the disinfectants along with the application of a long-duration disinfection treatment significantly exceeded typical medical practice in order to simulate long-term use. The effects of these treatments on the electrode's performance and surface properties were further evaluated through impedance and wetting angle measurements.

2.2. Investigation of Impedance Characteristics of the Capacitive Electrode

To characterize the frequency-dependent impedance response of the electrode and assess its stability under moisture exposure, two complementary methods were employed. The first method involved a conventional indirect measurement approach, in which the voltage drop across the device under test (DUT) was measured directly, and the current was calculated by monitoring the voltage drop across a precision reference resistor connected in series with the DUT. The data was complemented by electrochemical impedance spectroscopy (EIS) using a potentiostat–galvanostat system.

The low-frequency impedance behavior of the capacitive electrode, a classical indirect measurement method, was employed. In this configuration, the voltage drop across the capacitive electrode (DUT). The current is then calculated indirectly by monitoring the voltage drop across a precision reference resistor R_{ref} connected in series with the DUT. A simplified schematic of the method is shown in Figure 2.

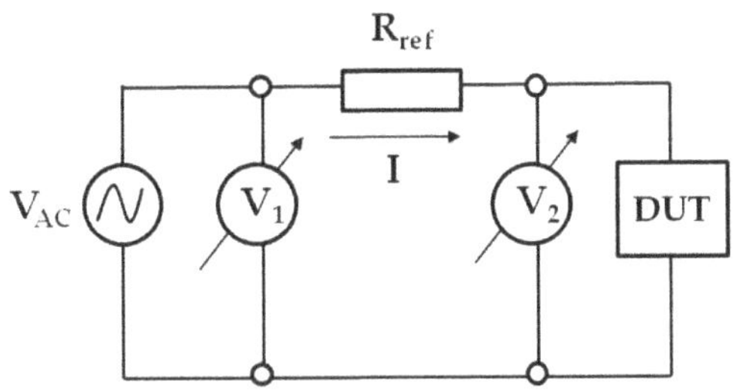

Figure 2. Simplified circuit for measuring the impedance of a DUT by evaluating two voltage drops.

An AC voltage source of relatively low frequency, labeled V_{AC}, drives the system. The complex impedance of the DUT can be calculated from the voltage drops V_1 and V_2, and the known reference resistor R_{ref}, using the following relation:

$$Z_{DUT} = \frac{V_2}{V_1 - V_2} R_{ref} \quad (1)$$

To determine the magnitude and phase angle of the impedance, the system is analyzed using phasor representation in polar coordinates [39,40]. The magnitude of the impedance and the phase shift φ between the voltage signals are determined via the law of cosines and trigonometric relationships as follows:

$$|Z_{DUT}| = \frac{V_2}{\sqrt{V_1^2 + V_2^2 - 2V_1^2V_2^2\cos\theta}} R_{ref} \tag{2}$$

$$\varphi = arctg\frac{-V_2\sin\theta}{V_1 - V_2\cos\theta} \tag{3}$$

The overall impedance phase angle α is obtained by subtracting φ from the measured phase difference θ, as follows:

$$\alpha = \theta - \varphi = \theta - arctg\frac{-V_2\sin\theta}{V_1 - V_2\cos\theta} \tag{4}$$

To extract the equivalent series resistance R_{ESR} and capacitance C from the complex impedance, the impedance is converted back to rectangular form as follows:

$$Z = R_{ESR} - \frac{j}{2\pi fC} \tag{5}$$

$$Z = |Z|\cos\alpha + j|Z|\sin\alpha, \tag{6}$$

where f is the frequency of the AC signal. From this, the real and imaginary components yield the following:

$$R_{ESR} = |Z|\cos\alpha \tag{7}$$

$$C = \frac{-1}{2\pi f|Z|\sin\alpha} \tag{8}$$

The experimental setup for this analysis is illustrated in Figure 3.

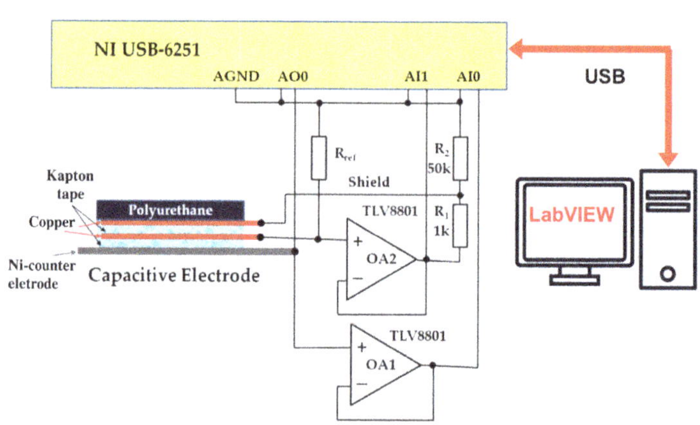

Figure 3. Block diagram of the experimental setup of a virtual system for impedance measurement.

The system is based on a National Instruments USB-6251 data acquisition (DAQ) system. To minimize the influence of the DAQ input impedance, operational amplifiers OA1 and OA2 (TLV8801, Texas Instruments) are used in the front end. Amplifier OA2 additionally provides shield-driving voltage to the capacitive electrode via a voltage divider

(R_1 and R_2). The reference resistor R_{ref} is switchable depending on the measurement frequency: 1 GΩ is used for the 0.1–10 Hz range, and 100 MΩ for the 10–100 Hz range.

To probe the extended frequency range (10^{-1}–10^5 Hz), additional measurements with electrochemical impedance spectroscopy were performed without external DC polarization, using a sinusoidal excitation voltage with an amplitude of 50 mV. Measurements were conducted with a PalmSens4 potentiostat–galvanostat, and equivalent circuit modeling was pe1rformed using PSTrace 5.9 software.

The tests were carried out on two identical capacitive electrodes, with 5 to 8 measurements for each condition. Electrodes were evaluated immediately after fabrication ("fresh"), after extended use involving repeated impedance measurements and ECG signal acquisition ("used"), and following contact with textiles soaked in two common disinfectant solutions. The disinfectant treatments included 30-min contact with ethyl alcohol and 30-min contact with tincture of iodine. In all cases, the textile separator and counter-electrode configuration were consistent with that described in Section 2.1.

These measurements aimed to assess the chemical and electrochemical stability of the Kapton-based electrode structure during prolonged exposure to sweat and repeated cleaning—conditions representative of wearable sensor reuse. Equivalent circuit models were used to extract impedance parameters such as interfacial resistance, capacitance of the dielectric barrier, and constant phase element behavior, which are discussed in detail in Section 3.1. and summarized in Supplementary Table S1.

2.3. ECG Signal Acquisition Using a Simulator and Practical Demonstration

To evaluate the functional performance of the capacitive electrode and illustrate feasibility and signal detectability in capturing biopotential signals, ECG tests were conducted using both a commercial patient simulator and healthy volunteers. Simulator-based ECG recordings were performed through the controlled cotton textile layer to demonstrate through-clothing measurement capability. For the volunteer tests, the electrodes were placed over the participants own garments as a demonstration of signal acquisition through everyday clothing.

The custom front-end readout circuit developed for this purpose utilizes charge-sensitive amplifiers at the input stage, incorporating T-network feedback structures as shown in Figure 4.

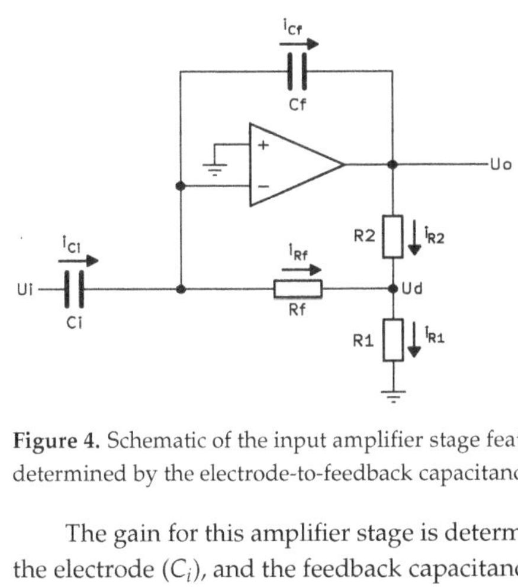

Figure 4. Schematic of the input amplifier stage featuring a T-shaped feedback network. The gain is determined by the electrode-to-feedback capacitance ratio.

The gain for this amplifier stage is determined through the ratio of the capacitance of the electrode (C_i), and the feedback capacitance (C_f). The feedback T-network consists of resistors R_f, R_1, and R_2 forming an equivalent resistance R_{eq} across the feedback path.

Starting with Kirchoff current law and Ohm's law:

$$i_{R_2} = i_{R_1} - i_{R_f} \tag{9}$$

$$i_{R_f} = -\frac{U_d}{R_f} \tag{10}$$

$$i_{R_1} = -\frac{U_d}{R_1} \tag{11}$$

$$U_d = -i_{R_f}.R_f \tag{12}$$

$$U_o = R_2.i_{R_2} + U_d \tag{13}$$

Substituting (10) and (11) into (9):

$$i_{R_2} = \frac{U_d\left(R_f + R_1\right)}{R_1 R_f} \tag{14}$$

Combining (12) and (14):

$$i_{R2} = -\frac{i_{Rf}(\mathrm{R_f} + \mathrm{R_1})}{\mathrm{R_1}} \tag{15}$$

Substituting into (13):

$$U_o = -i_{Rf}\left(\frac{R_2\left(R_f + R_1\right)}{R_1} + R_f\right) \tag{16}$$

Rearranging:

$$i_{R_f} = -\frac{U_o}{R_f\left(\frac{R_2}{R_1} + 1\right) + R_2} = -\frac{U_o}{R_{eq}} \tag{17}$$

Now, considering the currents through the capacitors:

$$i_{C_i} = C_i \frac{dU_i(t)}{dt}, \tag{18}$$

$$i_{C_f} = -C_f \frac{dU_{C_f}(t)}{dt} \tag{19}$$

$$i_{C_i} = i_{C_f} + i_{R_f} \tag{20}$$

Substituting (17)–(19) into (20), and taking Laplace transforms:

$$C_i s U_i = -C_f.s.U_o - \frac{U_o}{R_{eq}} \tag{21}$$

Thus, the gain U_o/U_i can be expressed as:

$$\frac{U_o}{U_i} = -\frac{s.C_i.R_{eq}}{s.C_f.R_{eq} - 1} \tag{22}$$

Finally, assuming $sC_fR_{eq} \gg 1$, the gain simplifies to:

$$\frac{U_o}{U_i} = \frac{C_i}{C_f} \tag{23}$$

This relationship shows that the amplifier gain depends on the ratio of the input and feedback capacitances, with minimal influence from the resistive elements under typical operating frequencies. In practice, however, this dependence of the gain on electrode capacitance represents a limiting factor for the calibration of cECG recorders and, consequently, for the accurate measurement of the amplitude components of the electrocardiogram.

For ECG signal generation, a PS2240 ECG Patient Simulator, dedicated to testing and calibration of ECG units, was used to output standardized ECG waveforms, including normal sinus rhythm and various arrhythmias, with adjustable amplitude and frequency. The simulator output was routed through the textile-interfaced capacitive electrode setup shown in Figure 1.

In addition to the simulator tests, a practical signal acquisition was carried out using the same electrode and readout system on two healthy volunteers (a 23-year-old male and a 22-year-old female). The participants were laboratory collaborators, and their electrocardiograms were recorded with personal informed consent and approval from the university ethics committee.

The setup used for these measurements is shown in Figure 5.

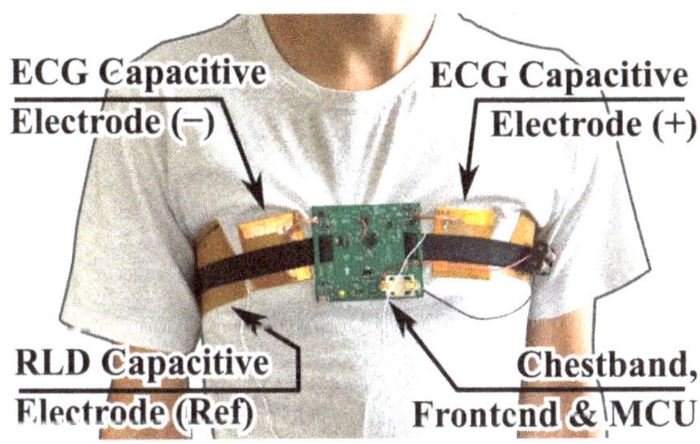

Figure 5. Setup for a practical demonstration of through-textile cECG acquisition employed in measurements on volunteers.

As illustrated, two capacitive electrodes were employed, pressed against the chest of the test subject via an elastic chest band, which also served as a support for the analog front end (described in later sections) and the readout electronics with microcontroller unit (MCU). A separate flexible electrode provided the reference right-leg drive (RLD) signal for common-mode interference suppression. The recorded ECG signals were acquired with the electrodes positioned over a cotton T-shirt, rather than through direct skin contact.

The volunteer measurements in this study represent a preliminary demonstration of applicability, aimed at validating the functionality of the proposed capacitive electrode structure and front-end configuration under realistic usage conditions and allowing for a side-by-side evaluation between simulator-generated and real biopotential signals.

3. Results and Discussion

3.1. Impedance Behavior of the Capacitive Electrode from 0.1 Hz to 100 Hz

To characterize the frequency-dependent behavior of the flexible capacitive electrode, impedance was measured across a wide frequency range (0.1–100 Hz) with the data presented in Figure 6.

The plot demonstrates a characteristic capacitive response, with impedance decreasing monotonically with increasing frequency. At low frequencies (below 10 Hz), the dry configuration exhibits impedance values above 1 GΩ, highlighting the high resistivity of the dielectric interface in the absence of moisture. Upon adding even a small amount of artificial sweat (8 μL cm^{-2}), the impedance drops by nearly an order of magnitude in the sub-10 Hz region, confirming that ionic conductivity plays a significant role in coupling efficiency.

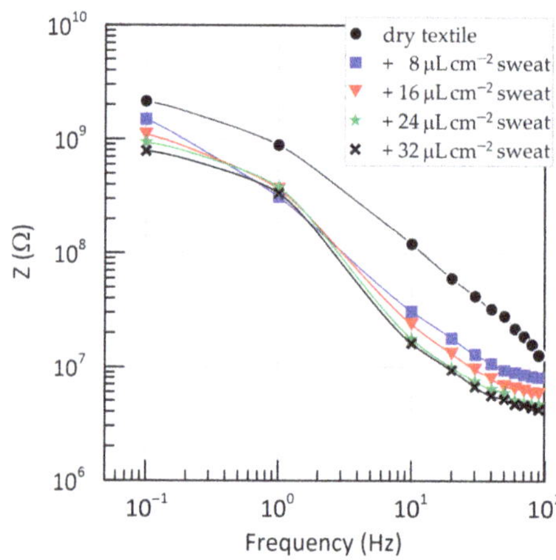

Figure 6. Impedance magnitude of the capacitive electrode as a function of frequency for different levels of artificial sweat added to the textile separator.

With further moisture addition (16–32 μL cm^{-2}), the impedance decreases slightly but progressively less, and the curves begin to overlap, especially above 10 Hz. This plateau indicates that the textile separator reaches moisture saturation at ~16 μL cm^{-2}, beyond which its conductive properties remain relatively unchanged.

Overall, these measurements verify the strong capacitive behavior of the electrode across all tested conditions and confirm that even small levels of perspiration can significantly reduce interface impedance. This reduction in impedance is a consequence of increased capacitance, which in turn facilitates the registration of low-frequency ECG components. At the same time, this effect may also introduce variability and potential distortion at very low frequencies.

3.2. Electrochemical Impedance Spectroscopy for Evaluating the Chemical Stability of the Electrode

The aim of the electrochemical tests was to evaluate whether the flexible electrode can be reliably used as a capacitive ECG electrode for long-term and repeated application. For consistent performance in such applications, the electrode must maintain high interfacial resistance that remains stable during prolonged exposure to chemically aggressive fluids, such as human sweat or disinfectant solutions. The high salt content and mildly acidic pH of sweat may compromise the barrier properties of the polyimide film, potentially allowing electrolyte penetration to the copper conductor and initiating undesirable oxidation processes. These could introduce parasitic noise in the acquired ECG signals. Additional concerns stem from typical disinfectants, which often contain organic solvents like ethanol or strong oxidizers such as iodine compounds and hydrogen peroxide, all of which could affect the integrity of the polyimide–copper interface.

Electrochemical impedance spectroscopy (EIS) results obtained from a "fresh" electrode, tested under varying degrees of wetting of the textile separator with artificial sweat, are shown in Figure 7. The data are presented as Nyquist and Bode plots for the sandwich configuration: test electrode/textile (dry or wetted with artificial sweat)/Ni-plated copper counter electrode. The Nyquist plots (Figure 7a) and the frequency dependence of the impedance magnitude (Figure 7b) both exhibit typical capacitive behavior, with experimental points forming nearly linear trajectories. However, the phase angle–frequency relationships (Figure 7c) reveal the presence of two time constants, indicating that a more complex equivalent circuit is needed beyond a single ideal capacitor.

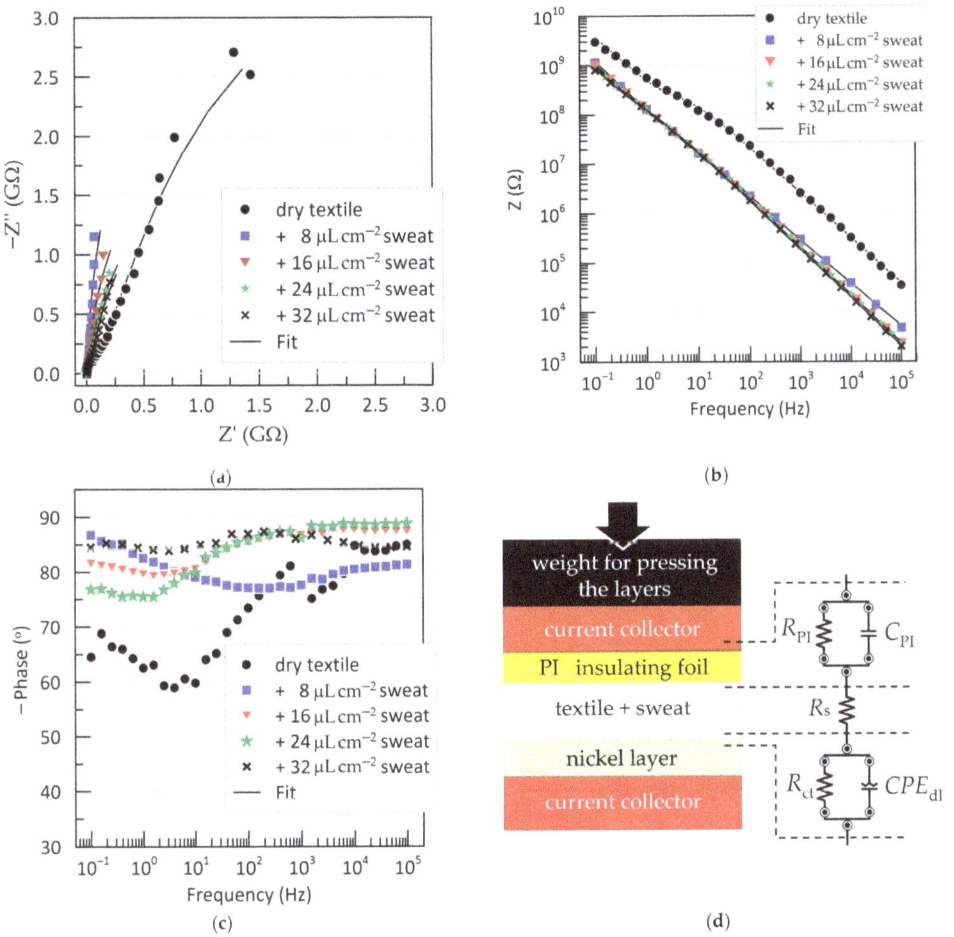

Figure 7. EIS results for the tested electrode using a textile separator with different quantities of artificial sweat: (**a**) Nyquist plots; (**b**) Bode impedance plots; (**c**) Bode phase angle plots; (**d**) equivalent circuit used to fit the EIS data. R_{PI}—resistance of the polyimide foil; C_{PI}—capacitance of the polyimide foil; R_s—resistance of the sweat-soaked textile separator; R_{ct}—charge transfer resistance; CPE_{dl}—constant phase element representing the electric double layer at the Ni/textile interface.

The remaining EIS measurements, performed after various treatments of the Kapton film, showed no significant deviations compared to the untreated "fresh" electrode (see Supplementary Figure S1).

In addition to the qualitative characterization of electrode behavior, quantitative interpretation of the EIS results requires fitting to an equivalent circuit. Various circuit models for capacitive ECG electrodes have been proposed in the literature [28,32,41–43]. However, their direct application is not suitable in our case, as these models describe the electrode-body interface, whereas our objective is to assess the chemical resistance of materials in contact with sweat. The simplest equivalent circuit of an electrochemical interface is the parallel resistor-capacitor (RC) model in series with the electrolyte resistance [41,44]. We adopted this model, with the final equivalent circuit shown in Figure 7d. It provided an excellent fit to the experimental data and consists of the resistance of the sweat-soaked textile separator (R_s), positioned between two RC blocks. The first RC branch ($R_{PI}C_{PI}$) corresponds to the Kapton@Cu/textile interface, while the second branch ($R_{ct}CPE_{dl}$) represents the textile/Ni interface. Instead of a pure capacitance, the electric double-layer capacitance at the metal-electrolyte interface is modeled using a constant phase element (CPE_{dl}), described by the following:

$$Z_{CPE} = \frac{1}{Y_0(i\omega)^n}, \tag{24}$$

as opposed to the ideal capacitor:

$$Z = \frac{1}{i\omega C_{dl}} \tag{25}$$

here, Y_0 is a pseudo-capacitance parameter (with units $\Omega^{-1}s^n = Fs^{n-1}$), and n is a dimensionless factor ranging from 0 to 1 that accounts for deviations from ideal capacitive behavior. The origin of the CPE element is typically attributed to surface phenomena on the metal electrode in contact with the electrolyte—such as surface roughness, passivation, and specific adsorption—which lead to non-uniform distributions of resistance, capacitance, and current density across the interface [44].

The extracted values for all equivalent circuit elements under different conditions—varied artificial sweat levels and surface treatments—are listed in Supplementary Table S1. The change in the interfacial capacitance C_{PI} (polyimide interface) is graphically presented in Figure 8.

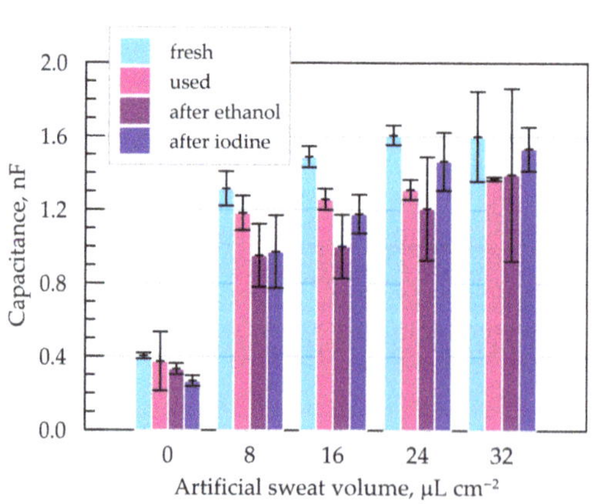

Figure 8. Influence of textile wetting with artificial sweat on the C_{PI} values determined via EIS.

As expected, the largest difference in C_{PI} occurs between the dry textile condition and all other sweat-wetted configurations. The dry system exhibited the lowest capacitance values, ranging from 0.2 to 0.6 nF. Even minimal wetting of the textile with an ionic medium (artificial sweat) caused a sharp increase in C_{PI} above 1 nF. Interestingly, the capacitance increases only moderately with further increases in sweat volume—from 8 μL cm^{-2} (semi-dry) to 32 μL cm^{-2} (fully soaked), with total variation around 0.3 nF. Beyond 16 μL cm^{-2}, C_{PI} remains effectively constant, suggesting saturation behavior.

EIS tests were also performed on the same electrodes after prolonged use, including repeated sweat exposure and mechanical bending. In these "used" samples, the C_{PI} values were slightly lower than those of the fresh electrodes, with the deviation becoming more noticeable at higher moisture levels. For dry textiles, the decrease was around 30 pF, while for sweat-soaked samples, C_{PI} was reduced by up to 0.3 nF.

Surface treatment with iodine did not significantly affect C_{PI}, with variations remaining within the margin of experimental error. Ethanol treatment resulted in a slight decrease in polyimide capacitance and increased uncertainty in the measurements. To better assess the extent of interaction between the electrode surface and the different fluids, contact angle measurements were also performed after 30 min of exposure to artificial sweat, ethanol, and iodine solutions. The results, shown in Supplementary Figure S2, indicate a marginal improvement in wettability following contact with iodine and sweat and a slight deterioration after ethanol exposure.

It should be emphasized that these surface treatments represent more aggressive and prolonged exposure than would be expected under typical disinfection conditions in wearable sensor applications. In conclusion, the Kapton-based capacitive electrodes demonstrate adequate mechanical robustness and chemical resistance to artificial sweat, ethanol, and iodine tincture, indicating their promise for long-term material durability in wearable sensor platforms.

3.3. ECG Signal Acquisition with the Capacitive Electrode System

The applicability of the developed capacitive electrodes for electrocardiographic (ECG) signal acquisition was explored through laboratory demonstrations using standardized signals from an ECG simulator and through experimental recordings from volunteers under real contact conditions. Figure 9 shows the block diagram of the front-end measurement system based on the widely used module AD8232 (IA), specially developed for ECG and other biopotential measurement applications. It is designed to extract, amplify, and filter small biopotential signals in the presence of noisy conditions, such as those created by motion or remote electrode placement. The integrated right-leg drive (RLD) amplifier further suppresses the power line interference.

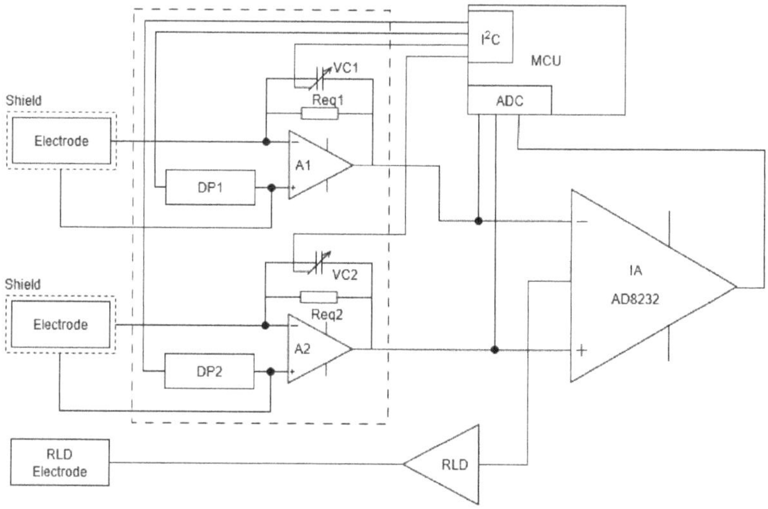

Figure 9. Block diagram of the developed front-end measurement system for capacitive ECG signal acquisition. The design includes charge-sensitive amplifiers, common-mode suppression, and baseline correction elements.

To minimize the influence of variable parameters of the capacitive electrodes, charge-sensitive amplifiers (A1 and A2) are included at the input stage. These amplifiers incorporate T-network feedback structures, as illustrated in Figure 4. As was mentioned above, the gain of these amplifiers is determined from the ratio of the input (electrode) and feedback capacitances. In practice, the gain in the proposed schematic solution is digitally adjustable via NCD2400MTR variable capacitors (VC1 and VC2) in the feedback loops of the amplifiers. This adjustment is managed by a PIC32MX150F128D microcontroller (MCU), which monitors the amplitude of the common-mode signal at the output of the IA and corrects (via I2C interface) the amplification of both amplifiers in a manner that reduces the imbalance in the input contours relating to common-mode signals.

This correction compensates for small variations in capacitance between the electrodes, especially when the textile interface is relatively dry. The gain adjustment remained within $\pm 5\%$ for properly mounted electrodes under realistic conditions, ensuring both signal fidelity and amplitude accuracy in controlled test environments.

Figure 10 presents the recorded waveforms from a PS2240 ECG simulator (BC Marketplace). The top trace corresponds to a standard ECG waveform with a 2 mV peak amplitude. The middle and bottom traces show signals with ST-segment elevation and depression, respectively. We chose to test the reproducibility of these cardiac pathologies because they are indicative of the low-frequency range of the amplifier, dependent on the capacitance of the electrodes and the input impedance of the amplifiers.

As can be seen, the simulator-induced waveform variations were distinguishable. The degree of elevation or depression shown in the plots cannot be regarded as absolutely reliable, since calibration of the measurement system sensitivity (mV/cm) is not possible. What can be stated, however, is that the signals exhibit characteristic changes in the ST-segment region, which should be interpreted only as indicative and requiring precise confirmation through conventional ECG diagnostics. A comparison of these measurements was also performed with those obtained using a traditional in-hospital patient monitor (AnyView P12). The corresponding control records are provided in the Supplementary Information (Figure S3) and demonstrate a good overlap in signal shape under simulator conditions with the waveforms acquired by the capacitive sensor setup. A more pronounced difference is observed in the baseline recovery of signals with ST-segment elevation, which can be attributed to the lower cutoff frequency (0.5 Hz) of the developed recording system.

The system also incorporates a digital baseline correction mechanism. This is achieved using MCU-controlled (via I2C interface) digital potentiometers CAT5171 (DP1 and DP2) that adjust the DC offset at the inputs of the buffer amplifiers based on real-time feedback. This approach optimizes the amplifier's dynamic range, allowing operation even under intensive baseline wander.

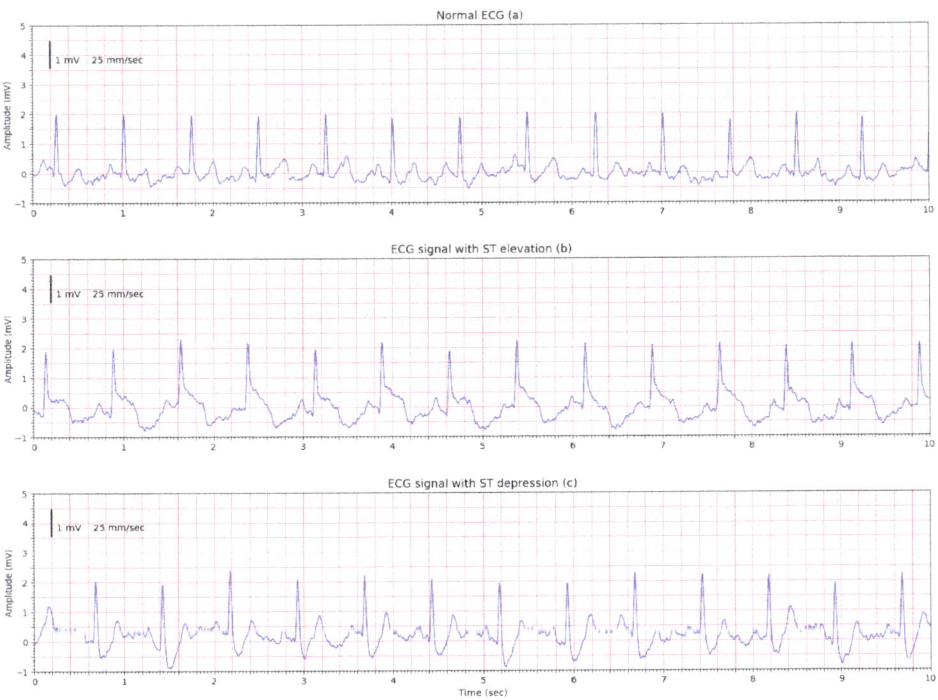

Figure 10. ECG signals recorded from the PS2240 simulator using the capacitive electrode system: (**a**) Normal ECG waveform; (**b**) ECG with ST-segment elevation (+300 μV); (**c**) ECG with ST-segment depression (−300 μV). The system reproduces characteristic waveform features of simulator outputs.

In addition to simulator-based testing, the system's ability to record real ECG signals was demonstrated using measurements from two healthy volunteers. Figure 11 shows a comparison between ECG signals acquired without symmetry correction (Figure 11a) and with active symmetry correction enabled (Figure 11b). The ECGs were acquired without changing the electrode positions, and the comparison confirms the effectiveness of the designed front-end circuitry in suppressing common-mode interference and improving signal quality.

This is confirmed in a second volunteer trial, where the improvement of the non-corrected signal (Figure 12a) is obvious, however, an intentional movement of the volunteer introduced a pronounced motion artifact visible in the middle of the signal (Figure 12b). The system's DC offset adjustment mechanism in the input amplifiers responded promptly, successfully compensating for the baseline shift. Despite the motion-induced disturbance, the system maintained a usable signal, suitable for demonstrating technical robustness.

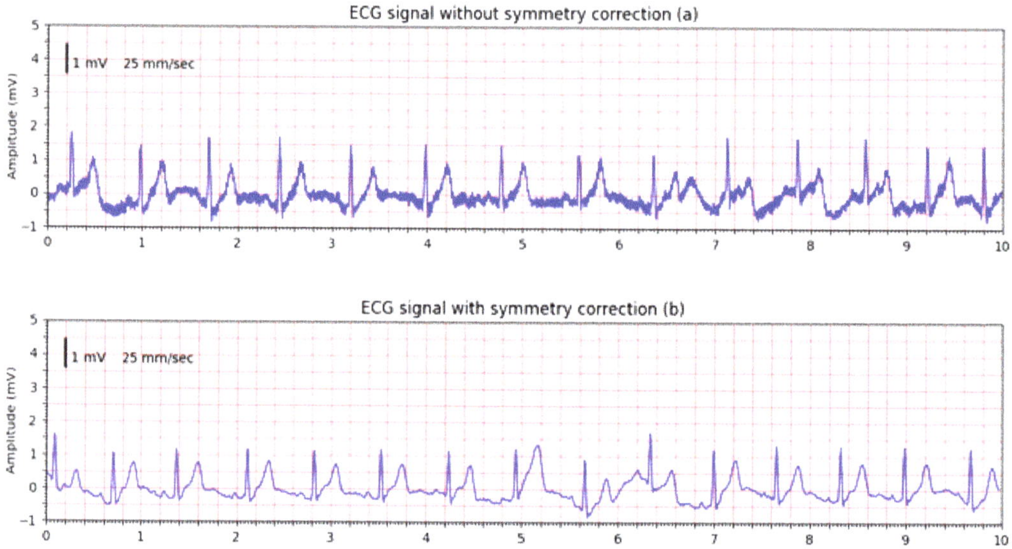

Figure 11. ECG signals recorded from volunteers: (**a**) ECG signal without symmetry correction; (**b**) ECG signal recorded with active feedback control.

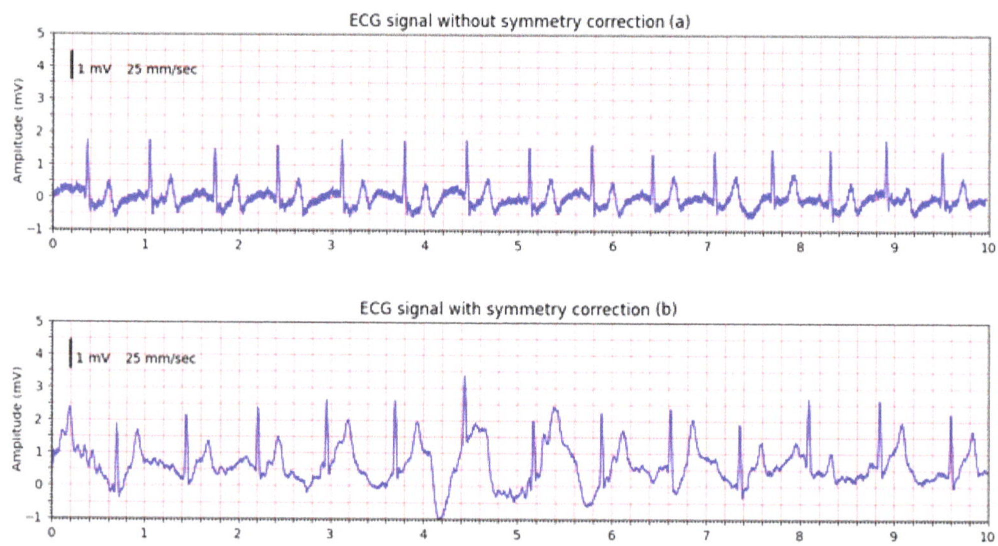

Figure 12. ECG signals recorded from a healthy volunteer: (**a**) ECG signal without symmetry correction; (**b**) ECG signal recorded with active feedback control, showing transient motion artifact. The system compensates for baseline drift in proof-of-concept tests.

A power spectral density (PSD) analysis was carried out on the data presented in Figures 11 and 12 using Hann windowing, 800-sample segments with 50% overlap. The resulting PSD spectra in the 0–80 Hz frequency range, obtained with and without symmetry correction, are shown in Figure 13. As can be seen, a pronounced suppression of the 50 Hz common-mode interference is achieved when symmetry correction is enabled, demonstrating the ability of the adapted front-end circuitry to reduce interference.

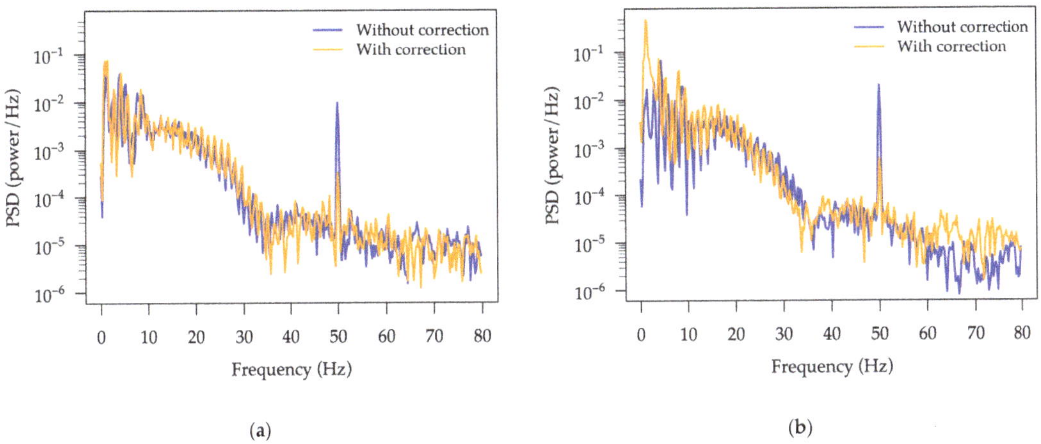

Figure 13. PSD plots for the ECG data presented in: (**a**) Figure 11; and (**b**) Figure 12. The clear suppression of the 50 Hz common-mode interference is notable when symmetry correction is active.

The system reacts so as to keep the signal within the measurable range and to allow reliable recognition of R peaks, enabling subsequent rhythm analysis of the signal, as illustrated in Figure 14.

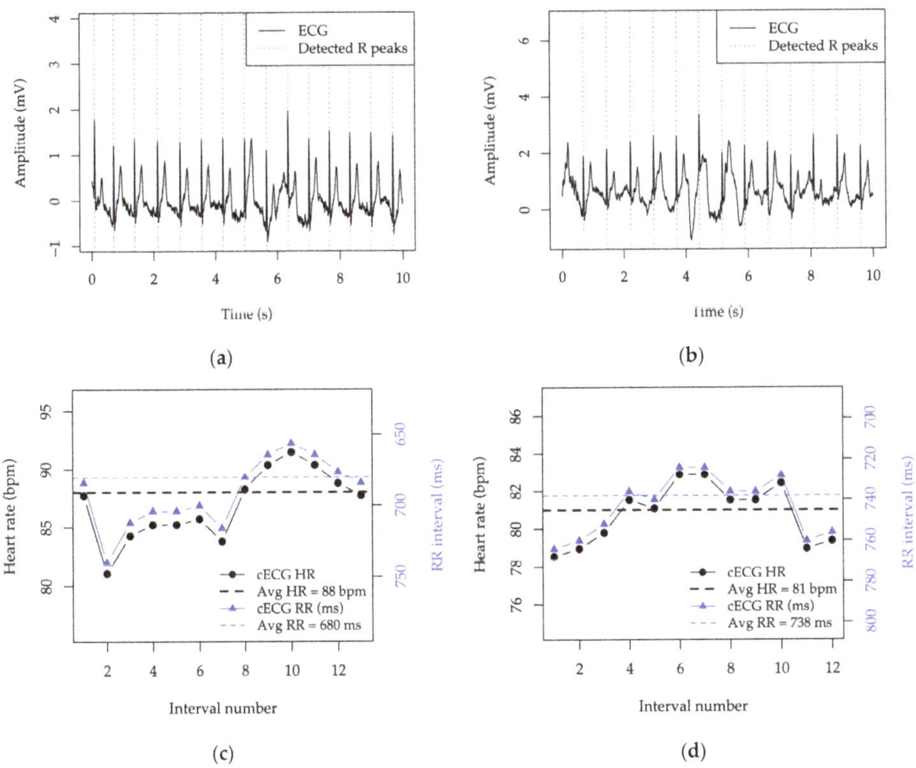

Figure 14. Demonstration of R-peak detection from the ECG signals and rhythm analysis based on the estimated R-R interval, based on the data shown in: (**a**,**c**) Figure 11; and (**b**,**d**) Figure 12.

4. Conclusions

This study demonstrates the viability of a flexible, cost-effective capacitive electrode for non-contact ECG signal acquisition, fabricated from readily available materials using a simple multilayer Kapton-insulated Cu foil structure on a flexible polyurethane structural support. Its performance was evaluated through a cotton textile interface, highlighting the electrode's ability to function reliably even when signals are acquired through clothing.

Electrochemical impedance spectroscopy (EIS) and broadband impedance measurements (10^{-1}–10^{5} Hz) confirmed that the electrode retains its capacitive behavior and chemical stability after repeated exposure to artificial sweat and common disinfectants such as ethanol and iodine tincture. A significant increase in interfacial capacitance was observed with moisture saturating beyond 16 μL cm^{-2}, which correlates with the sweat-retention capacity of the textile separator rather than the absolute sweat volume.

The custom-designed readout system, incorporating digitally adjustable gain and baseline correction, enabled robust ECG signal acquisition from both a simulator and human subjects. Simulator-based measurements reproduced characteristic waveform patterns. However, due to the absence of a guaranteed sensitivity criterion for the measurement system, low-frequency components such as ST-segment deviations can only be regarded as indicative of changes in cardiac activity, which would require confirmation through conventional clinical equipment. Volunteer recordings illustrated proof-of-concept feasibility under everyday clothing conditions.

Overall, the results indicate the potential of this low-cost, reusable electrode design as a technical platform for future wearable ECG research. While promising for unobtrusive signal monitoring through clothing, the inherent limitations of cECG, e.g., amplitude inaccuracy, low-frequency attenuation, variability between sessions, and strong susceptibility to motion artifacts, preclude clinical interpretation at this stage. A serious challenge remains the calibration of cECG recording systems: given the variable nature of the electrode–body coupling capacitance, calibration would need to be performed immediately prior to ECG acquisition. Further studies with larger cohorts, simultaneous reference ECG acquisition, and standardized signal quality assessment will be essential before any diagnostic use can be considered

Supplementary Materials: The following supporting information can be downloaded at: https://www.mdpi.com/article/10.3390/s25185856/s1, Table S1: Equivalent electrical circuit parameters from EIS fitting for all electrodes; Figure S1. EIS results for tested electrode after varied treatments; Figure S2. Variation in contact angles for water on the fresh Kapton surface, and after exposure to synthetic sweat, ethanol and iodine treatment; Figure S3. Control records of the PS2240 simulator generated ECG patterns, measured with an AnyView P12 patient monitor, corresponding to the data presented in Figure 10.

Author Contributions: Conceptualization, I.I. and G.T.N.; methodology, I.I., G.T.N., N.T. and B.T.; software, N.T. and G.T.N.; validation, I.I. and N.T.; formal analysis, I.I., G.T.N. and B.T.; investigation, I.I., G.T.N., N.T., B.T. and B.I.S.; resources, I.I., G.T.N. and B.T.; data curation, I.I., B.T., G.T.N. and B.I.S.; writing—original draft preparation, I.I., B.T. and G.T.N.; writing—review and editing, B.I.S., I.I. and B.T.; visualization, B.T., B.I.S. and I.I.; supervision, I.I.; project administration, I.I.; funding acquisition, I.I. All authors have read and agreed to the published version of the manuscript.

Funding: This study was financed by the European Union NextGenerationEU through the National Recovery and Resilience Plan of the Republic of Bulgaria, project No. BG-RRP-2.004-0005 "Improving research capacity and quality for international recognition and sustainability of the TU-Sofia".

Institutional Review Board Statement: This study was conducted in accordance with the Declaration of Helsinki and approved by the Ethics Committee of Technical University of Sofia (Approved by protocol №2, issued on 25 July 2025).

Informed Consent Statement: Informed consent was obtained from all subjects involved in the study.

Data Availability Statement: The data are not publicly available due to data privacy regulations. The data presented in this study are available on request from the corresponding author.

Acknowledgments: Part of the experimental investigations were performed on equipment obtained under the project BG16RFPR002-1.014-0006 "National Centre of Excellence Mechatronics and Clean Technologies".

Conflicts of Interest: The authors declare no conflicts of interest. The funders had no role in the design of the study, in the collection, analyses, or interpretation of data, in the writing of the manuscript, or in the decision to publish the results.

References

1. World Health Organization. Aging and Health. Available online: https://www.who.int/news-room/fact-sheets/detail/ageing-and-health (accessed on 16 June 2025).
2. Buyting, R.; Melville, S.; Chatur, H.; White, C.; Légaré, J.-F.; Lutchmedial, S.; Brunt, K.R. Virtual care with digital technologies for rural and remote Canadians living with cardiovascular disease. *CJC Open* **2021**, *4*, 133–147. [CrossRef]
3. Olmedo-Aguirre, J.O.; Reyes-Campos, J.; Alor-Hernández, G.; Machorro-Cano, I.; Rodríguez-Mazahua, L.; Sánchez-Cervantes, J.L. Remote Healthcare for Elderly People Using Wearables: A Review. *Biosensors* **2022**, *12*, 73. [CrossRef] [PubMed]
4. Yuan, Y.; Liu, B.; Li, H.; Li, M.; Song, Y.; Wang, R.; Wang, T.; Zhang, H. Flexible Wearable Sensors in Medical Monitoring. *Biosensors* **2022**, *12*, 1069. [CrossRef] [PubMed]
5. World Health Organization. Cardiovascular Diseases (CVDs). Available online: https://www.who.int/news-room/fact-sheets/detail/cardiovascular-diseases-(cvds) (accessed on 11 June 2025).
6. World Health Organization. Cardiovascular Diseases. Available online: https://www.who.int/health-topics/cardiovascular-diseases/#tab=tab_1 (accessed on 11 June 2025).
7. Cosoli, G.; Spinsante, S.; Scalise, L. Wrist-worn and chest-strap wearable devices: Systematic review on accuracy and metrological characteristics. *Measurement* **2020**, *159*, 107789. [CrossRef]
8. Majumder, S.; Chen, L.; Marinov, O.; Chen, C.; Mondal, T.; Deen, M. Noncontact Wearable Wireless ECG Systems for Long-Term Monitoring. *IEEE Rev. Biomed. Eng.* **2018**, *11*, 306–321. [CrossRef]
9. Fong, E.-M.; Chung, W.-Y. A Hygroscopic Sensor Electrode for Fast Stabilized Non-Contact ECG Signal Acquisition. *Sensors* **2015**, *15*, 19237–19250. [CrossRef]
10. Chi, Y.M.; Jung, T.-P.; Cauwenberghs, G. Dry-Contact and Noncontact Biopotential Electrodes: Methodological Review. *IEEE Rev. Biomed. Eng.* **2010**, *3*, 106–119. [CrossRef]
11. Avenel-Audran, M.; Goossens, A.; Zimerson, E.; Bruze, M. Contact dermatitis from electrocardiograph-monitoring electrodes: Role of p-tert-butylphenol-formaldehyde resin. *Contact Dermat.* **2003**, *48*, 108–111. [CrossRef]
12. Chen, Y.-H.; De Beeck, M.O.; Vanderheyden, L.; Carrette, E.; Mihajlović, V.; Vanstreels, K.; Grundlehner, B.; Gadeyne, S.; Boon, P.; Van Hoof, C. Soft, comfortable polymer dry electrodes for high quality ECG and EEG recording. *Sensors* **2014**, *14*, 23758–23780. [CrossRef]
13. Chlaihawi, A.A.; Narakathu, B.B.; Emamian, S.; Bazuin, B.J.; Atashbar, M.Z. Development of Printed and Flexible Dry ECG Electrodes. *Sens. Bio-Sens. Res.* **2018**, *20*, 9–15. [CrossRef]
14. Puurtinen, M.M.; Komulainen, S.M.; Kauppinen, P.K.; Malmivuo, J.A.V.; Hyttinen, J.A.K. Measurement of noise and impedance of dry and wet textile electrodes, and textile electrodes with hydrogel. In Proceedings of the 2006 International Conference of the IEEE Engineering in Medicine and Biology Society, New York, NY, USA; 2006; pp. 6012–6015. [CrossRef]
15. Meziane, N.; Yang, S.; Shokoueinejad, M.; Webster, J.G.; Attari, M.; Eren, H. Simultaneous comparison of 1 gel with 4 dry electrode types for electrocardiography. *Physiol. Meas.* **2015**, *36*, 513–529. [CrossRef] [PubMed]
16. Meziane, N.; Webster, J.G.; Attari, M.; Nimunkar, A.J. Dry electrodes for electrocardiography. *Physiol. Meas.* **2013**, *34*, R47–R69. [CrossRef] [PubMed]
17. Griss, P.; Enoksson, H.; Tolvanen-Laakso, P.; Merilainen, S.; Ollmar, G. Spiked biopotential electrodes. In Proceedings of the 13th International Conference on Micro Electro Mechanical Systems, Miyazaki, Japan, 23–27 January 2000; pp. 323–328. [CrossRef]
18. Salvo, P.; Raedt, R.; Carrette, E.; Schaubroeck, D.; Vanfleteren, J.; Cardon, L. A 3D printed dry electrode for ECG/EEG recording. *Sens. Actuators A Phys.* **2012**, *174*, 96–102. [CrossRef]
19. Baek, J.Y.; An, J.H.; Choi, J.M.; Park, K.S.; Lee, S.H. Flexible polymeric dry electrodes for the long-term monitoring of ECG. *Sens. Actuators A Phys.* **2008**, *143*, 423–429. [CrossRef]
20. Elango, P.; Dhanabalan, S.; Robel, M.; Sriram, S.; Bhaskaran, M. Dry electrode geometry optimization for wearable ECG devices. *Appl. Phys. Rev.* **2023**, *10*, 041408. [CrossRef]

21. Young, B. New standards for ECG equipment. *J. Electrocardiol.* **2019**, *57*, S1–S4. [CrossRef]
22. Torfs, T.; Chen, Y.-H.; Kim, H.; Yazicioglu, R.F. Noncontact ECG Recording System with Real Time Capacitance Measurement for Motion Artifact Reduction. *IEEE Trans. Biomed. Circuits Syst.* **2014**, *8*, 617–625. [CrossRef]
23. Portelli, A.J.; Nasuto, S.J. Design and Development of Non-Contact Bio-Potential Electrodes for Pervasive Health Monitoring Applications. *Biosensors* **2017**, *7*, 2. [CrossRef]
24. Chen, C.C.; Lin, S.Y.; Chang, W.Y. Novel Stable Capacitive Electrocardiogram Measurement System. *Sensors* **2021**, *21*, 3668. [CrossRef]
25. Spinelli, E.; Guerrero, F.; García, P.; Haberman, M. A simple and reproducible capacitive electrode. *Med. Eng. Phys.* **2016**, *38*, 286–289. [CrossRef]
26. Badarov, D.; Mihov, G.; Iliev, I. Development of analog front-end for capacitive ECG signal acquisition. In Proceedings of the IEEE International Scientific Conference Electronics (ET), Sozopol, Bulgaria, 15–17 September 2021; pp. 1–4. [CrossRef]
27. Haberman, M.A.; Spinelli, E.M.; García, P.A.; Guerrero, F.N. Capacitive Driven-Right-Leg Circuit Design. *Int. J. Biomed. Eng. Technol.* **2015**, *17*, 115–126. [CrossRef]
28. Khalili, M.; GholamHosseini, H.; Lowe, A.; Kuo, M.M. Motion artifacts in capacitive ECG monitoring systems: A review of existing models and reduction techniques. *Med. Biol. Eng. Comput.* **2024**, *62*, 3599–3622. [CrossRef] [PubMed]
29. Nagai, S.; Anzai, D.; Wang, J. Motion artefact removals for wearable ECG using stationary wavelet transform. *Healthc. Technol. Lett.* **2017**, *4*, 138–141. [CrossRef] [PubMed]
30. He, Z.; Min, H. Skin-induced motion artifact removal for ambulatory electrocardiography. *IEEE Sens. J.* **2022**, *22*, 15033–15043. [CrossRef]
31. Stanešić, A.; Klaić, L.; Cindrić, D.; Cifrek, M. Temporal Stability of Signal Quality in Non-Contact Biopotential Electrodes. *Sensors* **2025**, *25*, 3077. [CrossRef]
32. Lee, J.S.; Heo, J.; Lee, W.K.; Lim, Y.G.; Kim, Y.H.; Park, K.S. Flexible Capacitive Electrodes for Minimizing Motion Artifacts in Ambulatory Electrocardiograms. *Sensors* **2014**, *14*, 14732–14743. [CrossRef]
33. Tang, Y.; Chang, R.; Zhang, L.; Yan, F.; Ma, H.; Bu, X. Electrode Humidification Design for Artifact Reduction in Capacitive ECG Measurements. *Sensors* **2020**, *20*, 3449. [CrossRef]
34. Leicht, L.; Eilebrecht, B.; Weyer, S.; Leonhardt, S.; Teichmann, D. Closed-Loop Control of Humidification for Artifact Reduction in Capacitive ECG Measurements. *IEEE Trans. Biomed. Circuits Syst.* **2017**, *11*, 300–313. [CrossRef]
35. Cui, R.; Zheng, X.; Wang, K.; Chen, H.; Chen, W.; Chen, C.; Luo, J. Exploring Key Factors on Capacitive ECG Signal Quality Under Practical Considerations. *IEEE Trans. Instrum. Meas.* **2024**, *73*, 4011614. [CrossRef]
36. Toral, V.; Castillo, E.; Albretch, A.; Romero, F.J.; García, A.; Rodríguez, N.; Lugli, P.; Morales, D.P.; Rivadeneyra, A. Cost-effective printed electrodes based on emerging materials applied to biosignal acquisition. *IEEE Access* **2020**, *8*, 127789–127800. [CrossRef]
37. Naser, S.A.; Hameed, A.A.; Hussein, M.A. Corrosion behavior of some jewelries in artificial sweat. *AIP Conf. Proc.* **2020**, *2213*, 020030. [CrossRef]
38. Coull, N.A.; West, A.M.; Hodder, S.G.; Wheeler, P.; Havenith, G. Body mapping of regional sweat distribution in young and older males. *Eur. J. Appl. Physiol.* **2021**, *121*, 109–125. [CrossRef]
39. Svoboda, M.; Sotner, R.; Polak, L. Practical aspects of impedance measurement using operational amplifier and oscilloscope. In Proceedings of the 2022 32nd International Conference Radioelektronika (RADIOELEKTRONIKA), Kosice, Slovakia, 21–22 April 2022; pp. 1–4. [CrossRef]
40. Keysight Technologies. *Impedance Measurement Handbook: A Guide to Measurement Technology and Techniques*, 6th ed.; Keysight Technologies: Santa Rosa, CA, USA, 2020.
41. Polachan, K.; Chatterjee, B.; Weigand, S.; Sen, S. Human Body–Electrode Interfaces for Wide-Frequency Sensing and Communication: A Review. *Nanomaterials* **2021**, *11*, 2152. [CrossRef]
42. Wang, T.W.; Lin, S.F. Negative impedance capacitive electrode for ECG sensing through fabric layer. *IEEE Trans. Instrum. Meas.* **2020**, *70*, 4002308. [CrossRef]
43. Uguz, D.U.; Tufan, T.B.; Uzun, A.; Leonhardt, S.; Antink, C.H. Physiological motion artifacts in capacitive ECG: Ballistocardiographic impedance distortions. *IEEE Trans. Instrum. Meas.* **2020**, *69*, 3297–3307. [CrossRef]
44. Lazanas, A.C.; Prodromidis, M.I. Electrochemical impedance spectroscopy–a tutorial. *ACS Meas. Sci. Au* **2023**, *3*, 162–193. [CrossRef] [PubMed]

sensors

MDPI

Article

Data-Driven Strain Sensor Design Based on a Knowledge Graph Framework

Junmin Ke [1,2], Furong Liu [1,2,*], Guofeng Xu [1,2] and Ming Liu [1,2]

1 Key Laboratory of Trans-Scale Laser Manufacturing, Beijing University of Technology, Ministry of Education, Beijing 100124, China; liuming123@emails.bjut.edu.cn (M.L.)
2 School of Physics and Optoelectronic Engineering, Beijing University of Technology, Beijing 100124, China
* Correspondence: liufr@bjut.edu.cn

Abstract: Wearable flexible strain sensors require different performance depending on the application scenario. However, developing strain sensors based solely on experiments is time-consuming and often produces suboptimal results. This study utilized sensor knowledge to reduce knowledge redundancy and explore designs. A framework combining knowledge graphs and graph representational learning methods was proposed to identify targeted performance, decipher hidden information, and discover new designs. Unlike process-parameter-based machine learning methods, it used the relationship as semantic features to improve prediction precision (up to 0.81). Based on the proposed framework, a strain sensor was designed and tested, demonstrating a wide strain range (300%) and closely matching predicted performance. This predicted sensor performance outperforms similar materials. Overall, the present work is favorable to design constraints and paves the way for the long-awaited implementation of text-mining-based knowledge management for sensor systems, which will facilitate the intelligent sensor design process.

Keywords: knowledge graph; strain sensor; machine learning; material science

1. Introduction

Due to recent advancements in nanomaterials and nanotechnology, wearable electronic devices, particularly flexible and stretchable strain sensors, have garnered significant interest across various fields, such as healthcare and human motion detection, sports and physical training, and smart robotics [1,2]. In this process, these devices can be directly attached to the body to collect signal information, which can be utilized for various physical activities and health states. Currently, 5G communication technology has matured, and strain sensors are anticipated to have extensive applications in the field of motion perception systems.

Generally, a reliable implementation for detecting strain signals involves hybridizing a network composed of conductive materials (metal nanowires, carbon-based materials, etc.) with stretchable substrates (cotton, yarn, etc.). The indicators of strain sensor performance include gauge factor, strain range, long-term stability, and response time. The gauge factor and strain range are the two most critical factors among them, as they have a direct correlation with the application scenarios of strain sensors [3]. The performance of a strain sensor is significantly influenced by the choice of materials and their respective structures. Currently, viable research is divided into two categories. The first category involves designing sensor structures, such as segregated conductive networks and polymeric conductive composites [4–7]. The second category focuses on incorporating advanced nanomaterials such as 0D carbon black, metal nanoparticles, and graphene, as well as flexible substrates such as rubber [8–12] to improve sensors' performance. For the design of strain sensors, most of the work to date has been focused on improving performance with a high sensitivity, strain range, and cycle ability, etc. This leads to extensive trial-and-error experiments to optimize the process parameters, which involves high costs. The data-driven method is no doubt a new research paradigm for designing high-performance strain sensors, which have already been used in some fields. Recent research has concentrated on incorporating machine learning techniques (ML) into sensor design processes to forecast sensor characteristics [13–20]. Wang et al. [21] developed a silent speech recognition system based on a linear discriminant analysis (LDA) algorithm. Ravenscroft et al. [18] implemented a selection of machine learning algorithms, including artificial neural networks (ANN), random forests (RF), and k-nearest neighbor (KNN), for categorizing input signals. The above studies rely on ML methods to recognize the motion information. In particular, using ML techniques to facilitate the sensor design process has seldom been investigated. An alternative method involves mathematical statistics or sensor modeling tools to optimize sensor characteristics [22]. Remarkably, a recent study utilized active learning technology to automate the design of sensors based on composites [23]. Although these methods have their advantages, the intricate relationships within the sensor data and the influence of various semantic information on the performance are still unclear. Sensor design may consider factors of conductive materials, flexible substrates, manufacturing methods, and structures to optimize performance. The correlation among the above items, which are regarded as semantic features in the ML field, is crucial for discovering the underlying information. However, utilizing semantic features, especially the relationship feature, into a single regression model (based on text feature) poses a significant challenge in facilitating sensor designs. Moreover, current predictions usually have a unidirectional nature, concentrating exclusively on either the positive or negative aspects of the design process. Consequently, the current ML regression methods lack the ability to describe sensor knowledge.

Therefore, there is a need to develop a new framework that harnesses machine learning techniques for accurate predictions and recommendations based on semantic features while efficiently managing sensor knowledge, particularly in the era of rapid data growth. Effective knowledge interpretation necessitates a thorough understanding of the sensor. In this context, knowledge graphs (KGs) can facilitate the creation of unified standard representations for data fusion by representing knowledge in the form of entities and relationships. It can aggregate the relationships of multiple aspects via semantic associations. By extracting semantic features, it can serve as a foundation for the development of robust modeling. Despite their effectiveness in managing and visualizing complex relationships and characteristics in the literature [24,25], knowledge graphs are still underutilized for innovative design discovery. Since representational learning methods can extract multi-dimensional information as inputs, it has been proposed to apply representational learning methods to deep feature processing for prediction and the discovery of new things. For example, Nie et al. utilized word embeddings to extract semantics features for literature mining of materials [25]. However, the disadvantage of this method is that the extracted features lack structural information, which is difficult to apply to the task of sensors. Therefore, a machine learning framework based on the combination of knowledge graphs and representational learning was further developed for the prediction of strain sensors. This framework's efficiency is described as follows. (1) Storing both textual and data information in an organized manner and using algorithms to convert it into scientific knowledge. (2) Efficiently conducting reasoning analyses on a trained neural network after learning its representation, which involves tasks such as mining sensor designs and predicting performance. (3) The finding design is relatively consistent with the model and shows excellent sensing performance (300% of strain range). Meanwhile, the durability of the sensor (above 1000 cycles under 10% strain) is excellent under the stretching–releasing process. Compared to extreme strain cases (small stain range (<50%) or low sensitivity (<20)), the performance enables real-time monitoring of in-plane strains between 30% and 300%. (4) Ultimately, once established, the AI framework becomes accessible, as scientific knowledge from previous publications can serve as a guide or platform for the rapid design of multifunctional devices in the future.

2. Materials and Methods

The purpose of this work is to create a new framework (KGAI) based on the knowledge graph for properties prediction and design exploration. Graph construction, representation learning, and knowledge reasoning are the three primary modules of the KGAI architecture (Figure 1) in the context of data-driven AI analyses.

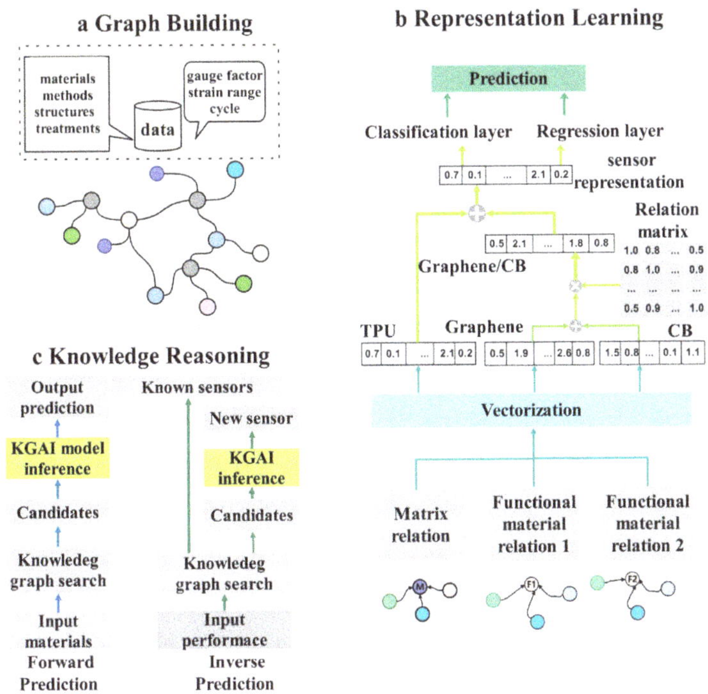

Figure 1. Illustration of a KGAI architecture developed with three procedures for (**a**) knowledge graph building of strain sensors, (**b**) representation learning of the knowledge graph (feature engineering), (**c**) knowledge reasoning of sensor design (performance prediction, sensor design query, and exploration).

2.1. Knowledge Graph Building of the Strain Sensors

Firstly, 200 fabrication methods of strain sensors based on piezoresistive mechanisms were collected from the literature. Secondly, the nodes of the graph were extracted by keyword-matching rules and manual extraction. These nodes include materials (flexible and conductive), manufacturing methods, sensor structure, and performance (strain range, measurement factor, and period). Among them, flexible materials included cotton, ecoflex, thermoplastic polyurethane (TPU), fibers, polymethylsiloxane (PDMS), nylon, polyurethane, rubber, and fabrics. Conductive materials included carbon nanotubes (CNTs), MXene, carbon black (CB), graphene, gold nanoparticles (AuNPs), silver nanowires (AgNWs), reduced graphene oxide (rGO), multi-carbon nanotubes (MWCNTs), PEDOT: PSS, cellulose nanofibers (CNFs), polyvinylidene fluoride (PVDF), cellulose nanocrystals (CNCs), carbon nanoparticles, polyacrylonitrile, and liquid metals. Then, the node data were standardized (SS1). After that, the triples were built using the above nodes as entities (Figure S2). In a triple, there is a head node, a relationship node, and a tail node, where the relationship between the head and tail nodes can be learned. The relationship can be divided into the following aspects: flexible matrix, functional material, structure, method, strain range, measurement factor, and cyclic stability. The number of constructed triples and entities is 978 and 2000, respectively. Finally, all triples were written to the graph database (nebula graph database).

2.2. Representation Learning

The representative learning module in Figure 1b converts semantic relationships from text strings into low-dimensional dense vectors, quantifying data within the knowledge graph using matrix calculations. Generally, generating embedding is a key procedure that affects the overall performance of the method. Given the size of the database, four lightweight algorithms (TranSE [26], DistMult [27], HOLE [28], and Node2vec [29]) are utilized. Meanwhile, two general indicators of Mean Reciprocal Rank (MRR) and Hit@n (namely, the proportion of correct entities ranked in the top 1/3/5/10) [26–28] are then applied to compare the performance of the algorithms above. The specific definitions of these two factors are given in detail in the Supplementary Materials. The aim is to learn representations of entities and relations that best explain a dataset by minimizing the logistic loss. The training epoch is 10,000. The optimizer is stochastic gradient descent, and the learning rate is 0.01. The L2_regularization is used for avoiding over-fitting and is set to 0.01.

2.3. Knowledge Reasoning

In the last stage of knowledge reasoning, functions such as prediction, mining, and detection are enabled using the embedding vectors established in the previous module. To achieve the property prediction and new findings, machine learning techniques that learn from a training dataset are utilized to make accurate predictions on unseen samples. Three robust and nonlinear methods of extreme gradient boosting (XGB), support vector machine (SVM), and multi-layer perception (MLP) are employed to compare performance and identify the best model for determining the probability of combinations of flexible substrates and functional materials. Furthermore, to make a distinction between positive and false designs, the probability threshold of classification is determined after extracting the embedding of the entity. The classification threshold is set above 0.74, according to the probability range of positive material combinations of sensors in the literature (Table S2).

2.4. Fabrication of a Designed Strain Sensor

According to the knowledge reasoning, the TPU/graphene/CB system was first found with a high correlation (0.93). Meanwhile, TPU with conductive composites had good merit in balancing the sensitivity and strain range as compared to the traditional design [6,15,16,30–33] of PDMS, rubber, cotton, etc./graphene/CB; therefore, it was chosen in the present study. Subsequently, a commonly used electro-spinning method was used to create the TPU composite film.

First, 2 g of TPU was dissolved in a dimethylformamide (DMF)/tetrahydrofuran (THF) (Macklin, Shanghai, China) mixture solution (volume ratio of 1:1) for 6 h by magnetic stirring to form a 20 wt% solution. Subsequently, the collecting distance was set to 20 cm, a flow rate of 3 mL h^{-1}, and a high voltage of 20 KV was supported. The syringe tip was 0.65 mm, and a rotation drum as the receiving device was set at 100 rpm. After 3 h, the TPU fibrous membranes were prepared. Next, 2 g of CB and graphene were ultrasonically dispersed in the solution (sodium dodecyl benzene sulfonate and deionized water) (Macklin, Shanghai, China) at 300 w ultrasonic power. Finally, the obtained TPU fibrous mats were placed in the suspension three times at the same work power. After three washing operations and drying at 70 °C for 90 min, the fibrous membrane was successfully prepared. The layer structure was obtained via the rolling-up process.

2.5. Characterization

The surface morphology of carbon-based TPU fibers was observed with a scanning electron microscope (JEOL 7001F, JEOL Ltd., Tokyo, Japan). Samples were subjected to electron microscopy after gold spraying. The electrical and mechanical properties of the fibers were measured by monitoring their resistance to the cyclic tensile tests in real-time with an electrochemical workstation (Ametek P4000, Ametek Inc., Berwyn, IL, USA) and Micro Tester (Instron 5948, Instron, Shanghai, China).

3. Results and Discussion

3.1. The Effect of the KGAI Representation Method

It is found that the algorithm of HOLE has the highest values of MRR (~0.899) and Hit@n (over 0.8) compared to the other three (Figure 2a,b). In order to consider the connections between multicomponent attributes, the same attribute relations are extracted through added embeddings, termed improved representations. For example, the combination of MXene and carbon nanotubes is examined both as a whole and as an individual, illuminating connections between various material combinations. Nevertheless, the original representation (OR) fails to take into account the relationship between materials.

To clarify the advantage of improved representation (IR), different combinations of flexible matrices and functional materials were tested. Figure 2c,d displays two-dimensional coordinates of various material combinations in the embedding space of the IR and OR methods. The distribution indicates that different functional materials with the same flexible substrate are more closely clustered together. The statement suggests that the design vector generated by the knowledge representation algorithm contains semantic information about sensor design. Additional findings from KMeans clustering indicate that a stronger correlation exists between flexible matrices and functional materials, leading to an increased likelihood of sensor design formation. Moreover, Table S3 provides a detailed list of typical combinations of matrix and functional material with a short distance from the improved method, which is consistent with previous experiments. Furthermore, the distance between the flexible substrate and functional materials in the new combinations is comparatively greater than in prior designs. The effective integration of existing material combinations between data indicates that the enhanced method is a reasonable representation of the sensor knowledge. Currently, the majority of studies [16–20] only focus on material properties as a feature to accomplish the prediction task. However, the relationship between materials in sensor design is vital, especially in the hydrogel design, which represents the interfacial property.

A heat map, which was reported to extract the correlation among different vectors [23], was also applied to demonstrate the relevance of designs. The correlations between 8 flexible matrixes and 8 functional materials were then extracted, and the results are plotted in Figure 2e. The block's color transition from pale yellow to dark yellow signifies a stronger correlation between them. Materials such as cotton, TPU, PDMS, and yarn exhibit significant correlations in their matrix structure, which is consistent with earlier research [34,35]. Previous research [36] has established a strong correlation between MXene and flexible matrices (PDMS and cotton). Furthermore, graphene/CB combinations with cotton and PDMS substrates follow the literature [37]. The observed correlation between TPU and graphene/CB is a new design due to comparable production processes. Despite the lack of direct reports for PEDOT/PSS combinations, the KGAI method's correlation learning revealed potential for yarn and graphene/cellulose nano-crystalline (CNC) designs. The involvement of domain experts with a thorough understanding of the nuances and interactions of material factors is required to interpret these complexities. In conclusion, the information-to-knowledge route (KGAI method) is capable of searching for existing designs.

In order to evaluate the prediction error of the KGAI method, a standard radar graph [23] is presented in Figure 2f. The false positive stands for the number of wrong predictions of designs in the graph database, while the true negative represents undetected designs. The KGAI method demonstrates robust strain sensor design capabilities, with fewer than 4 wrong predictions out of 64 samples.

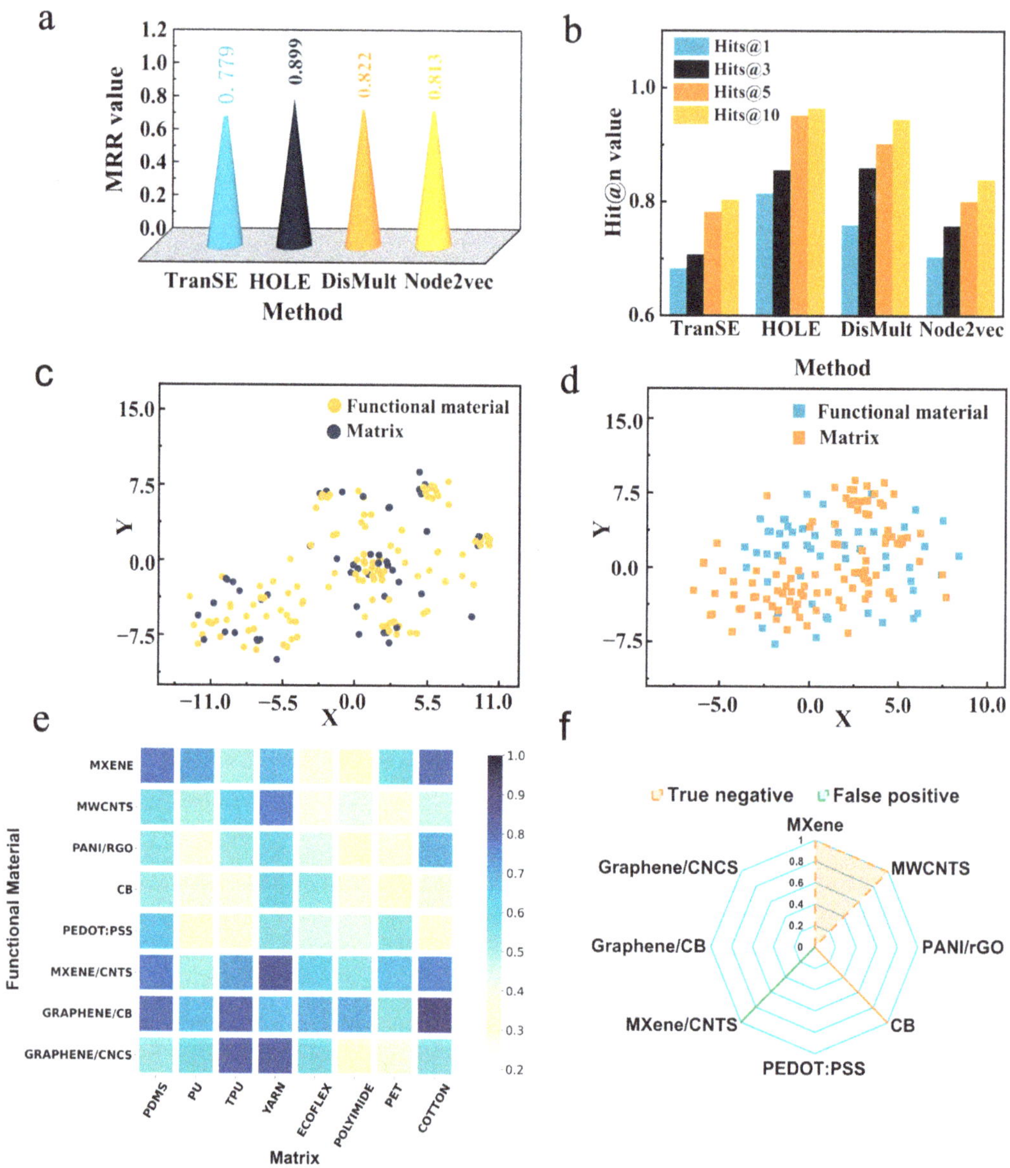

Figure 2. The evaluation of the KGAI method. (**a**,**b**) Values of MRR and Hit@n metrics of knowledge graph construction. (**c**,**d**) Cluster atlas of different combinations of representation methods (improved method with considering the correlation between functional materials and hole method without considering the correlation between functional materials) via unsupervised learning after dimensionality reduction. (**e**) The heat map of functional materials and flexible matrix. The colors scale with the values of the cosine similarity between embeddings. The dark color of the squares means a strong correlation. (**f**) The value of false positive (value = 1, wrong prediction in database) and true negative of designs (value = 3, undetected designs in database).

3.2. Navigation Model of Prediction

In order to assess prediction performance for the classification task (Figure 3a), typically used evaluation indicators like precision, recall, and F1-score were chosen. To evaluate the feasibility of the design, the classification task was executed. Ultimately, the multi-layer perceptron method performed better than SVM and XGB techniques and was selected as the study's navigation model (Figure 3b). The MLP method's accuracy and precision are shown in Figure 3c for ten distinct test sets, with corresponding values of 0.81 and 0.78, respectively. Figure 3d,e indicates that the predicted labels are basically consistent with the real labels.

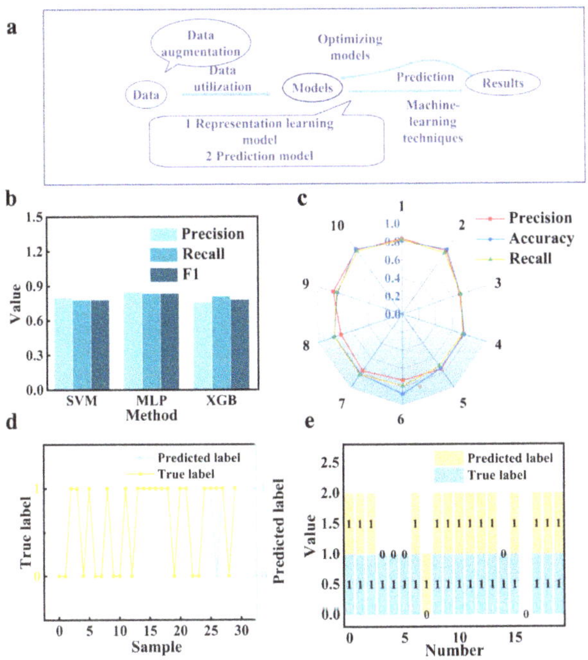

Figure 3. (**a**) The design process of the whole method. (**b**) The indicators (Precision, Accuracy, and Recall) of three methods for classification tasks by 10-fold cross-validation. (**c**) The indicators of the MLP method in ten different test datasets. (**d**,**e**) The predicted labels and true labels of train and test samples (one and four misclassified samples, respectively).

In this study, the prediction of performance is also considered and constructed, in addition to the possibility of predicting sensor designs. The regression task (XGB and MLP methods) utilizes the mean absolute error (MAE) and mean squared error (MSE) as measures of performance. Unlike accuracy, a smaller MAE value indicates a higher level of prediction accuracy in the decision-making process. Thus, the regression model is derived by training the multi-layer perception network using cross-validation to prevent over-fitting. The mean absolute error (MAE) is used as a metric to assess performance (Figure 4a,b). It is evident that the MAE and MSE of the MLP method are relatively smaller than SVM and XGB methods, owing to the learning and nonlinear approximation capabilities of the neural network. The final MAE of the MLP method on the training data is 0.98, and on the test data, it is 1.73. Additionally, the predictions of the data are illustrated in Figure 4c,d, showcasing relative fitting results. The results demonstrate that the feature generated by the relationship-based representation method is an effective way to predict properties.

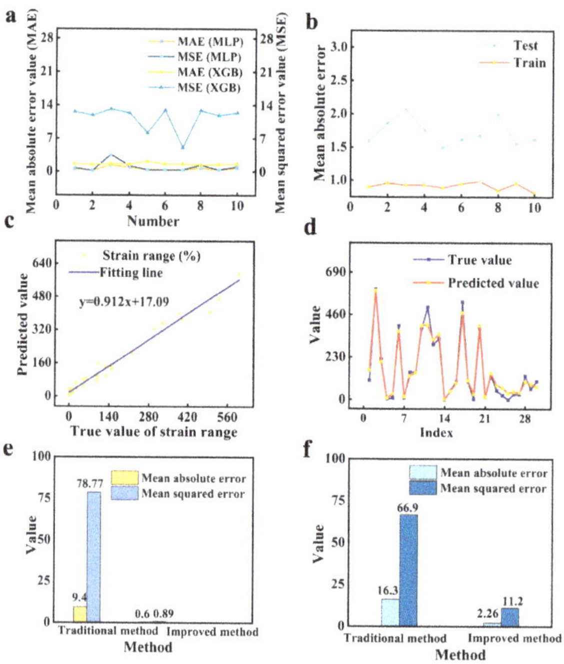

Figure 4. (**a**) The indicators of the XGB and MLP method for performance prediction. (**b**) The mean absolute error of 10-fold cross-validation of the test dataset. (**c**) The model prediction value and the true value for train samples. The line represents a prediction formula for the strain range. (**d**) The model prediction and the true value of test samples for strain range. (**e**,**f**) Evaluations of the design prediction. (**e**) The error prediction performance of the traditional method (only using the text feature without the correlation) and our KGAI method (using the graph feature containing the text feature and relationship feature) in the training process. (**f**) The error prediction performance of the traditional method and our KGAI method on the test dataset.

To further demonstrate the advantages of the relationship-based representation method (KGAI), the traditional method directly using text features [24] (e.g., the word2vec method without considering relationships) is used to complete regression tasks for comparisons (partial data in Table S2). Figure 4e demonstrates that the KGAI method has reduced the maximum absolute value error by nearly 8 times and the maximum square value error by nearly 5 times in comparison to the traditional method (Figure 4f). The primary reason for the excellent results is the utilization of deeper-level information hidden in the data, enhancing representations of the interaction between materials. The findings suggest that using ergodic sensor attribute knowledge as an entity and using relationships between attributes as features can improve the precision of traditional text-based machine learning methods, hence facilitating strain sensor mining and design tasks.

Most machine learning work of strain sensors in the literature either focuses on signal processing for resistance readout or lacks a comprehensive framework that combines knowledge graph-based data fusion with machine learning techniques for modeling. However, the graph embedding framework under study focuses on the correlation between performance and design, bridging the gap in the strain sensor field. This approach specifically addresses the issue of ignoring intra-aspect relationships, which is often present in medical modeling. The proposed method has obvious advantages: enhanced generalization, simplicity, and efficiency. The algorithm converts triples into low-dimensional and dense vectors to achieve efficient design reuse. A potential limitation of the present work is the inconsistent performance of most reports, restricting the capability of the machine learning model. Future data augmentation is warranted to enhance predictive power. This work is a preliminary demonstration of the application of the knowledge graph method for strain sensor fabrication. Future research could explore model optimization for the fabrication process.

3.3. Entity Prediction

The knowledge reasoning process, as shown in Table 1, is capable of executing two-way tasks, including (1) sensor performance prediction and (2) recommendation of feasible sensor designs. Since the construction of a knowledge graph consists of key characteristics of sensors (referred to as entity), these two-way predictions can be regarded as entity predictions. The performance of strain sensors is evaluated by two crucial parameters: the gauge factor (GF) and strain range (l). GF is defined by the slope of the relative resistance change versus the strain range change.

Table 1. The query is implemented in the strain sensor graph.

Code: Graph Query Implemented to Access Results
Query rules: Match (v:matrix{name: "PDMS"})\WHERE v. matrix.name == "PDMS"\ RETURN v;
Results = graph. run(match_str)

Table 2 presents a forward design for predicting performance based on the inputs of the flexible matrix and functional materials. The performance of rGO is significantly affected by the type of fabric it is coupled with, such as yarn and cotton, even if they are manufactured in a similar manner [38,39]. In an inverse design, Table 3 displays the recommended materials and structures for achieving an appropriate design with extremely high-performance results (e.g., 1 gauge factor, 120% strain range, and 500% strain range are set in advance). Moreover, the performance of various material combinations and structures can be captured. For example, Figure 5 demonstrates the trend of performance and structure strategies. Further, the balance between strain range and sensitivity of PDMS should be considered. In this regard, the utility of the KGAI framework enables rapid tracking of the current trend in strain sensors.

Table 2. Task 1 (Forward design to predict performance).

Input	Verification	
Flexible substrate: yarn Functional materials: rGO Method: dip-coating	Gauge factor:50 Strain range: 90%	Reference [38]
Flexible substrate: cotton Functional materials: rGO Method: dip-coating	Gauge factor: 4 Strain range: 11.6%	Reference [39]

Table 3. Task 2 (Inverse design to predict materials and structure).

Input	Verification	
Functional material: carbon nanotubes (CNTs) Gauge factor: 1.24 Strain range: 120%	Substrate: Flax fabric	Reference [40]
Substrate: Polyaniline Gauge factor: 1 Strain range: 500% Strain range: 240%	Functional material: Silver nanowires	Reference [41]

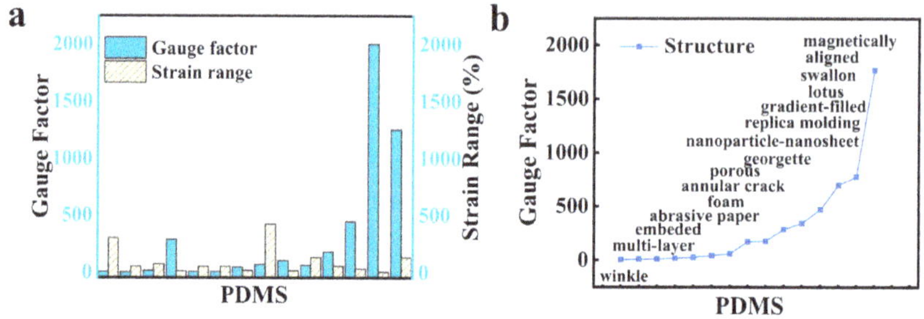

Figure 5. The trend of flexible substrate (PDMS) in the knowledge graph. (**a**) The sensor performance of strain range and gauge factor. (**b**) The trend of gauge factor under different structures of PDMS.

3.4. Automatic Strain Sensor Design

Due to the high correlation of materials in Section 3.1, TPU was first chosen as the matrix in this work due to its merits such as biocompatibility, breathability, flexibility, and toughness. The KGAI method found a mixed graphene/CB design due to their excellent correlation. The classification of this combination (TPU/graphene/CB) obtained a high probability (above 0.93), signifying the feasibility of this combination. Regarding the manufacturing method, electro-spinning technology was predicted to be used. The machine learning model applied to predict this combination had a good strain range (above 240%). Moreover, the verification in a later section shows that the combination of TPU and graphene/CB exhibits a strain range that is relatively consistent with the model prediction.

3.5. Characteristics of the Designed Strain Sensor

The fabrication process of the strain sensor is shown in the Method section. Figure 6 demonstrates the characteristics of flexibility, ultra-thin structure, and surface morphology. The multi-layer network from the top surface of the film was observed. Additionally, it was found that carbon-based materials were evenly distributed on the fibers, attributed to the loose structure of the TPU substrate, which may facilitate good contact between functional materials and substrates, resulting in excellent stability [42].

In order to verify practical performance, the mechanical properties of the strain sensor were first evaluated by a tensile test, as shown in Figure 7a. It is evident that over a 300% strain range was obtained, conforming to the prediction. Except for the strain range, the relative resistance change corresponding to the strain range is shown in Figure 7b, which demonstrated a good gauge factor (110 at 80% strain). When comparing spherical CB nanoparticles, CB and graphene are more likely to form a complete conductive network, thereby enhancing electrical conductivity. Table 4 summarizes the comparisons of performance (such as strain range, gauge factor, and durability) to compare the performance of the designed sensor. It can be seen that the designed sensor has a balanced performance between strain range (300%) and gauge factor (110 at 80% strain). The excellent performance can be attributed to its internal network structure. The fibers underwent straightening and elongating, leading to deformation of the internal network structure, resulting in resistance changes. Meanwhile, the micro-crack that is generated during the stretching process helps enhance the performance [21]. Figure 7c reveals the sensor 's excellent durability under the stretching–releasing process (10% of strain). The insert picture in Figure 7c shows the response time (110 ms) and recovery time (112 ms). To further demonstrate the stability of the strain sensor, different tensile rates (40 mm/min and 20 mm/min) were applied to the strain sensor, presenting repeatable stress (Figure 7d,e). The first cycle exhibits a relatively large variation in stress, most likely due to the permanent destruction of the conductive network, which aligns with the previously reported sensor [22]. Stress in subsequent cycles was reduced compared to the first cycle and almost overlapped, suggesting stable destruction and reconstruction of the conductive network. The response of the strain sensor during stretch–release cycles in various strain ranges (30%, 60%, and 80%) is depicted in Figure 7f. Different strains exhibit the same response patterns, suggesting a continuous reaction to cyclic loading. Furthermore, during the stretching and releasing process, the resistance may increase proportionally to the tensile stress and recover naturally without any external intervention. This phenomenon can be attributed to the partial breakdown of electrically conductive pathways and the gradual increase in separation between carbon-based materials, both leading to an increase in resistance. Furthermore, the sensing signals remain stable, and the alteration in stress is only determined by strain, indicating the reliability of the sensor in stress recognition.

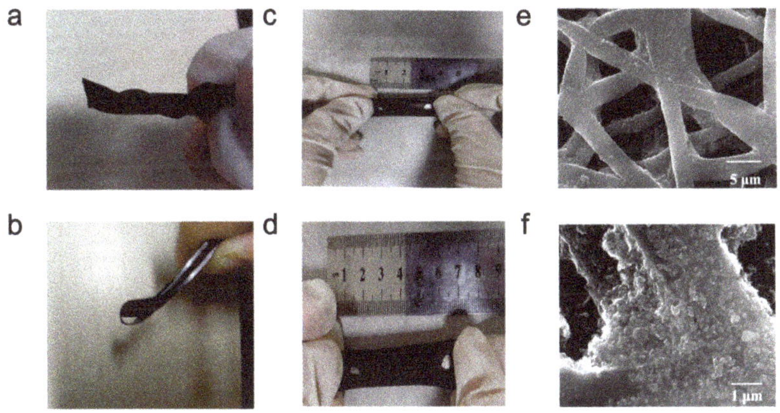

Figure 6. Photographs of the film showing excellent flexibility when (**a**) twisted, (**b**) bent, and (**c**,**d**) stretched. (**e**,**f**) Top views of the composite film under the SEM observation with different scales (5 μm and 1 μm).

Table 4. The summarization of the reported sensor performance.

Ref.	Main Materials	Strain Range	Sensitivity	Repeatability	Response/Recovery Times	Limit Detection
[1]	CNTs ink/PU	350%	2.7	1000	~	~
[6]	CNTs/CNF/PDMS/TPU	217.5%	12.7	800	~	~
[7]	Silver fillers/LM ink	170%	~	5000		
[8]	CNTs/latex tube	200%	91.1	4000	290 ms/310 ms	0.1%
[15]	CB/PDMS	250%	12	1000	~	~
[16]	CNTs/CB/PDMS	80%	7.7	10,000	100 ms/110 ms	0.04%
[20]	CNTs/LiCl/elastic core-spun yarn	100%	1.35	1000	300 ms	~
[30]	Graphene/PU	160%	86.86	100	~	~
[31]	LM/TPU	548%	6	1000	~	~
[32]	CNTs/TPU	140%	~	1250	~	~
[33]	PAN/graphene/TPU	2%	1700	300	~	~
[37]	CNTs@carbon black/PDMS	80%	7.7	10,000		2%
[40]	CNTs/fabric	128%	4.73	~	~	~
[42]	MXene/TPU/PAN	80%	9.69	1750	140.6 ms	0.1%
[43]	Conductive ink/rubber	50%	12.14	5000	71.43 ms/178.49 ms	~
[44]	CNTs/cotton	300%	21.85	~	~	~
This work	Graphene/CB/TPU	300%	110	1000	110 ms/112 ms	~

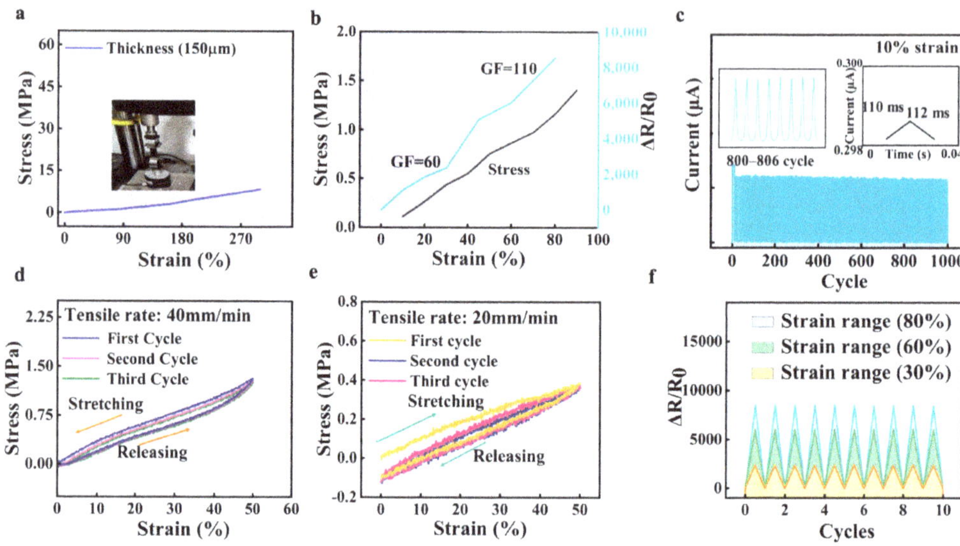

Figure 7. Main metrics of the strain sensor. (**a**) The stress under different strain ranges. (**b**) The relative resistance–strain variation curves and stress–strain curves of strain sensors. (**c**) The durability of the sensor. The insert pictures are the subgraph of the cycle (800–806) and response/recovery times of the sensor. (**d**,**e**) The three cycles of the stretching process with a tensile rate of 40 mm/min and 20 mm/min, respectively. (**f**) The cycle stability of relative resistances under different strains (35%, 45%, and 80%).

3.6. Demonstration of a Human Action Monitoring Application

According to the above results, the prepared composite sensor has excellent properties such as extensive strain range, flexibility, and good sensitivity, showing great prospects in wearable devices. Consequently, it was used to detect human motion across several levels, including swallowing, speaking, hand bending, and knee movements. The small strain of swallowing was first tested by attaching the sensor to the neck, followed by testing the speaking process of two words ("MXene" and "science") (Figure 8a,b). Different changes in relative resistance during these activities caused by vocal cord vibrations were captured, indicating sensitivity to tiny motions. Figure 8c exhibits that the sensor responded to hand curvature with a consistent correlation between the bending angle and resistance change. To monitor larger joint movements, measurements were performed while sitting and standing by mounting the sensor on the knee (Figure 8d). The resistance-time waveform presented a clear and periodic response, showing promising potential for human-friendly rehabilitation. The pressure characteristic was also observed according to different forces (Figure S3).

As discussed above, the designed strain sensor has the potential for health applications. The high-quality design is mainly attributed to the constructed knowledge graph for strain sensors. On the surface, the sensor design data consists of discrete symbols that fulfill diverse requirements. However, these discrete symbols represent the semantic correlation between design elements, such as dimension mixtures. The relationship between designs (material, structure, property) of strain sensors is unveiled in the graph database, enabling data-driven mining. With more designs of films or sensors and appropriate data-processing methods, the model could be continuously updated, leading to improvement [45]. It is feasible to design multifunctional films, such as frost resistance, UV blocking, electromagnetic interference shielding, and so on.

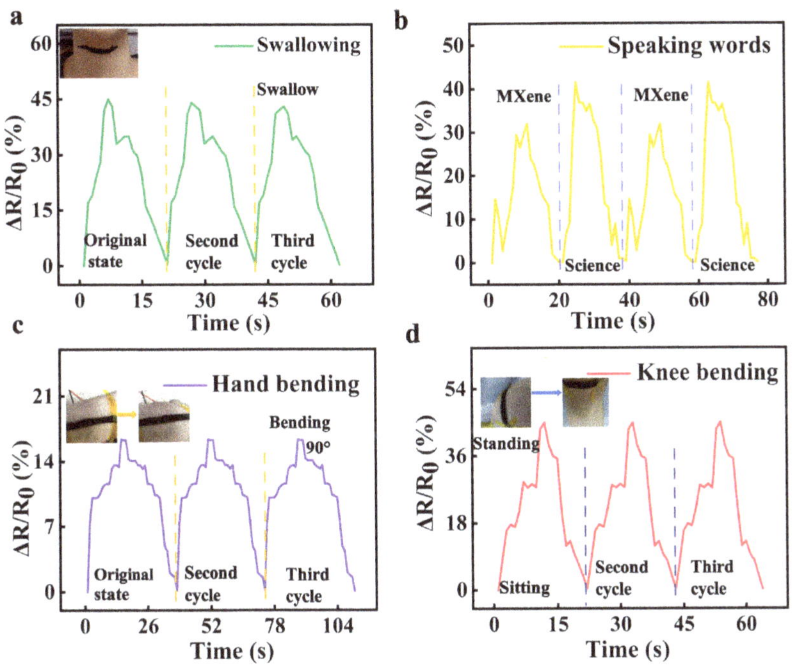

Figure 8. Applications of the strain sensor to detect human motions. (**a**,**b**) Motions of swallowing and speaking. (**c**) The resistance changes with the bending of the hand. (**d**) The resistance changes of knee motions (standing and sitting state).

4. Conclusions

This work develops a framework that combines knowledge graphs and graph representation learning methods to reveal the interconnectedness of strain sensor design. The established knowledge graph has a high MRR (mean reciprocal rank), with a value greater than 0.89, allowing for effective retrieval and querying of information. Furthermore, by establishing an MLP regression model, the feasibility of the design and strain range prediction are demonstrated. The strain sensor created using this framework has a prediction accuracy of 0.93, demonstrating its excellent performance (strain range of 300%) and the validity of this relational semantic prediction method in strain sensor design. The efficiency of this method not only opens up the possibility of rapid fabrication of strain sensors but it is also expected to extend to other material science design challenges by guiding algorithms to learn and perform specific tasks. For precise prediction, some promising directions could be considered in future work. (1) The current knowledge graph database could be expanded to encompass a wider variety of strain sensor data. This will involve the aggregation of data from diverse sources, thereby improving the model's generalizability and applicability. Furthermore, the creation of industry-standard flexible sensor datasets is imperative to establish a benchmark. For example, the low detection limit, response time, biocompatibility, and cost should be properly considered. (2) The construction of a dynamic knowledge graph may help continuously optimize design and prediction models, such as incremental learning algorithms (online sequential extreme learning machine), online learning algorithms (adaptive linear neuron), and neural link prediction algorithms (neural collaborative filtering or matrix factorization techniques). This dynamic nature will allow the system to automatically absorb new data and knowledge to continuously optimize designs and predictive models. (3) Future research could explore approaches such as ensemble learning and deep learning to further improve prediction accuracy. (4) The real-time feedback and adaptive design are also important for the design process. By integrating machining learning models with real-time sensor data, an adaptive design process can be achieved where the model can automatically adjust design parameters based on new input data. This method's application potential will be continually explored as sensor technology advances and application scenarios expand (including aerospace, medical devices, and smart manufacturing), and its value is expected to be realized in a broader range of fields in the future.

Supplementary Materials: The following supporting information can be downloaded at: https://www.mdpi.com/article/10.3390/s24175484/s1, Figure S1. The introduction of confusion matrix. Figure S2. The example of the partial knowledge graph. Figure S3. The cluster altas of different combinations of TranSE and DisMult methods. Table S1. The full names of materials. Table S2. The probability of combinations in the literature. Table S3. The coordinate of some functional materials and substrates, refs. [22,46–63].

Author Contributions: Conceptualization, J.K.; Methodology, J.K.; Validation, J.K.; Formal analysis, J.K.; Investigation, F.L. and G.X.; Writing—original draft, J.K.; Writing—review and editing, F.L.; Visualization, M.L.; Supervision, G.X. and M.L. All authors have read and agreed to the published version of the manuscript.

Funding: This research is supported by the Beijing Natural Science Foundation-Municipal Education Committee Joint Funding Project (Project No. KZ201910005004).

Institutional Review Board Statement: Not applicable.

Informed Consent Statement: Not applicable.

Data Availability Statement: The original contributions presented in the study are included in the article/supplementary material, further inquiries can be directed to the corresponding authors.

Conflicts of Interest: The authors declare no conflicts of interest.

References

1. Yin, Y.; Guo, C.; Li, W.; Liu, H.; Mu, Q. A super-elastic wearable strain sensor based on PU/CNTs yarns for human-motion detection. *ACS Appl. Mater. Interfaces* **2024**, *50*, 102017. [CrossRef]
2. Dong, H.; Sun, J.; Liu, X. Highly sensitive and stretchable MXene/CNTs/TPU composite strain sensor with bilayer conductive structure for human motion detection. *ACS Appl. Mater. Interfaces* **2022**, *13*, 15504–15516. [CrossRef]
3. Guo, H.; Chu, Z.; Fu, L.; Lv, Y.; Liu, X.; Fan, X.; Zhang, W. Thickness-induced gradient micro-wrinkle PDMS/MXene/rGO wearable strain sensor with high sensitivity and stretchability for human motion detection. *Chem. Eng. J.* **2024**, *495*, 153684. [CrossRef]
4. Zhang, D.; Sun, H.; Huang, M.; Su, M.; Ma, Y.; Shi, M.; Liu, H. Construction of "island-bridge" microstructured conductive coating for enhanced impedance response of organohydrogel strain sensor. *Chem. Eng. J.* **2024**, *496*, 153752. [CrossRef]
5. Tan, P.; Wang, H.; Xiao, F. Solution-processable, soft, self-adhesive, and conductive polymer composites for soft electronics. *Nat. Commun.* **2022**, *13*, 358. [CrossRef]
6. Lin, T.; Li, S.; Hu, Y. Ultrastretchable and adhesive agarose/Ti3C2Tx-crosslinked-polyacrylamide double-network hydrogel for strain sensor. *Carbohydr. Polym.* **2022**, *290*, 119506. [CrossRef]
7. Xue, S.S.; Tang, Z.H.; Zhu, W.B. Stretchable and ultrasensitive strain sensor from carbon nanotube-based composite with significantly enhanced electrical and sensing properties by tailoring segregated conductive networks. *Compos. Commun.* **2022**, *29*, 100987. [CrossRef]
8. Yuan, Z.; Li, H.; Duan, Z.; Huang, Q.; Zhang, M.; Zhang, H. High sensitivity, wide range and waterproof strain sensor with inner surface sensing layer for motion detection and gesture reconstruction. *Sens. Actuators A* **2024**, *369*, 115202. [CrossRef]
9. Yin, J.; Lu, C.; Li, C.; Yu, Z.; Shen, C.; Yang, Y.; Zhang, Y. A UV-filtering, environmentally stable, healable and recyclable ionic hydrogel towards multifunctional flexible strain sensor. *Compos. Part B Eng.* **2022**, *230*, 109528. [CrossRef]
10. Luo, L.Y.R.; Park, H.; Jin, S.W. High sensitivity and wide sensing range flexible strain sensors based on pre-stretching and layer-by-layer self-assembly. *Adv. Funct. Mater.* **2024**, *49*, 104385. [CrossRef]
11. Mei, S.; Xu, B.; Wan, J.; Chen, J. Preparation of CNT/CNF/PDMS/TPU Nanofiber-Based Conductive Films Based on centrifugal spinning method for strain sensors. *Sensors* **2024**, *24*, 4026. [CrossRef] [PubMed]
12. Wu, P.; He, H. Integrating high-performance flexible wires with strain sensors for wearable human motion detection. *Sensors* **2024**, *24*, 4795. [CrossRef] [PubMed]
13. Huang, Q.; Jiang, Y.; Duan, Z.; Wu, Y.; Yuan, Z.; Zhang, M.; Tai, H. Electrochemical self-powered strain sensor for static and dynamic strain detections. *Nano Energy* **2023**, *118*, 108997. [CrossRef]

14. Lei, M.; Feng, K.; Ding, S.; Wang, M.; Dai, Z.; Liu, R.; Gao, Y.; Zhou, Y.; Xu, Q.; Zhou, B. Breathable and waterproof electronic skin with three-dimensional architecture for pressure and strain sensing in nonoverlapping mode. *ACS Nano* **2022**, *16*, 12620–12634. [CrossRef]
15. Dai, Z.; Ding, S.; Lei, M.; Li, S.; Xu, Y.; Zhou, Y.; Zhou, B. A superhydrophobic and anti-corrosion strain sensor for robust underwater applications. *J. Mater. Chem. A* **2021**, *9*, 15282–15293. [CrossRef]
16. Wang, C.; Wang, X.; Li, Q.; Tao, X. Recognizing and predicting muscular fatigue of biceps brachii in motion with novel fabric strain sensors based on machine learning. *Biomed. Signal Process. Control* **2024**, *96*, 106647. [CrossRef]
17. Lee, J.; Jin, J. A novel method to design and evaluate artificial neural network for thin film thickness measurement traceable to the length standard. *Sci. Rep.* **2022**, *12*, 2212. [CrossRef]
18. Ravenscroft, D.; Prattis, I.; Kandukuri, T.; Samad, Y.A.; Mallia, G.; Occhipinti, L.G. Machine learning methods for automatic silent speech recognition using a wearable graphene strain gauge sensor. *Sensors* **2022**, *22*, 299. [CrossRef] [PubMed]
19. Hu, M.; He, P.; Zhao, W.; Zeng, X.; He, J.; Chen, Y.; Yang, J. Machine learning-enabled intelligent gesture recognition and communication system using printed strain sensors. *ACS Appl. Mater. Interfaces* **2023**, *15*, 51360–51369. [CrossRef]
20. Yang, C.; Zhang, D.; Wang, D.; Luan, H.; Chen, X.; Yan, W. In situ polymerized MXene/polypyrrole/hydroxyethyl cellulose-based flexible strain sensor enabled by machine learning for handwriting recognition. *ACS Appl. Mater. Interfaces* **2023**, *15*, 5811–5821. [CrossRef]
21. Wang, Y.; Tang, T.; Xu, Y.; Bai, Y.; Yin, L.; Li, G.; Huang, Y. All-weather, natural silent speech recognition via machine-learning-assisted tattoo-like electronics. *npj Flex. Electron.* **2021**, *5*, 20. [CrossRef]
22. Yang, H.; Ding, S.; Wang, J. Computational design of ultra-robust strain sensors for soft robot perception and autonomy. *Nat. Commun.* **2024**, *15*, 1636. [CrossRef] [PubMed]
23. Yang, H.; Li, J.; Lim, K.Z. Automatic strain sensor design via active learning and data augmentation for soft machines. *Nat. Mach. Intell.* **2022**, *4*, 84–94. [CrossRef]
24. Chatterjee, A.; Nardi, C.; Oberije, C. Knowledge graphs for COVID-19: An exploratory review of the current landscape. *J. Pers. Med.* **2021**, *11*, 300. [CrossRef] [PubMed]
25. Nie, Z.; Zheng, S.; Liu, Y. Automating materials exploration with a semantic knowledge graph for Li-ion battery cathodes. *Adv. Funct. Mater.* **2022**, *32*, 2201437. [CrossRef]
26. Bordes, A.; Usunier, N.; Garcia-Duran, A. Translating embeddings for modeling multi-relational data. In Proceedings of the Advances in Neural Information Processing Systems, Lake Tahoe, NV, USA, 5–8 December 2013; Volume 26.
27. Yang, B.; Yih, W.; He, X. Embedding entities and relations for learning and inference in knowledge bases. *arXiv* **2014**, arXiv:1412.6575. [CrossRef]
28. Nickel, M.; Rosasco, L.; Poggio, T. Holographic embeddings of knowledge graphs. In Proceedings of the AAAI Conference on Artificial Intelligence, Phoenix, AZ, USA, 12–17 February 2016; Volume 30. [CrossRef]
29. Grover, A.; Leskovec, J. node2vec: Scalable feature learning for networks. In Proceedings of the 22nd ACM SIGKDD International Conference on Knowledge Discovery and Data Mining, San Francisco, CA, USA, 13–17 August 2016; pp. 855–864. [CrossRef]
30. Tian, X.; Chan, K.; Hua, T. Wearable strain sensors enabled by integrating one-dimensional polydopamine-enhanced graphene/polyurethane sensing fibers into textile structures. *J. Mater. Sci.* **2020**, *55*, 17266–17283. [CrossRef]
31. Uzabakiriho, P.C.; Wang, M.; Wang, K. High-strength and extensible electrospun yarn for wearable electronics. *ACS Appl. Mater. Interfaces* **2022**, *14*, 46068–46076. [CrossRef]
32. Tang, J.; Wu, Y.; Ma, S. Flexible strain sensor based on CNT/TPU composite nanofiber yarn for smart sports bandage. *Compos. Part B Eng.* **2022**, *232*, 109605. [CrossRef]
33. Chen, S.; Zhang, J.; Zhang, Q.; Cai, G.; Xu, A.; Yan, S. Highly Stretchable and Durable Electrospinning Polyurethane Nanofiber Composite Yarn for Electronic Devices. *Fibers Polym.* **2022**, *23*, 352–359. [CrossRef]
34. Yang, K.; Yin, F.; Xia, D. A highly flexible and multifunctional strain sensor based on a network-structured MXene/polyurethane mat with ultra-high sensitivity and a broad sensing range. *Nanoscale* **2019**, *11*, 9949–9957. [CrossRef]
35. Sun, S.; Liu, Y.; Chang, X.; Jiang, Y.; Wang, D.; Tang, C.; Gao, Y. A wearable, waterproof, and highly sensitive strain sensor based on three-dimensional graphene/carbon black/Ni sponge for wirelessly monitoring human motions. *J. Mater. Chem. C* **2020**, *8*, 2074–2085. [CrossRef]
36. Wu, G.; Yang, Z.; Zhang, Z. High-performance stretchable fibrous supercapacitors and flexible strain sensors based on CNTs/MXene-TPU hybrid fibers. *Electrochim. Acta* **2021**, *395*, 139141. [CrossRef]
37. Souri, H.; Bhattacharyya, D. Highly sensitive, stretchable, and wearable strain sensors using fragmented conductive cotton fabric. *J. Mater. Chem. C* **2018**, *6*, 10524–10531. [CrossRef]
38. Niu, B.; Hua, T.; Hu, H. A highly durable textile-based sensor as a human-worn material interface for long-term multiple mechanical deformation sensing. *J. Mater. Chem. C* **2019**, *7*, 14651–14663. [CrossRef]
39. Liu, Q.Q.; Zhang, Y.; Li, A. Reduced graphene oxide-coated carbonized cotton fabric wearable strain sensors with ultralow detection limit. *J. Mater. Sci. Mater. Electron.* **2020**, *31*, 17233–17248. [CrossRef]
40. Bi, S.; Hou, L.; Lu, Y. An integrated wearable strain, temperature, and humidity sensor for multifunctional monitoring. *Compos. Part A Appl. Sci. Manuf.* **2021**, *149*, 106504. [CrossRef]
41. Zhang, Y.; He, P.; Luo, M. Highly stretchable polymer/silver nanowires composite sensor for human health monitoring. *Nano Res.* **2020**, *13*, 919–926. [CrossRef]

42. Wang, H.; Jia, Z.; Li, Z.; Ma, S.; Zhang, W.; Chen, Y.; Luo, Y.; Jia, D.; Zhong, B.; Razal, J.M.; et al. Constructing conductive titanium carbide nanosheet (MXene) network on polyurethane/polyacrylonitrile fiber framework for flexible strain sensor. *J. Colloid Interface Sci.* **2021**, *584*, 1–10. [CrossRef]
43. Huang, Q.; Jiang, Y.; Duan, Z.; Yuan, Z.; Wu, Y.; Peng, J.; Xu, Y.; Li, H.; He, H.; Tai, H. A Finger Motion Monitoring Glove for Hand Rehabilitation Training and Assessment Based on Gesture Recognition. *IEEE Sens. J.* **2023**, *23*, 13789–13796. [CrossRef]
44. Zeng, Z.; Hao, B.; Li, D. Large-scale production of wearable, dyeable and durable spandex/CNT/cotton core-sheath yarn for wearable strain sensors. *Compos. Part A Appl. Sci. Manuf.* **2021**, *149*, 106520. [CrossRef]
45. Xu, B.; Ye, F.; Chen, R. A wide sensing range and high sensitivity flexible strain sensor based on carbon nanotubes and MXene. *Ceram. Int.* **2022**, *7*, 48. [CrossRef]
46. Wang, Y.; Li, W.; Li, C. Fabrication of ultra-high working range strain sensor using carboxyl CNTs coated electrospun TPU assisted with dopamine. *Appl. Surf. Sci.* **2021**, *566*, 150705. [CrossRef]
47. Chen, Q.; Xiang, D.; Wang, L. Facile fabrication and performance of robust polymer/carbon nanotube coated spandex fibers for strain sensing. *Compos. Part A Appl. Sci. Manuf.* **2018**, *112*, 186–196. [CrossRef]
48. Wang, Z.; Zhang, K.; Liu, Y. Modified MXene-doped conductive organosilicon elastomer with high stretchability, toughness, and self-healability for strain sensors. *Compos. Struct.* **2022**, *282*, 115071. [CrossRef]
49. Huang, T.; He, P.; Wang, R. Porous fibers composed of polymer nanoball decorated graphene for wearable and highly sensitive strain sensors. *Adv. Funct. Mater.* **2019**, *29*, 1903732. [CrossRef]
50. Bu, Y.; Shen, T.; Yang, W.; Yang, S.; Zhao, Y.; Liu, H.; Shen, C. Ultrasensitive strain sensor based on superhydrophobic microcracked conductive Ti3C2Tx MXene/paper for human-motion monitoring and E-skin. *Sci. Bull.* **2021**, *66*, 1849–1857. [CrossRef]
51. Wu, X.; Han, Y.; Zhang, X.; Lu, C. Highly sensitive, stretchable, and wash-durable strain sensor based on ultrathin conductive layer@polyurethane yarn for tiny motion monitoring. *ACS Appl. Mater. Interfaces.* **2016**, *8*, 9936–9945. [CrossRef]
52. Akhtar, I.; Chang, S.-H. Highly aligned carbon nanotubes and their sensor applications. *Nanoscale.* **2020**, *12*, 21447–21458. [CrossRef]
53. Om, J.; Heo, J.S.; Kim, M. Highly sensitive textile-based strain sensors using poly(3,4-ethylenedioxythiophene): Polystyrene sulfonate/silver nanowire-coated nylon threads with poly-L-lysine surface modification. *RSC Adv.* **2017**, *7*, 53373–53378. [CrossRef]
54. Qin, R.; Hu, M.; Li, X.; Yan, L.; Wu, C.; Liu, J.; Huang, W. A highly sensitive piezoresistive sensor based on MXenes and polyvinyl butyral with a wide detection limit and low power consumption. *Nanoscale* **2020**, *12*, 17715–17724. [CrossRef] [PubMed]
55. Cheng, Y.; Wang, R.; Sun, J.; Gao, L. A stretchable and highly sensitive graphene-based fiber for sensing tensile strain, bending, and torsion. *Adv. Mater.* **2015**, *27*, 7365–7371. [CrossRef] [PubMed]
56. Wang, R.; Jiang, N.; Su, J.; Yin, Q.; Zhang, Y.; Liu, Z.; Liu, Z. A bi-sheath fiber sensor for giant tensile and torsional displacements. *Adv. Funct. Mater.* **2017**, *27*, 1702134. [CrossRef]
57. Hu, J.; Wu, Y.; Yang, Q.; Zhou, Q.; Hui, L.; Liu, Z.; Ding, D. One-pot freezing-thawing preparation of cellulose nanofibrils reinforced polyvinyl alcohol based ionic hydrogel strain sensor for human motion monitoring. *Carbohydr. Polym.* **2022**, *275*, 118697. [CrossRef]
58. Wang, S.L.; Xu, X.; Han, Z.; Li, H.; Wang, Q.; Yao, B. Highly stretchable liquid-metal based strain sensor with high sensitivity for human activity monitoring. *Mater. Lett.* **2022**, *308*, 131277. [CrossRef]
59. Soomro, A.M.; Khalid, M.A.U.; Shah, I.; wan Kim, S.; Kim, Y.S.; Choi, K.H. Highly stable soft strain sensor based on Gly-KCl filled sinusoidal fluidic channel for wearable and water-proof robotic applications. *Smart Mater. Struct.* **2020**, *29*, 025011. [CrossRef]
60. Wu, L.; Xu, C.; Fan, M.; Tang, P.; Zhang, R.; Yang, S.; Bin, Y. Lotus root structure-inspired Ti_3C_2-MXene-based flexible and wearable strain sensor with ultra-high sensitivity and wide sensing range. *Compos. Part A Appl. Sci. Manuf.* **2022**, *152*, 106702. [CrossRef]
61. Cai, Y.; Shen, J.; Ge, G.; Zhang, Y.; Jin, W.; Huang, W.; Dong, X. Stretchable Ti_3C_2Tx MXene/carbon nanotube composite based strain sensor with ultrahigh sensitivity and tunable sensing range. *ACS Nano.* **2018**, *12*, 56–62. [CrossRef] [PubMed]
62. Zhang, P.; Chen, Y.; Li, Y.; Zhang, Y.; Zhang, J.; Huang, L. A flexible strain sensor based on the porous structure of a carbon black/carbon nanotube conducting network for human motion detection. *Sensors* **2020**, *20*, 1154. [CrossRef]
63. Tan, C.; Dong, Z.; Li, Y.; Zhao, H.; Huang, X.; Zhou, Z.; Sun, B. A high performance wearable strain sensor with advanced thermal management for motion monitoring. *Nat. Commun.* **2020**, *11*, 3530. [CrossRef]

sensors

MDPI

Communication

A Unique Time-Reversal Algorithm-Enabled Flexible Ultrasound Transducer with a Controllable Acoustic Field

Lu Jia [1,2], Yingzhan Yan [1,2], Jing Xu [1,2] and Yuan Gao [1,2,*]

1 Information Science Academy, China Electronics Technology Group Corporation, Beijing 100142, China; jialu199408@163.com (L.J.); yyz712@163.com (Y.Y.); xujing_1999@126.com (J.X.)
2 National Key Laboratory of Integrated Circuits and Microsystems, Beijing 100142, China
* Correspondence: ygaoah@connect.ust.hk

Abstract: Flexible ultrasonic devices represent a feasible technology for providing timely signal detection and even a non-invasive disease treatment for the human brain. However, the deformation of the devices is always accompanied by a change in the acoustic field, making it hard for accurate focusing. Herein, we report a stable and flexible transducer. This device can generate a high-intensity acoustic signal with a controllable acoustic field even when the device is bent. The key is to use a low-impedance piezoelectric material and an island-bridge device structure, as well as to design a unique time-reversal algorithm to correct the deviation of signals after transcranial propagation. To provide an in-depth study of the acoustic field of flexible devices, we also analyze the effects of mechanical deformation and structural parameters on the corresponding acoustic response.

Keywords: flexible ultrasonic devices; phased-array transducer; time-reversal algorithm; controllable acoustic field

1. Introduction

The fast-growing interest in the Internet of Things, artificial intelligence, soft robotics, and personal health care has promoted the development of flexible and wearable electronic devices [1–9]. So far, flexible sensing devices that can detect optical signals [1], strain [2,3], and electrochemical signals [6] are mostly in the form of wearable accessories or skin patches. These devices can only obtain biomedical information from human skin, superficial blood vessels, or bodily fluids, but they can hardly collect information from deep organs (e.g., the detection of brain signals). The design of implantable sensing devices can enable deep signal detection from the human body [10,11]. However, these implant devices must fulfill stringent requirements for excellent biocompatibility, and the implant surgery is mostly painful. Particularly when detecting brain signals, the implant surgery may lead to secondary infections during device implantation and removal, which seriously harm the human body.

Ultrasonic devices appear to be promising candidates to address this challenge because acoustic waves can penetrate crucial internal organs and tissues and remain harmless to the human body if the signal intensity is not too high. As such, the use of ultrasonic devices can provide timely signal detection [12–15] and even non-invasive disease treatment of the human brain [16–20]. In order to obtain stable detection of signals, high-performance and flexible ultrasonic devices, which can conformally contact and closely adhere to the rigid or irregular surfaces of organs, are highly sought after. To date, almost all studies of flexible ultrasonic devices have focused on either the synthesis of flexible ultrasonic materials or the design of flexible device structures. Among them, the fabrication of a phased-array transducer using piezoelectric materials [21,22] (e.g., polyvinylidene fluoride and its copolymers) and an "island-bridge" device structure [14,23–26] is a promising approach to develop ultrasonic devices with both high performance and prominent flexibility. These devices have been successfully applied in biomedical fields such as real-time monitoring of blood flow [27], deep tissue sensing [12,15], cardiac pacing [28], treatment of brain diseases [19,20], etc.

It is noted that, for diseases such as epilepsy and Parkinson's, sudden or long-term neuroregulatory therapy is often needed. The reported ultrasonic neuroregulatory systems can help provide timely and effective intervention and treatment for these diseases [18,19]. However, a common issue with these devices resides in the reflection or absorption of acoustic signals by the organs of animals/humans (especially the skull), leading to a drastic decrease in signal intensity. Additionally, the deformation of the devices is always accompanied by changes in the acoustic field, making it hard to precisely monitor the propagation direction of the acoustic signals.

Herein, we report a flexible phased-array transducer, which can generate a high-intensity acoustic signal with traceable signal propagation even when the device is bent. The key is to use a thin and low-impedance piezoelectric composite as the transducer material and to design a unique algorithm to correct the deviation of the signals after transcranial propagation. To provide an in-depth study of the acoustic field of the flexible devices, we also analyze the effects of mechanical deformation and structural parameters on the corresponding acoustic response using FEA.

2. Design and Simulation

To guide the design and fabrication of desirable ultrasonic devices, we first simulate the emission acoustic field of the transducer with a single pixel or multiple pixels using Field II in MATLAB R2021b. To reduce the boundary effects of the signals during the simulation, a sinusoidal signal modulated with a Hanning window is used to stimulate the transducer. The pixel possesses a square geometry with a side length of 2 mm. The distance between neighboring pixels is 0.5 mm. Figure 1a–e illustrate the steady-state emission acoustic field of a single-pixel transducer at various frequencies. With the increase in frequency, the intensity of the acoustic signal is enhanced in the forward direction and decreased in the lateral direction, indicating an improved propagation direction of the acoustic field. The corresponding acoustic field of a dual-pixel transducer is shown in Figure 1f–j. Although the intensity of the acoustic signals is enhanced with the increase in frequency, the converged acoustic beam gradually splits into two separate beams at a frequency of 6 MHz. This result differs from that of a single-pixel transducer, which can be attributed to the superposition of signals from different pixels.

The conformal contact between the flexible transducer and the surface of the human brain enables better acoustic signal control and detection. In this scenario, the desirable transducers should possess a high tolerance to deformation, while maintaining a stable acoustic field. We thus further simulate the acoustic field of the dual-pixel transducer when bent at a radius of 30 mm, as shown in Figure 1k–o. The flexed transducer shows a relatively clear main lobe at frequencies of 1 MHz and 2 MHz. When compared to the flat transducer, the bent design of the device has a negligible effect on its acoustic field at low frequencies. As such, we believe that the transducer made with an array structure and stimulated at low frequencies (e.g., 2 MHz) can be tolerant to deformation and show a stable acoustic field.

The acoustic field of a transducer can also be affected by the arrangement of its pixels. To verify this statement, 25 pixels are aligned in a linear array, a square array, and a multirow array, as shown in Figure 2a–c. For the linear and square arrays, each pixel is electrically connected with serpentine metal wire and can work individually. To simplify the simulation model, a multirow array consisting of five individual lines is used, with each line comprising five pixels that connect closely w/o using metal wires. All the arrays possess the same focusing point located below the central pixel with a distance of 30 mm. We then compare the echo imaging of artificial phantoms generated by these flexible transducers at three points, including 0, 25 mm, 0, 30 mm, and 0, 35 mm, as shown in Figure 2d–g. According to the simulation results, the echo imaging generated by the square array and the multirow array is similar at all three observed points. Both arrays exhibit better focused acoustic signals than the linear array, as demonstrated by the smaller spot size at the focusing point. In addition, the square array and multirow array show a higher signal intensity (~10 times) than that of the linear array at the three observed points under the same conditions, which further proves that a desirable acoustic field of the transducer can be achieved with the design of either a square or multirow array. Considering the practical application of brain disease detection or treatment, the multirow array design requires fewer flexible interconnections among each pixel than the square array, benefiting the fabrication of transducers with smaller future sizes. In this scenario, we regard the multirow array as the optimal transducer design in this work.

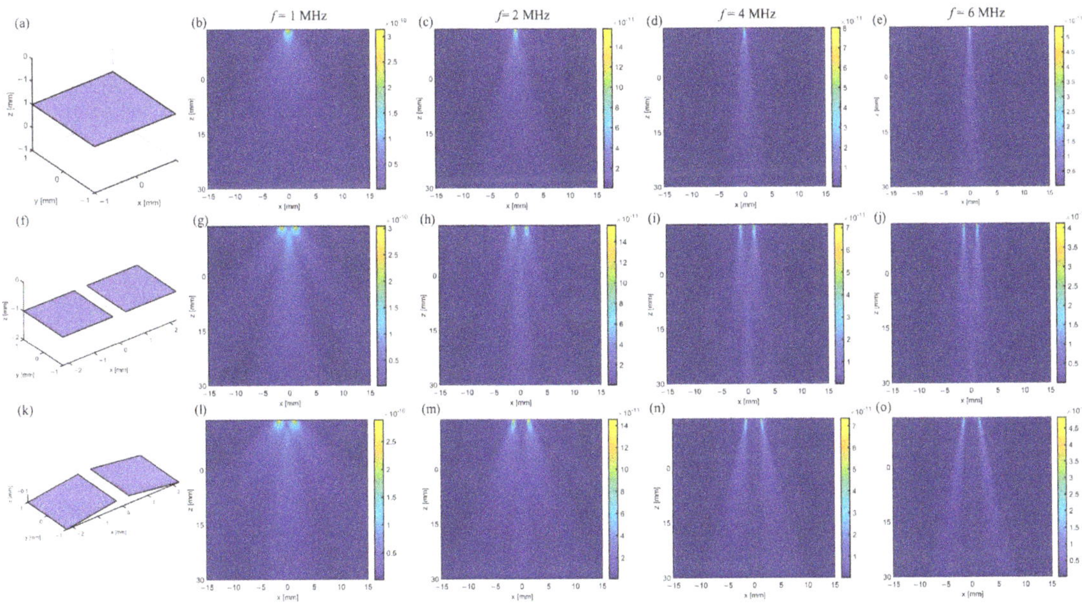

Figure 1. Steady-state emission acoustic field of the flexible transducer under different conditions. (**a**) A single-pixel transducer. The pixel possesses a square geometry with a side length of 2 mm. (**b–e**) The emission acoustic field of a single-pixel transducer at the frequencies of 1 MHz, 2 MHz, 4 MHz, and 6 MHz, respectively. (**f**) A dual-pixel transducer w/o bending. The distance between neighboring pixels is 0.5 mm. (**g–j**) The emission acoustic field of a dual-pixel transducer at the frequencies of 1 MHz, 2 MHz, 4 MHz, and 6 MHz, respectively. (**k**) A flexed dual-pixel transducer with the bending radius of 30 mm. (**i–o**) The emission acoustic field of a flexed dual-pixel transducer with the bending radius of 30 mm at the frequencies of 1 MHz, 2 MHz, 4 MHz, and 6 MHz, respectively.

We also simulate the evolution of the acoustic field when the multirow array is bent under different bending conditions. Under flat conditions, the acoustic waves generated by different pixels propagate vertically and result in a converged acoustic beam with a uniform energy distribution along the propagation direction, as shown in Figure 3a. When the array is bent, the acoustic waves can also focus on a certain point right below the center of the array, and the distance from the focal point to the array can be reduced with an increase in bending curvature, as shown in Figure 3b,c. It should be noted that at a large bending curvature, e.g., 25 mm^{-1}, the energy of the acoustic wave attenuates after the focal point because of the large diffusion angle, which negatively affects the propagation distance of the acoustic waves, as shown in Figure 3d.

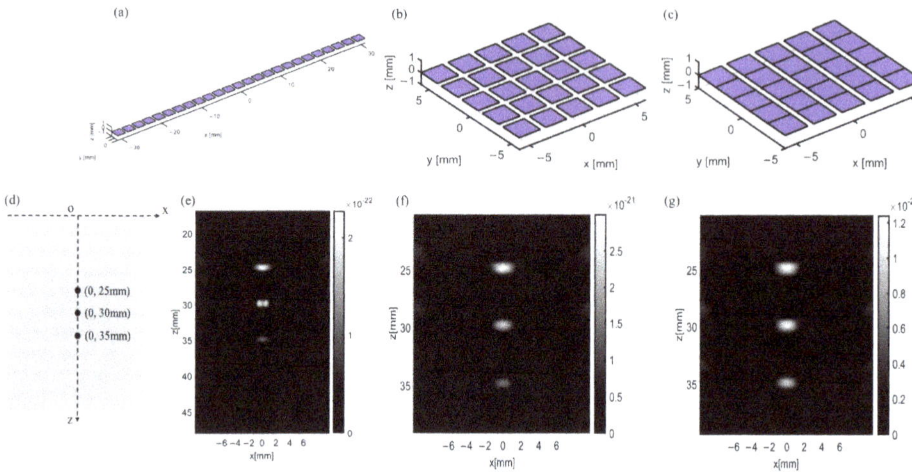

Figure 2. Different arrangement of 25 pixels in the phased-array transducers and their corresponding echo imaging. (**a**) A linear array. (**b**) A square array. Each pixel in the square array can work individually. (**c**) A multirow array. It consists of five individual lines and each line is made of five close-connected pixels. (**d–g**) Echo imaging is generated by a linear array, a square array, and a multirow array at three observed points.

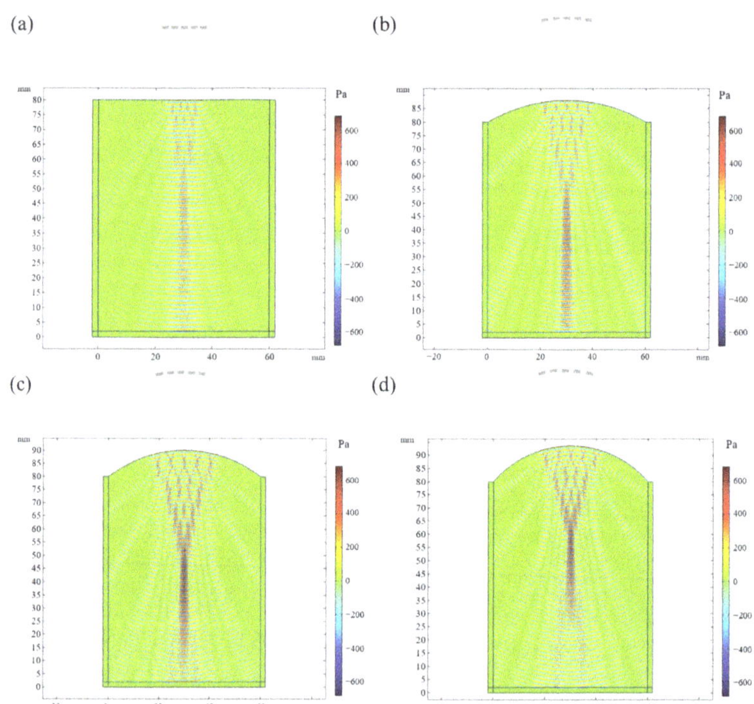

Figure 3. The evolution of acoustic field when the phased-array transducer is bent under different bending conditions. (**a**) Flat condition w/o bending. (**b**) Bending at a curvature of 16.7 mm^{-1}. (**c**) Bending at a curvature of 20 mm^{-1}. (**d**) Bending at a curvature of 25 mm^{-1}.

Based on the above simulation results, we then design a flexible transducer. In this transducer, each pixel possesses a geometry size of 2 mm × 2 mm × 1 mm and is bridged by the serpentine interconnects, as shown in Figure 4. A bulk 1–3 piezoelectric composite with a thickness of 1 mm and a high resonant frequency of 2 MHz is used as the active material because of its multiple appealing properties. Firstly, the 1–3 piezoelectric composite exhibits a lower sound impedance of 15 MRal than traditional PZT materials (30 MRal), which is in favor of a better matched impedance with the organs of the animal/human tissues (e.g., 2 MRal for the skulls). The matched impedance can reduce the signal reflection between the devices and organs, hence ensuring a high energy retention of the acoustic signals. Additionally, benefiting from the unique material structure, 1–3 piezoelectric composite materials exhibit a suppressed transverse vibration when subjected to external stimulation so as to achieve better control of the acoustic signal.

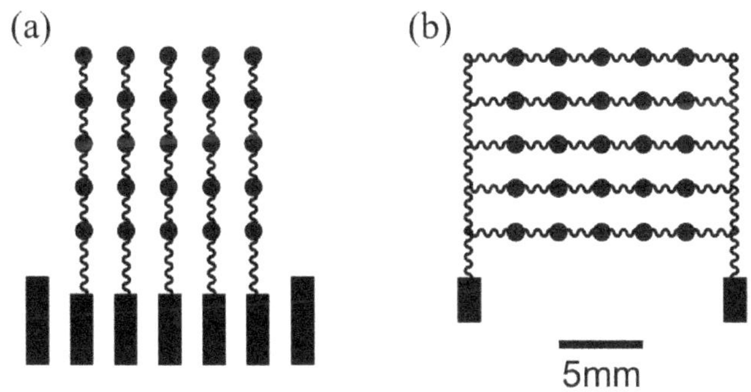

Figure 4. Schematic image of the designed flexible phased-array transducer. (**a**) Top electrodes. (**b**) Bottom electrodes.

3. Fabrication Process and Performance Testing

The fabrication of a phased-array transducer and the resulting device image are shown in Figures 5 and 6. Typically, a layer of polymethyl methacrylate (PMMA) is spin-coated on a silicon wafer, followed by the coating of another polyimide (PI) layer on the PMMA. In this scheme, PI serves as a protective layer, which can relieve the strain mismatch among different layers. To achieve flexible interconnects, a layer of copper (Cu) is evaporated on the PI and etched into targeted patterns. Afterwards, the excess PI is removed via reactive ion etching (RIE). The top or bottom serpentine Cu electrodes are achieved after transferring the patterned Cu layer onto a thin PDMS film. The 1–3 piezoelectric composite, serving as the pixel of the transducer, is sandwiched between the top and bottom Cu electrodes and further bonded by the silver paste. In this device, ACF lines are used as the readout circuit to collect the signals from individual pixels. At last, another polydimethylsiloxane (PDMS) layer is used to encapsulate the whole device, ensuring the good electrical properties and flexibility of the device.

The good reliability of the phased-array transducer can be reflected by the performance consistency of its channel. Typically, the resonant frequency and the intensity of the acoustic signals are two important performance parameters for transducers. Figure 7 displays the setup used for testing the acoustic performance of the as-fabricated transducer. During the test, a sinusoidal wave is generated by a generator and amplified by a power amplifier, which can be used as the stimuli signal. The response signal produced by the transducer is collected by a highly sensitive hydrophone and then can be visualized via an oscilloscope. To mimic the environment of the skull, the acoustic performance of the phased-array transducer is tested in a tank filled with water.

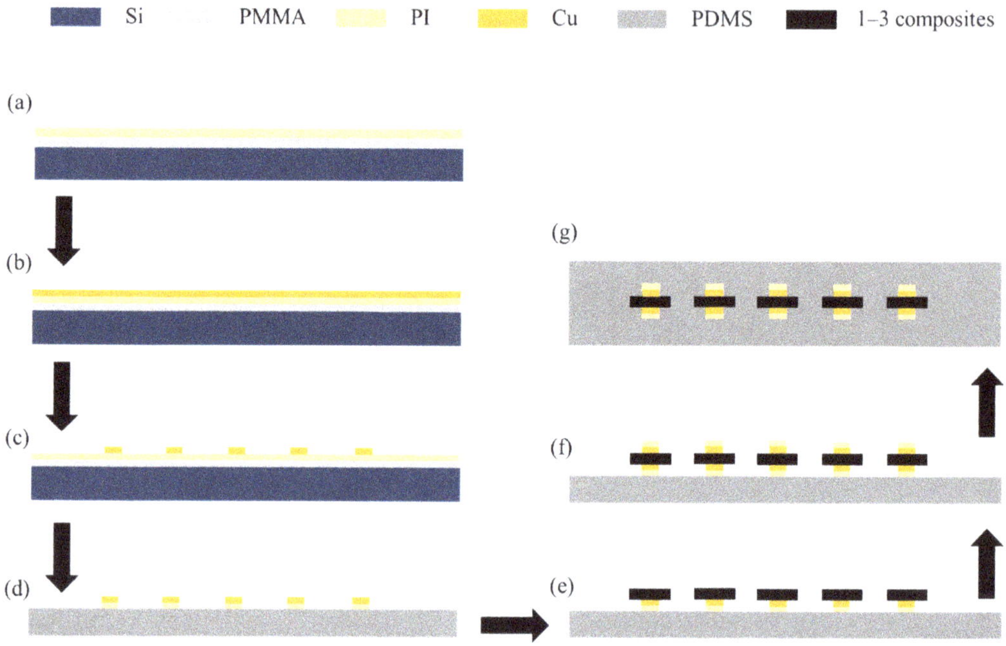

Figure 5. Fabrication of a phased-array transducer. (**a**) Deposition of polymethyl methacrylate (PMMA) and polyimide (PI) onto silicon substrates via spin coating. (**b**) Deposition of copper (Cu) thin films onto PI via evaporation. (**c**,**d**) Fabrication of patterned Cu electrodes via reactive ion etching (RIE) and the removal of excess PI. (**e**,**f**) Connection of each piezoelectric composite pixel to the bottom and top Cu electrode with silver paste. (**g**) Encapsulation of transducer with polydimethylsiloxane (PDMS).

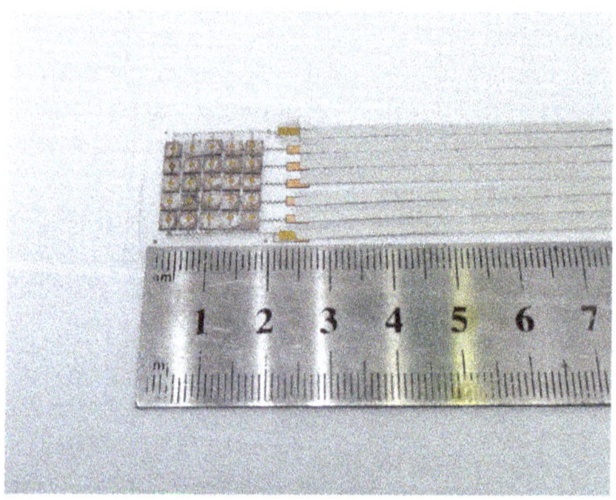

Figure 6. Image of the phased-array transducer.

Five channels are tested using an impedance analyzer. The resonance frequency of each channel exhibits a similar value of around 2 MHz, as shown in Figure 8. The sound pressure shows a linear relationship with the intensity of the acoustic signals, which can be calculated using the following formula:

$$P = I\varepsilon \quad (1)$$

where P represents the effective sound pressure, ε is the sensitivity of the hydrophone at 2 MHz, and I is the intensity of the signal. According to Table 1, each pixel renders a similar sound pressure of ~36 kPa. Both the similar resonant frequency and the sound pressure indicate that each pixel possesses a good consistency of acoustic performance.

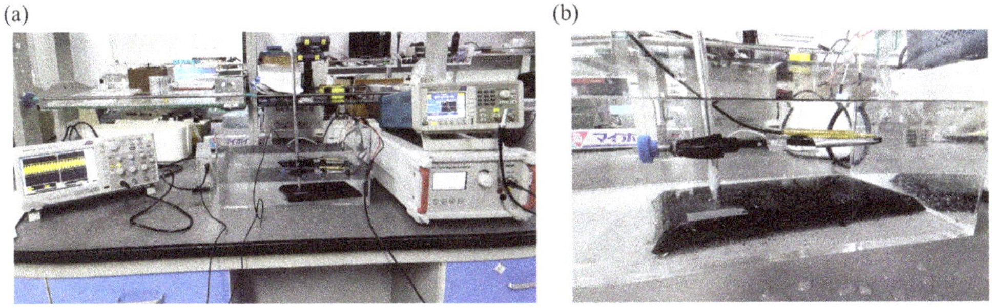

Figure 7. Setup used for testing the acoustic performance of phased-array transducer. (**a**) Image of the whole testing setup. (**b**) Zoomed-in image of phased-array transducer under test.

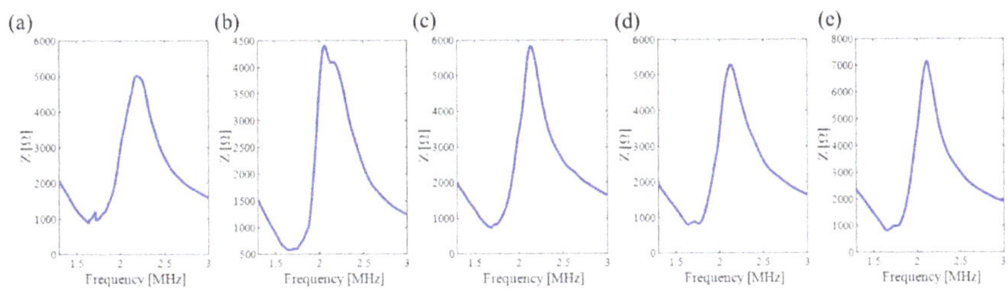

Figure 8. Test of resonant frequency of each channel. (**a**) Channel 1. (**b**) Channel 2. (**c**) Channel 3. (**d**) Channel 4. (**e**) Channel 5.

Table 1. Intensity of the acoustic signal for one pixel in each channel.

Excitation Voltage \ Channel	1	2	3	4	5
10 V	35.46 kPa	36.76 kPa	35.86 kPa	37.23 kPa	36.79 kPa

4. Transcranial Focused Ultrasound Simulation

It is noted that the transducer should be able to generate a high-intensity acoustic signal, which can overcome the high blood–brain barrier permeability. That is to say, the acoustic intensity should be higher than 150 kPa if the transducer is used for brain signal detection and treatment in mice [29]. Considering that the phased-array transducer consists of 25 pixels, the device can theoretically generate an acoustic signal with an overall signal intensity of ~900 kPa.

For brain signal detection or disease treatment, the focus (where the disease resides inside the skull) and the position of a transducer should be predefined, so that the acoustic intensity and the propagation can be regulated accordingly with a specific algorithm. The use of a phased-array transducer with multiple pixels can further enhance the accuracy of acoustic signal detection and regulation as well as provide good flexibility. However, the levels of deformation for each pixel in a phased-array transducer are different, which generates a time difference for ultrasound propagation, thus making it hard for accurate focusing.

To tackle this issue, we propose a proof-of-concept algorithm, namely the Virtual Point Source Time Reversal Algorithm (VPSTRA), which can compensate for time differences in ultrasound transmission through the skull. The time-reversal method is one of the most widely used methods in brain-focusing therapy [30–32]. With the development of computing technology, high-precision numerical simulation methods for acoustic fields can be used to simulate the time-reversal process based on a "virtual source" on a computer, achieving truly non-destructive and precise focusing of transcranial ultrasound. Compared to traditional algorithms, this method can overcome the non-uniformity of skull bones and the difficulty in ultrasound focusing caused by the deformation of pixels in the phased-array transducer. The design principle of VPSTRA is shown in Figure 9a and can be illustrated as follows: Firstly, we define a virtual point source (marked with a red dot) inside the skull and ensure that each pixel of the transducer can receive signals from this point. Since the time for the signal that propagates from the virtual point to each pixel is different, we thus eliminate the time differences by using the time-reversal algorithm. Finally, we use the corrected signals as the stimuli for each pixel so as to achieve focal reformation at the virtual point.

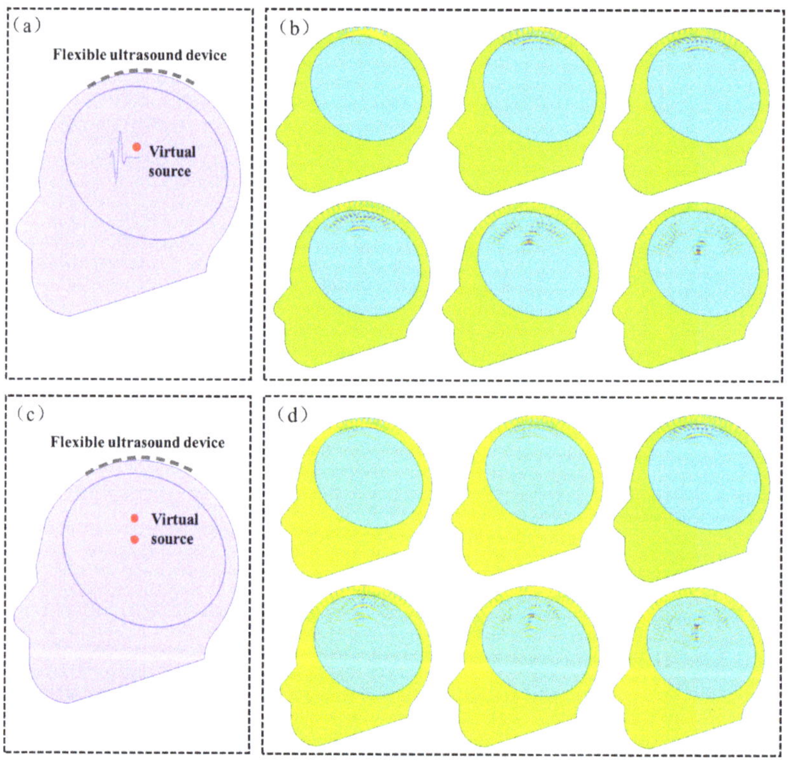

Figure 9. Simulation study for the acoustic field at the interface between the skull and the phased-array transducer in single and two virtual points. (**a**,**b**) Geometric model and snapshots of the acoustic field at different times in a single virtual point. (**c**,**d**) Geometric model and snapshots of the acoustic field at different times in two virtual points.

To demonstrate the feasibility of VPSTRA, we simulate the change in the acoustic field at the interface between skulls and the transducers for both a single virtual point and two virtual points, shown in Figure 9. The acoustic speeds in the skull and other parts are set as 3360 m s^{-1} and 1500 m s^{-1}, respectively. The densities of the skull region and other parts are 1658 kg m^{-3} and 1000 kg m^{-3}, respectively. Using the corrected signals as stimuli, the acoustic field of the phased-array transducer shows a good focal formation both in single and two virtual source points in Figure 9.

5. Conclusions

This paper reported a stable and flexible transducer consisting of 25 pixels, in which a low-impedance piezoelectric material and an island-bridge device structure were used. Ultrasonic emission experiments were carried out in water to research the ultrasonic output of a flexible device. The measured intensity of each single element was around 36 kPa. A unique Virtual Point Source Time Reversal Algorithm (VPSTRA) was proposed to correct the deviation in the signals after transcranial propagation. The flexible phased-array transducer design, together with VPSTRA, can simultaneously reduce the future size of the transducer, avoid the use of hard materials, and resolve the difficulty in device fixation, which greatly facilitates the application of ultrasound methods in the treatment of brain diseases.

Author Contributions: Conceptualization, L.J. and Y.G.; Funding acquisition, Y.G.; Methodology, L.J.; Project administration, J.X.; Resources, Y.Y.; Software, L.J.; Writing—original draft, L.J.; Writing—review and editing, L.J. and Y.G. All authors have read and agreed to the published version of the manuscript.

Funding: This research received no external funding.

Institutional Review Board Statement: Not applicable.

Informed Consent Statement: Not applicable.

Data Availability Statement: The data presented in this study are available on request from the corresponding author.

Conflicts of Interest: All authors were employed by the Information Science Academy, China Electronics Technology Group Corporation. All authors declare that the research was conducted in the absence of any commercial or financial relationships that could be construed as a potential conflict of interest.

References

1. Zhang, X.; Wang, C.; Zheng, T.; Wu, H.; Wu, Q.; Wang, Y. Wearable Optical Fiber Sensors in Medical Monitoring Applications: A Review. *Sensors* **2023**, *23*, 6671. [CrossRef] [PubMed]
2. Ji, S.; Wan, C.; Wang, T.; Li, Q.; Chen, G.; Wang, J.; Liu, Z.; Yang, H.; Liu, X.; Chen, X. Water-Resistant Conformal Hybrid Electrodes for Aquatic Endurable Electrocardiographic Monitoring. *Adv. Mater.* **2020**, *32*, e2001496. [CrossRef]
3. Javaid, A.; Zulfiqar, M.H.; Saleem, M.S.; Khan, M.A.; Zubair, M.; Massoud, Y.; Mehmood, M.Q. Skin-Mount Textile-Based Flexible Strain Sensors for Physiotherapy. *Adv. Eng. Mater.* **2024**, *2301138*, 2301138. [CrossRef]
4. Gao, Y.; Hu, H.; Chang, J.; Huang, Q.; Zhuang, Q.; Li, P.; Zheng, Z. Realizing High-Energy and Stable Wire-Type Batteries with Flexible Lithium—Metal Composite Yarns. *Adv. Energy Mater.* **2021**, *11*, 2101809. [CrossRef]
5. Wang, S.; Gao, Y.; Huang, Q.; Guo, X.; Yang, A.; Zhang, Y.; Zhuang, Q.; Chen, D.; Chen, L.; Ju, X.; et al. Inkjet-Printed Xerogel Scaffolds Enabled Room-Temperature Fabrication of High-Quality Metal Electrodes for Flexible Electronics. *Adv. Funct. Mater.* **2022**, *32*, 2203730. [CrossRef]
6. Zhao, Y.; Jin, K.Q.; Li, J.D.; Sheng, K.K.; Huang, W.H.; Liu, Y.L. Flexible and Stretchable Electrochemical Sensors for Biological Monitoring. *Adv. Mater.* **2023**, *2305917*, 2305917. [CrossRef] [PubMed]
7. Gao, Y.; Guo, Q.; Zhang, Q.; Cui, Y.; Zheng, Z. Fibrous Materials for Flexible Li—S Battery. *Adv. Energy Mater.* **2020**, *11*, 2002580. [CrossRef]
8. Fan, Y.; Wang, C.; Sun, J.; Peng, X.; Tian, H.; Li, X.; Chen, X.; Chen, X.; Shao, J. Electric-Driven Flexible-Roller Nanoimprint Lithography on the Stress-Sensitive Warped Wafer. *Int. J. Extrem. Manuf.* **2023**, *5*, 035101. [CrossRef]
9. Gao, Y.; Xie, C.; Zheng, Z. Textile Composite Electrodes for Flexible Batteries and Supercapacitors: Opportunities and Challenges. *Adv. Energy Mater.* **2021**, *11*, 2002838. [CrossRef]
10. Carpentier, A.; Canney, M.; Vignot, A.; Reina, V.; Beccaria, K.; Horodyckid, C.; Karachi, C.; Leclercq, D.; Lafon, C.; Chapelon, J.Y.; et al. Clinical Trial of Blood-Brain Barrier Disruption by Pulsed Ultrasound. *Sci. Transl. Med.* **2016**, *8*, 343re2. [CrossRef]
11. Blesa, J.; Pineda-Pardo, J.A.; Inoue, K.I.; Gasca-Salas, C.; Balzano, T.; López-González Del Rey, N.; Reinares-Sebastián, A.; Esteban-García, N.; Rodríguez-Rojas, R.; Márquez, R.; et al. BBB Opening with Focused Ultrasound in Nonhuman Primates and Parkinson's Disease Patients: Targeted AAV Vector Delivery and PET Imaging. *Sci. Adv.* **2023**, *9*, eadf4888. [CrossRef] [PubMed]
12. Lin, M.; Hu, H.; Zhou, S.; Xu, S. Soft Wearable Devices for Deep-Tissue Sensing. *Nat. Rev. Mater.* **2022**, *7*, 850–869. [CrossRef]
13. Wang, C.; Li, X.; Hu, H.; Zhang, L.; Huang, Z.; Lin, M.; Zhang, Z.; Yin, Z.; Huang, B.; Gong, H.; et al. Monitoring of the Central Blood Pressure Waveform via a Conformal Ultrasonic Device. *Nat. Biomed. Eng.* **2018**, *2*, 687–695. [CrossRef]

14. Wang, C.; Chen, X.; Wang, L.; Makihata, M.; Liu, H.; Zhou, T.; Zhao, X. Imaging of Diverse Organs. *Science* **2022**, *377*, 517–523. [CrossRef] [PubMed]
15. Wang, C.; Qi, B.; Lin, M.; Zhang, Z.; Makihata, M.; Liu, B.; Zhou, S.; Huang, Y.H.; Hu, H.; Gu, Y.; et al. Continuous Monitoring of Deep-Tissue Haemodynamics with Stretchable Ultrasonic Phased Arrays. *Nat. Biomed. Eng.* **2021**, *5*, 749–758. [CrossRef] [PubMed]
16. Lee, J.H.; Cho, I.J.; Ko, K.; Yoon, E.S.; Park, H.H.; Kim, T.S. Flexible Piezoelectric Micromachined Ultrasonic Transducer (PMUT) for Application in Brain Stimulation. *Microsyst. Technol.* **2017**, *23*, 2321–2328. [CrossRef]
17. Liu, W.; Chen, W.; Zhu, C.; Wu, D. Design and Micromachining of a Stretchable Two-Dimensional Ultrasonic Array. *Micro Nano Eng.* **2021**, *13*, 100096. [CrossRef]
18. Zhou, H.; Niu, L.; Xia, X.; Lin, Z.; Liu, X.; Su, M.; Guo, R.; Meng, L.; Zheng, H. Wearable Ultrasound Improves Motor Function in an MPTP Mouse Model of Parkinson's Disease. *IEEE Trans. Biomed. Eng.* **2019**, *66*, 3006–3013. [CrossRef]
19. Zhou, H.; Meng, L.; Xia, X.; Lin, Z.; Zhou, W.; Pang, N.; Bian, T.; Yuan, T.; Niu, L.; Zheng, H. Transcranial Ultrasound Stimulation Suppresses Neuroinflammation in a Chronic Mouse Model of Parkinson's Disease. *IEEE Trans. Biomed. Eng.* **2021**, *68*, 3375–3387. [CrossRef] [PubMed]
20. Zhou, S.; Gao, X.; Park, G.; Yang, X.; Qi, B.; Lin, M.; Huang, H.; Bian, Y.; Hu, H.; Chen, X.; et al. Transcranial Volumetric Imaging Using a Conformal Ultrasound Patch. *Nature* **2024**, *629*, 810–818. [CrossRef]
21. Liu, W.; Wu, D. Low Temperature Adhesive Bonding-Based Fabrication of an Air-Borne Flexible Piezoelectric Micromachined Ultrasonic Transducer. *Sensors* **2020**, *20*, 3333. [CrossRef] [PubMed]
22. Liu, W.; Zhu, C.; Wu, D. Flexible Piezoelectric Micro Ultrasonic Transducer Array Integrated on Various Flexible Substrates. *Sens. Actuators A Phys.* **2021**, *317*, 112476. [CrossRef]
23. Yang, Y.; Tian, H.; Yan, B.; Sun, H.; Wu, C.; Shu, Y.; Wang, L.G.; Ren, T.L. A Flexible Piezoelectric Micromachined Ultrasound Transducer. *RSC Adv.* **2013**, *3*, 24900–24905. [CrossRef]
24. Lyu, W.; Ma, Y.; Chen, S.; Li, H.; Wang, P.; Chen, Y.; Feng, X. Flexible Ultrasonic Patch for Accelerating Chronic Wound Healing. *Adv. Healthc. Mater.* **2021**, *10*, 2100785. [CrossRef]
25. Powell, D.J.; Ward, G.H. A Performance Appraisal of Flexible Array Structures Using a Facet Ensemble Scattering Technique. *Proc. IEEE Ultrason. Symp.* **1991**, 753–756. [CrossRef]
26. Wang, Y.F.; Ren, T.L.; Yang, Y.; Chen, H.; Zhou, C.J.; Wang, L.G.; Liu, L.T. High-Density PMUT Array for 3-D Ultrasonic Imaging Based on Reverse-Bonding Structure. In Proceedings of the 2011 IEEE 24th International Conference on Micro Electro Mechanical Systems, Cancun, Mexico, 23–27 January 2011; pp. 1035–1038. [CrossRef]
27. Wang, F.; Jin, P.; Feng, Y.; Fu, J.; Wang, P.; Liu, X.; Zhang, Y.; Ma, Y.; Yang, Y.; Yang, A.; et al. Flexible Doppler Ultrasound Device for the Monitoring of Blood Flow Velocity. *Sci. Adv.* **2021**, *7*, eabi9283. [CrossRef]
28. Jin, P.; Fu, J.; Wang, F.; Zhang, Y.; Wang, P.; Liu, X.; Jiao, Y.; Li, H.; Chen, Y.; Ma, Y.; et al. A Flexible, Stretchable System for Simultaneous Acoustic Energy Transfer and Communication. *Sci. Adv.* **2021**, *7*, eabg2507. [CrossRef]
29. Arif, W.M.; Elsinga, P.H.; Gasca-Salas, C.; Versluis, M.; Martínez-Fernández, R.; Dierckx, R.A.J.O.; Borra, R.J.H.; Luurtsema, G. Focused Ultrasound for Opening Blood-Brain Barrier and Drug Delivery Monitored with Positron Emission Tomography. *J. Control. Release* **2020**, *324*, 303–316. [CrossRef]
30. Pinton, G.; Aubry, J.F.; Fink, M.; Tanter, M. Effects of Nonlinear Ultrasound Propagation on High Intensity Brain Therapy. *Med. Phys.* **2011**, *38*, 1207–1216. [CrossRef]
31. Leduc, N.; Okita, K.; Sugiyama, K.; Takagi, S.; Matsumoto, Y. Focus Control in HIFU Therapy Assisted by Time-Reversal Simulation with an Iterative Procedure for Hot Spot Elimination. *J. Biomech. Sci. Eng.* **2012**, *7*, 43–56. [CrossRef]
32. Zheng, Y.; Yang, Y.; Zhang, Q.; Jiang, D.; Tu, J.; Zhang, D.; Duan, H. Ultrasonic Methods for Brain Imaging: Techniques and Implications. *IEEE Trans. Biomed. Eng.* **2022**, *69*, 3526–3537. [CrossRef] [PubMed]

sensors

MDPI

Review

Wearable and Flexible Sensor Devices: Recent Advances in Designs, Fabrication Methods, and Applications

Shahid Muhammad Ali [1,2,*], Sima Noghanian [3], Zia Ullah Khan [4], Saeed Alzahrani [5], Saad Alharbi [6], Mohammad Alhartomi [5] and Ruwaybih Alsulami [7]

1 Department of Engineering and Technology, School of Computing and Engineering, University of Huddersfield, Queensgate, Huddersfield HD1 3DH, UK
2 Engineering Department, The City of Liverpool College, Liverpool L3 6BN, UK
3 CommScope Ruckus Wireless, 350 W Java Dr, Sunnyvale, CA 94089, USA; sima_noghanian@ieee.org
4 National Physical Laboratory, Electromagnetic & Electrochemical Technologies Department, Teddington TW11 0LW, UK; zia.khan@npl.co.uk
5 Department of Electrical Engineering, University of Tabuk, Tabuk 71491, Saudi Arabia; saeedalzahrani@ut.edu.sa (S.A.); malhartomi@ut.edu.sa (M.A.)
6 King Abdulaziz City for Science and Technology, Riyadh 11442, Saudi Arabia; saalharbi@kacst.gov.sa
7 Department of Electrical Engineering, Umm Al-Qura University Makkah, Mecca 24382, Saudi Arabia; rrsulami@uqu.edu.sa
* Correspondence: shahid.ali@liv-coll.ac.uk or shahid_afridi333@yahoo.com

Abstract: The development of wearable sensor devices brings significant benefits to patients by offering real-time healthcare via wireless body area networks (WBANs). These wearable devices have gained significant traction due to advantageous features, including their lightweight nature, comfortable feel, stretchability, flexibility, low power consumption, and cost-effectiveness. Wearable devices play a pivotal role in healthcare, defence, sports, health monitoring, disease detection, and subject tracking. However, the irregular nature of the human body poses a significant challenge in the design of such wearable systems. This manuscript provides a comprehensive review of recent advancements in wearable and flexible smart sensor devices that can support the next generation of such sensor devices. Further, the development of direct ink writing (DIW) and direct writing (DW) methods has revolutionised new high-resolution integrated smart structures, enabling the design of next-generation soft, flexible, and stretchable wearable sensor devices. Recognising the importance of keeping academia and industry informed about cutting-edge technology and time-efficient fabrication tools, this manuscript also provides a thorough overview of the latest progress in various fabrication methods for wearable sensor devices utilised in WBAN and their evaluation using body phantoms. An overview of emerging challenges and future research directions is also discussed in the conclusion.

Keywords: wearable devices; DIW and DW fabrication methods; WBAN

1. Introduction

Wireless body area networks (WBANs) are used in monitoring physical activities, navigation and searching, healthcare, and defence, as shown in Figure 1. WBAN relies on wearable devices to provide a possible solution to clinical diagnostics using physical, chemical, and biological sensors to monitor vital body signals in real-time. These devices can be integrated into wearable materials such as glasses, jewellery, face masks, wristbands, watches, fitness bands, tattoos, bandages, patches, and wearable logos that can be used in both non-invasive and invasive applications [1]. Flexible and stretchable devices are available that can adapt to any surface without breaking and have the capability to elongate or deform under stress without losing functionality. They are usually made from an elastic; flexible substrate that has active sensing components bonded to it and can be employed in numerous applications. Recently, a self-powered and repairable humidity sensor constructed from silk nanofibre (SNF) film advanced the development of wearable technology by offering novel approaches to the domains of intelligent control, health monitoring, and next-generation wearable technology, including self-powered triboelectric nanogenerator (TENG) sensors utilising polyethylene oxide and copper-oxide nanofibres to monitor humidity as well as ethanol in simulated exhalation [2,3]. A range of wearable devices has been developed incorporating different sensing applications. These devices utilise systems-on-chips (SoCs) to combine the sensing and processing of bio signals [4]. The most current developments in skin-worn, tattoo-based wearable electrochemical devices, electrolyte and metabolite sensors, biofuel cells, batteries, and nanomaterial-based wearable skin devices were investigated [5]. The application of liquid metal or its alloys as printing inks holds great potential in the realm of flexible electronics. It presents a convenient and direct approach to fabricating electronic components by simply writing them out using these liquid materials. This innovative method involves incubating the liquids to create the desired electronic structures [6]. Within the context of the DIW method, there is a comprehensive evaluation of crucial components for wearable electronic devices. These components include nanogenerators for energy generation, lithium–ion batteries for energy storage, and strain sensors for managing energy consumption. The focus of this examination revolves around assessing their performance metrics and exploring various fabrication techniques [7]. Shoe-based sensor systems can handle various vertical ground reaction forces (vGRFs), motion intent learning, plantar pressure, and fall detection, and they are useful in the calculation of the centre of mass displacement while walking, as well as pedestrian navigation and tracking [8]. A loop-based sensing antenna was designed using NinjaFlex material on a 3D hemispherical-shaped substrate [9]. The printing method was fused deposition modelling (FDM). An ultra-wideband (UWB) monopole-sensing antenna was designed on a flexible substrate and printed using the fuse-based filament method [10]. Another paper reports a new radio frequency identification (RFID) tag was printed on a cotton fabric using graphene ink via an nScrypt 3D-DW dispensing system [11]. The recent progress in flexible electronic systems, which utilises transistor-based circuitry and active-matrix technology built on plastics or textile substrates, has been extensively reviewed [12]. The focus lies on fabrication techniques using conventional materials that are commonly used in electronics. These flexible systems have paved the way for the development of various wearable sensor devices, including tattoo-based and biofluidic prototypes. These sensors have found applications in healthcare, particularly in WBAN systems [13].

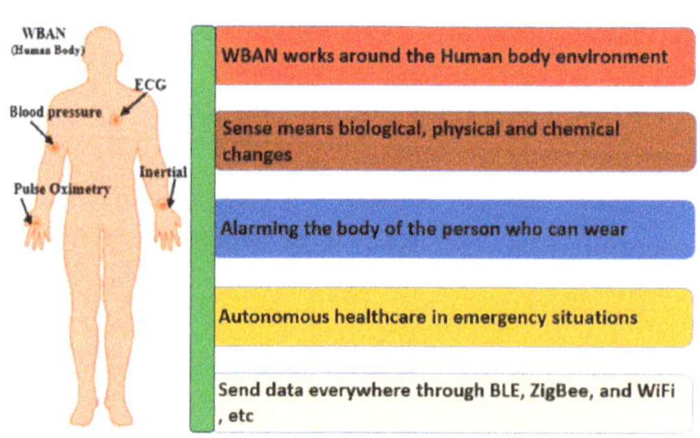

Figure 1. Examples of technologies used in WBAN systems.

However, there has been a significant surge in the development of sensing systems for industrial electronics. These sensing devices have undergone design, prototyping, and testing processes in both laboratory and real-life scenarios [14]. To cater to different electrical, mechanical, and thermal requirements, a wide range of flexible sensor devices has been developed using materials such as rigid, flexible, liquid, and single-crystal silicon, as well as a variety of polymers and nanomaterials, such as polydimethylsiloxane (PDMS), polyethylene terephthalate (PET), polyimide (PI), and polyethylene–naphthalate (PEN). Several flexible sensor devices have been designed using conductive polymer types of nanocomposite materials, such as poly (3,4-ethylene dioxythiophene), polystyrene sulfonate (PEDOT: PSS), and polyaniline. In addition, carbon-based allotropes have been employed as processed materials to design wearable and flexible sensor devices [15]. A new stretching-insensitive bending sensor was designed and tested in [16]. A thin, porous PDMS sponge with a CNT network coating (CCPPS), which is flexible and bending-insensitive, was integrated with two electrodes to monitor the human body [17]. To support or use soft active materials such as carbon, metal, ion, or liquid metals, researchers have reported using flexible/stretchable non-active materials, which are discussed in [18], such as a small contact-lens platform that can be used for various human–machine interactions (HMI) [19]. Several human movements are sensed by shoe-mounted wearable sensors, which may be used for various purposes, including activity detection and energy expenditure estimation [20]. A disposable, ultrathin multi-layer adhesive patch with electrodes and interconnects for skin interfacing based on an Ag-In-Ga-filled elastomer was digitally produced and reported [21]. Nanomaterials have made it possible to develop tactile sensors that are well-suited for applications involving body parts such as skin. However, these sensors often encounter issues with a noisy environment and low power signals, which may contain contaminated data. The presence of noise can hinder the accurate collection of relevant information. Considering this, researchers have suggested the adoption of signal-processing methods based on neural networks (NNs) to improve the signal-to-noise ratio (SNR) and improve signal accuracy. Such approaches can help mitigate the impact of noise and enhance the reliability and accuracy of the collected data [22]. An advancement in machine learning combined with wearable technologies has revolutionised medical monitoring and human–machine interaction (HMI) by leveraging the following four key bio signals: electrocardiograms (ECGs), electromyography (EMG), electroencephalography (EEG), and photoplethysmography (PPG). ECG, when integrated with machine learning, enables accurate disease detection, heartbeat classification, and real-time cardiac monitoring, enhancing cardiac healthcare and identifying conditions like sleep apnea and stroke risk. EMG-driven machine learning has transformed human–machine interaction (HMI) applications, such as virtual reality interfaces and prosthetics, while also aiding in the diagnosis of neuromuscular diseases. EEG, powered by machine learning, facilitates brain–computer interface development and cognitive state analysis, benefiting psychology and neurology. Meanwhile, PPG, with its non-invasive nature and enhanced signal processing, supports the diagnosis of cardiovascular conditions and real-time health monitoring [19,22]. With the development of biodegradable materials, there is now a great chance to revolutionise healthcare technologies using sensors that will naturally break down after use. With the recent development of superhydrophobic wearable sensors, it is now evident that wearable devices with superhydrophobic functions have significant potential for diverse applications [23]. MXene ink (transition metal carbides, nitrides, and carbonitrides) has penetrated such diverse fields of research as wonder materials. The most attractive features of their applications show great promise in energy storage conversion, harvesting, and sensing [24].

E-skins with multimodal sensing capabilities have been extensively reviewed, highlighting their potential for diverse applications, including health monitoring, intelligent prosthetics, HMIs, and robotic systems. Further, these multimodal e-skin systems are classified into three modes, as follows: (1) integration of multiple physical sensors, (2) integration of physical and electrophysiological sensors, and (3) integration of physical and chemical sensors. These systems primarily mimic human skin functions by detecting physical stimuli like force, strain, vibration, temperature, and humidity while also integrating additional sensing modalities, such as ultra-violet (UV) light sensors, to create next-generation smart skins [5]. These advancements have shed light on the future direction of next-generation wearable sensor devices. This evolving trend encompasses various areas, including the applications of sixth generation (6G) radio frequency (RF) in sensing, imaging, and healthcare. These applications encompass a broad spectrum, ranging from wireless networks employed in neural implants and drug delivery systems to guaranteeing safety concerns related to exposure to microwaves. The exploration of these avenues paves the way for innovative developments in wearable technology [25]. In wearable applications, specific absorption rate (SAR) measures the rate at which the human body absorbs energy from electromagnetic fields (EMF), with regulatory limits set to prevent harmful exposure. In the U.S., the Federal Communications Commission (FCC) has set the SAR limit at 1.6 W/kg, averaging over 1 g of tissue for the head, and 4.0 W/kg, averaging over 10 g for extremities. The International Commission on Non-Ionizing Radiation Protection (ICNIRP) has slightly different guidelines, with 2.0 W/kg for head exposure in the EU, Japan, and China. In overall wearable and flexible sensor devices, accurate SAR measurement and adherence to these regulatory limits are essential for ensuring user safety, particularly with newer high-frequency applications. Updating safety guidelines to account for prolonged exposure and new frequency ranges is increasingly important as these devices become more integrated into daily life. To minimise the effects of microwave wave radiation, various methods such as reducing direct exposure, maintaining distance from radiation sources, using protective layers like shielding, and period structures can be used [25,26].

In WBAN, additive manufacturing (AM) methods have undergone significant advancements over the past few decades, catering to diverse applications such as medical devices and orthopaedic equipment [27]. These advancements have opened new possibilities for the manufacturing and utilisation of wearable sensor devices within the WBAN domain using various inks or liquids. For example, DW is an additive manufacturing (AM) method that directly transfers onto a substrate material, also called digital writing or digital printing, while DIW, called the direct-write assembly, is an extrusion-based 3D AM printing method in which liquid or semisolid colloidal inks are dispensed via small nozzles under controlled pressure and flow rates. These methods are fabrication methods that are used to produce micro devices, printed circuit board (PCB) electronics, optics biosensing devices including biocompatible scaffolds for tissue engineering, and high-performance energy storage systems [27,28]. In [29], the research investigated a liquid-based eutectic gallium–indium (EGaIn) to design sensing devices in WBAN. Intelligent sensing based on commercial Wi-Fi devices is an emerging technology for next-generation healthcare monitoring with great potential for future development due to its low cost and availability. However, an integration of TENG technology presents a promising solution for wearable health devices and disease diagnostics, as it allows for real-time, portable monitoring without the need for external power sources. Key characteristics of the sensors created in this manner include their biocompatibility, scalability, lightweight design, flexibility, and wearability. Additionally, connection with a triboelectric source has made it possible to develop a self-powered, wearable, portable sensor. Smart sensors are useful because they can be created rapidly and affordably, worn conveniently, have a long lifespan, react swiftly, and may be used anywhere and at any time. They are also resistant to water and other factors [30,31]. However, this manuscript aims to address these gaps by considering the following key requirements:

1. Exploration of the latest developments in flexible, rigid, liquid, and next-generation devices.
2. Examination of advanced DIW and DW fabrication methods.
3. Integration of body phantom models to account for the impacts of the human body and development of more accurate and realistic human body phantom models.

By incorporating these requirements, this review paper offers a unique view that simultaneously considers three dynamic developments that have not been extensively studied together. Many researchers are currently involved in solving all these challenges. For example, several different types of wearable and flexible sensor devices have already been discussed and have shown a very reasonable performance in WBAN systems such as flexible and stretchable, self-powered and degradable humidity sensors, biofluidic, Mxene ink, liquid, and next-generation miniaturised wearable and flexible sensor devices [8,13,15,17]. State-of-the-art reviews based on advancements in DW- and DIW-based and emerging fabricated systems such as 3D to 6D have also been studied [27,28]. The human body methods such as on-body body phantom models for evaluation have already been drawn with the aim to provide researchers with a holistic understanding of devices and their impact on complex human body situations and provide advancements in design and manufacturing methods. Therefore, the development of realistic simulation and emulation platforms is essential for evaluating these techniques in their initial phase of development. In recent years, the research community has shown a keen interest in the development of more accurate and realistic phantoms for different sensing techniques based on microwaves, magnetic resonance imaging (MRI), optics, and acoustic optics [11–14,30]. Given this, it is fruitful to retrospectively examine the recent progress in the field of wearable and flexible sensor devices, including their design, fabrication, and on-body applications in WBAN systems. This manuscript is arranged as follows: It begins with an introduction, followed by Section 2, which describes wearable and wireless communication devices. Section 3 provides the latest comprehensive overview of DIW and DW fabrication techniques for wearable devices, featuring examples and detailed descriptions. Section 4 focuses on the impact on the human body to evaluate the performance of wearable devices through simulation and measurement. Lastly, Section 5 concludes the review and offers recommendations for future directions.

2. Wireless Communication Technologies

Wireless communication systems offer a versatile solution for various communication needs, providing a viable alternative to expensive and inflexible cable network systems. These wireless communication systems can be categorised into different types based on their range and transmission speed, including short-range, cellular, and long-range networks. Furthermore, they include outdoor and indoor positioning systems, further highlighting their diverse applications and capabilities [32,33]. Indoor environments pose a significant challenge for navigation systems, as they account for up to 80% of human activities. While reliable and widely used outdoor navigation systems like "Global Positioning Systems (GPS), the Global Navigation Satellite System (GLONASS), and the Bei Dou Navigation Satellite System (BDS)" offer precise location services, their accuracy diminishes indoors due to signal losses caused by buildings, the multipath effect, and time-delay issues. Consequently, the demand for positioning technologies falls short of meeting the requirements of an indoor system. To address this, an Indoor Positioning System (IPS) was developed which continuously and in real-time provides accurate locations of people or objects within enclosed spaces. IPSs serve multiple purposes, including object detection and tracking, assisting the daily routines of the elderly and disabled, and enabling medical and vital signs monitoring during emergencies. Indoor positioning services are crucial in various public areas, as they assist visitors in finding their desired destinations and offer navigation support for individuals who are blind or have impaired mobility. Moreover, hospitals may use IPSs for effective patient tracking [34]. Indoor positioning systems, including hardware platforms and localisation algorithms, have received great attention. Several sensing technologies, such as RFID, Wi-Fi, acoustic signals, and Bluetooth, have emerged as viable options. Different parameters can be used to create various subcategories within these IPSs. One common categorisation is based on the types of sensors employed. These include camera-based, infrared-based, tactile, polar, sound-based, Wi-Fi and WLAN, RFID systems, UWB, Assistant GNS, pseudo-based satellite systems, as well as other RF-based devices such as ZigBee, Bluetooth, digital television, cellular-based networks, radar, inertial-based navigation, and magnetic-based systems. This diverse range of sensor technologies contributes to the wide array of options available for indoor positioning and tracking purposes. Table 1 illustrates the available wireless technologies that are used in the literature for several applications.

Table 1. Overview of available wireless wearable technologies.

Tech/References	Frequency (GHz)	Range (m)	Power (mW)	Data Rate (kbps)
RFID [35,36]	0.1356, 0.86–0.96	0–3	200	640
Bluetooth [37–40]	2.4–2.5	1–100	2.5–100	1–3
Bluetooth low energy [41,42]	2.4–2.5	1–100	10	1
ZigBee [43–45]	2.4–2.5	10–100	35	250
Wi-Fi [46,47]	2.4–2.5	150–200	1 W	54
UWB [48,49]	3.1–10.6	3–10	250	53–480
Adaptive network topology (ANT) [50]	2.4–2.5	30	0.01–1	20–60
Medical implant communication system (MICS) [51]	0.402–0.405	2	25	200–800
Infrared data association (IrDA) [52]	0.00038	0.1	4–10	1×10^6
Near field communication (NFC) [53]	0.1356	0.05	0.15 mW	424

In [35,36], researchers evaluated power consumption and the quality of the data exchange in WBAN systems. RFID plays a significant role in numerous applications within short-range wireless networking. Although not as commonly used as RFID, Near Field Communication (NFC) is compatible with smartphones and tablets. In addition to Wi-Fi for short-range high-speed communication, cellular wireless systems like GSM (Global System for Mobile Communications), GPRS (General Packet Radio Service), and 3G, 4G, and 5G (third-, fourth-, and fifth generation) wireless technology for mobile telecommunication enable long-range wireless connectivity. GPRS-based systems are employed in various cellular wireless communication networks. Various researchers employed a design based on Wi-Fi and GPRS to monitor multiple wireless communication applications, achieving minimal data loss. For long-range networks, LPWANs (low-power wide-area networks) have been specifically designed to be compatible with the Internet of Things (IoT) devices. LPWANs offer low data rates and energy consumption, starting from 2G and extending to 5G networks [46–48]. The principle of free hardware has recently been implemented in various wireless applications to control and analyse data. However, these technologies have not yet been fully utilised in healthcare, even though they are gaining traction due to their compatibility with Bluetooth, Wi-Fi, ethernet, LoRa (long-range), and Lora WANs (long-range wide-area networks). One example is ESP8266EX, which boasts a combination of affordability, energy efficiency, compactness, and dependable performance [54].

Smart Wearable Devices

The global phenomenon of population aging, coupled with concerns surrounding family planning and its impact on birth rates, has led to significant implications for the socio-economic healthcare system. The cost of medicine, medical devices, and hospital care continues to rise. To address these challenges, wearable biosensors have emerged as non-invasive smart gadgets that enable continuous real-time patient monitoring. These technologies serve as effective tools for disease management, promoting healthier behaviours, and facilitating early identification of risk factors, diagnosis, and treatment [55,56]. Sensing electronic devices attached to soft tissue and worn as skin patches have recently been developed to establish a new interface for monitoring applications, as shown in Figure 2a [57]. Skin patches, when placed discreetly beneath clothing, can capture highly precise data without being influenced by human movement. These patches are capable of accurately monitoring vital signals like temperature, tension, sweat, and heart rate when affixed to the surface of the skin [58]. Sweat holds significant importance as a crucial bodily fluid due to its composition of essential elements such as electrolytes, small chemicals, and proteins. In recent years, the development of sweat-analysis devices has enabled the detection of various components present in sweat. These devices utilise sensing techniques to conduct a comprehensive analysis of sweat composition [59].

It is crucial to monitor changes in skin temperature during the initial phases of diagnosis and treatment [60]. Recently, a group of researchers developed a flexible stress-monitoring patch with a small area of skin contact. This is an important factor since it reduces the effects of the patch on the human body. In [61], it is reported that a group of scientists created a photonic crystal for non-invasive glucose monitoring of tear fluid made of colloidal particles dissolved in the hydrogel in a face-centre cubic structure. Tear fluid had 100 mol/L or more of glucose, which was detected by photonic-based glucose-sensing materials. The detection threshold for tear fluid was 1 mol/L. A mouthguard, which uses built-in wireless electronics to test salivary uric acid, is a recent addition to wearable devices in WBAN systems [62]. In another research investigation, a mouthguard was used to monitor and detect salivary glucose [63]. To measure blood pressure (BP) without a cuff, a portable patch sensor with an adjustable flexible piezo-resistive sensor (FPS), as well as epidermal ECG sensors, has been created [64]. Numerous studies have investigated wearable flexible and fluidic devices used to measure vital signals [65]. Wearable devices have found applications beyond human health and well-being, extending to animal health in the field of pet care and animal husbandry, etc [66]. Autonomously functioning wearable sensor systems have the capability to integrate point-of-care functionality with mobile connectivity. These systems comprise a combination of active and passive components, as well as sensing devices. These autonomously functioning wearable sensor systems support ongoing evaluation for disease prevention, early reduction of health risk factors, and maintenance of optimal, lifetime health quality [66,67]. However, modern wearable devices perform high-quality measurements as compared to conventional medical devices. The first-generation wearables predominantly focused on biophysical devices designed to monitor and track metrics related to physical activity, such as heart rate and body temperature. These devices were integrated with various wearables such as watches, shoes, and headphones [68]. The initial wave of wearables gained widespread adoption, leading to a shift in focus toward second-generation devices. The second-generation wearables encompass non-invasive or minimally invasive biochemical and multi-functional designs. Second-generation devices encompass a range of options, including invasive microneedles, injectable devices, and compact alternatives such as skin patches, tattoos, tooth-mounted sensors, and contact lenses. These advancements offer diverse solutions for monitoring and enhancing various aspects of human health and well-being [69]. A group of second-generation devices utilise biofluids and employ biorecognition components to convert analytes into measurable signals. These devices are typically laboratory-based and include examples such as the Freestyle Libre for glucose monitoring and sweat-based patches for Gx detection. Numerous wearable biochemical and biophysical sensor devices have been utilised for wellness applications, including the detection of COVID-19 [70]. In [71], a novel nanostructured smart tactile sensing device was constructed using 3D printing with an integrated electronic component. A tactile sensing device was improved using neural networks (NN) on a digital unit. Various telehealth devices are employed to identify an illness of an individual. By utilising four essential bio signals such as ECG, EMG, EEG, and PPG, machine learning and wearable technology have transformed medical monitoring and HMI. When machine learning is combined with ECG, it can be used to accurately detect diseases, classify heartbeats, and provide real-time cardiac monitoring. This improves cardiac care and helps identify disorders like stroke risk and sleep apnea. EMG-driven machine learning has revolutionised HMI applications and helped diagnose neuromuscular illnesses. Examples of these applications include virtual-reality interfaces and prosthetics. Machine learning-powered EEG helps with cognitive state analysis and brain–computer interface development, which is beneficial in neurology and psychology. PPG, on the other hand, facilitates real-time health monitoring and the identification of cardiovascular disorders due to its non-invasive nature and improved data processing. Therefore, machine learning has revolutionised wearable and HMI applications and helped in diagnosis [19,22,72].

Researchers integrated a contact lens device based on a multifunctional ultrathin MoS2 transistor, as shown in Figure 2b [73]. In [74], a hard-structure sensor, along with circuit chips, was incorporated into the lens-type substrate and contact-tear fluid method via a microfluidic-based sensing channel. An ultrathin-type sensor was fixed onto the lenses and came into contact with the tear fluid, which provided detection without interfering with blinking. Soft electronic devices offer an excellent interface with the human body's skin, enabling real-time monitoring of vital signals [75]. Researchers investigated smart glasses and head-mounted computers capable of displaying information [76]. They employed nanomaterials with mechanical structures and stretchable polymers to develop electronic devices that are both flexible and stretchable, enabling seamless integration with living organisms. These devices are capable of detecting parameters such as temperature, ion concentrations, and electrocardiogram (ECG or EKG) signals, in vivo or in vitro [77]. Several transdermal patches were used for pharmaceutical and chemical agents through the human body's skin [78]. A single-walled carbon nanotube (SWCNT) was designed to monitor glucose oxidase–Nafion composites for sensing [79]. Recently, hybrid devices based on metal, dielectric, and metallic polymers were proposed, and their performance was simulated for various frequency bands to design wearable and flexible devices. A group of researchers recently investigated a self-powered device that can easily harvest ambient energy to run a sensor, removing the need for external power [80]. Piezo-electric nano-generators (PENGs) and tribo-electric nano-generators (TENGs) have predominantly been employed in the development of self-powered sensor devices. These wearable sensor devices play a crucial role in measuring vital signals such as respiratory rate, temperature, and blood-oxygen levels, which are essential in accurately diagnosing and monitoring COVID-19 [81]. High-performance wearable electrical devices require fibrous material with high strength and considerable stretchability to provide durability and stability. An ultra-thin, robust, and extensible conducting microfibre is constructed for designing fibrous-type mechanical sensors for wearable applications. These sensing devices demonstrated high sensitivity in detecting strains, with a high resolution and a large detection range simultaneously [82].

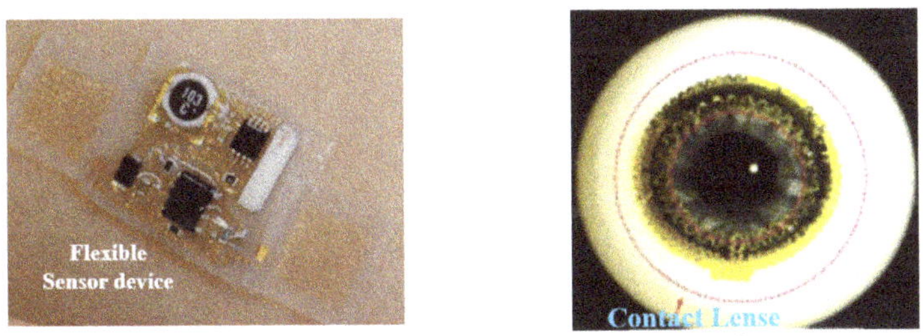

(a) (b)

Figure 2. Wearable electronic devices (**a**) soft tissue [57], and (**b**) contact lens [74].

In [83], researchers evaluated traditional polysomnography measurement tools to study sleep. Some of these tools lack comfort and impose high additional costs. As an alternative, several innovative designs, such as jewellery and body-mounted sensing devices, have been developed as non-invasive options to measure vital signals during sleep. They provide a more comfortable and cost-effective solution for monitoring sleep-related activities and signals [84]. Sleep disorders may be controlled through advanced technologies that make early diagnosis possible [85]. MXenes ink is a highly conductive material [86], as highlighted, and applications have been found in various flexible and stretchable devices. In [87], a highly integrated multipurpose sensor device for intracorporeal applications was proposed. This intracorporeal system incorporated temperature, capacitive-type pressure, and electrochemical-based oxygen sensors onto a single chip, along with their corresponding interface application-specific integrated circuits (ASICs). A low-cost UWB-based landslide monitoring device, named Wi-GIM, was introduced in [88]. This device utilises UWB technology to determine the relative distances between monitoring points and effectively track ground deformations associated with landslides. A novel conformable sensor interface was created in [89]. This sensor was designed to be easily attached and detached from face masks. Smartwatches have become a well-established tool for continuous activity and fitness tracking. Longitudinal tracking using smartwatches is being utilised in clinical applications, including remote monitoring of Parkinson's disease (PD), which is the most prevalent neurodegenerative disease [90]. In [91], researchers investigated biocompatible and biodegradable materials for flexible and stretchable devices without producing any environmental problems. As reported in [92], a laser-engraved wearable sensor was used for the sensitive detection of uric acid and tyrosine in sweat, as shown in Figure 3a. Extensive research has been conducted on intricate measurements, including ECG. For instance, the Apple watch screen has been investigated to detect atrial fibrillation simply through the touch of a finger [93]. Soft polymers have been explored as electrodes and active layers to produce flexible and stretchable electronic devices for wearable applications [94]. Various HMI-based (human–machine interface) vision assistance devices are now possible thanks to advancements in micro-electronics and nanofabrication for wearable and flexible materials. These applications include integrating sensors, circuits, and components into a small contact-lens design. Its interaction with apps for displaying information, detecting eye movements, recovering vision, and identifying specific biomarkers in tear fluid is made possible through various smart contact-lens materials, device designs, and componentry [95]. Google's Fitbit-based tracker system examines blood-oxygen levels [96]. Recently, a four-layer sensor device was used for the healthcare system, which is compatible with 5G networks [97].

In another study [98], researchers proposed a sensor network that harnesses 5G-wireless networks for healthcare and safety applications. The combination of 5G-wireless technology and machine-to-machine (M2M) communications enables the continuous monitoring of chronic diseases. Thus, M2M systems are seamlessly integrated with 5G-enabled devices to provide ongoing monitoring for chronic diseases. The design of an e-health system tailored for chronic patients was explored by studying the infrastructure and protocols of 5G networks [99]. Continuous data transfer and sensing applications were facilitated using 5G-cellular channels as highlighted [100]. The integration of IoT, machine learning, data analytics, and 5G and beyond-5G (B5G) networks have opened new ways for innovative diagnostic methods that have the potential to lower hospital and medical costs. Furthermore, 5G technology brings several advantages to robotic surgery care by providing faster download rates, video calls, and gaming capabilities [101]. Advancements in flexible and stretchable technology have also enabled the utilisation of implantable systems in various areas, including the deep brain, intravascular space, intracardiac space, and even the interior structure of a single cell [102]. Several new sensing devices have been designed; for example, a degradable, tuneable graphene-based antenna and a band stop filter/antenna sensor have been reported [103]. In [104], a 3D curved-shaped antenna was reported, featuring a curved layout and a non-radiating edge gap-coupled configuration, suitable for diverse applications. Furthermore, a lightweight and scalable flexible integrated circuit phased array system has been designed which can be conveniently stored or rolled in a limited space, as shown in Figure 3b [105]. In recent developments, a multi-material approach has emerged as a viable solution to enhance the safety of wireless, skin-based interfaced bioelectronic sensors. This approach, depicted in Figure 3c, utilises a combination of different materials to ensure improved safety and robust functionality of wearable and flexible devices [106]. Utilising biodegradable, implantable, and wearable sensors for healthcare could change the future of disease management in healthcare toward preventative, predictive, and personalised management of diseases. Furthermore, the manufacturing of biodegradable electronics and sensors that naturally dissolve under ambient conditions could mitigate electronic waste challenges. The degradation rates of materials and dissolution behaviour of bioresorbable devices in different biological systems can vary significantly. Furthermore, developing a high-performance and miniaturised power supply, circuits, and chips on biodegradable substrates and wireless transmission is required to develop a fully biodegradable platform. Among several options, biodegradable sensors (BD-SENs) have proven to be the best option and are widely used in the food and medical industries. Moreover, BD-SENs are widely used in implanted medical devices, which disintegrate into the body through various degradation mechanisms once the intended function is completed. The use of these implantable BD-SENs helps to avoid the need for additional sensor removal operations. Implantable sensors with controlled degradation mechanisms will also aid in determining how long the device should operate inside the body. The issue of electronic waste has been addressed by the design of these environmentally friendly and sustainable sensors, which also serve as a green substitute for traditional non-biodegradable sensors. Degradable piezoelectric biomaterials have recently attracted more attention as potential medicinal applications. They provide a self-sufficient solution using biomechanical energy harvesting. This promising path toward the development of cutting-edge, transient biomedical devices that support both improved patient care and real-time monitoring is made possible by the combination of degradable sensors with flexible, piezoelectric materials. Future implantable sensors need to use as little power as possible, for example, getting power from the body's movement, heat, or other in situ energy sources such as biofluid [91,103]. In [107], flexible MEMS-type sensor devices were proposed. For example, MEMS technologies have been used to produce such flexible electronics, they need equipment for vacuum deposition, photolithography, and wet and dry etching, as shown in Figure 3d.

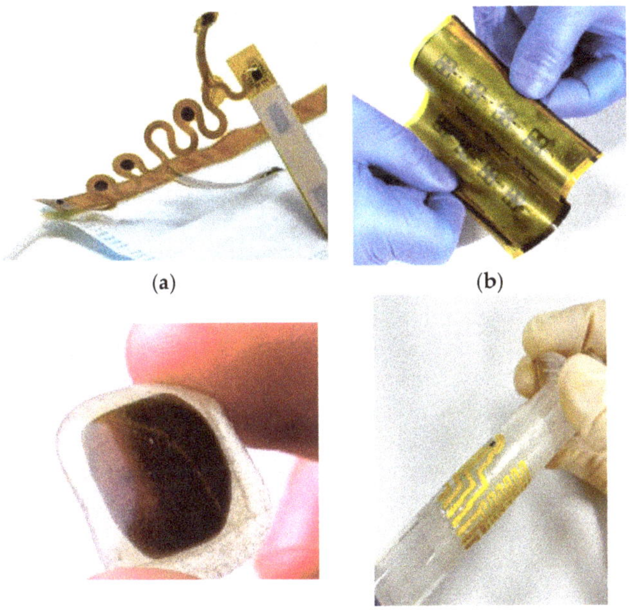

(a) (b) (c) (d)

Figure 3. Smart devices: (**a**) laser-engraved [92], (**b**) flexible [105], (**c**) multi-materials devices [106], and (**d**) MEMS sensor [107].

Various advanced flexible electrochemical sensor devices have been used for various applications, such as in fitness and biomedicine [108]. With significant advancements in wearable devices, there has been an exploration of near-infrared spectroscopy (NIRS) electronics, which hold potential for numerous healthcare applications [109]. Several unique shapes of wearable body-worn sensing devices are proposed in the literature. In general, there are two types of sensing and monitoring systems: contact-based and contactless wearable sensor devices. They include optical and RF signal-based devices. Vital signals are frequently monitored by wearable sensors; nevertheless, there are drawbacks to this technology, including the high costs, dependence on user compliance, and the potential for disease transmission if not adequately disinfected. In situations where hygiene is a concern, like hospitals, contactless systems, especially those that use RF signals in the range of 30 kHz to 300 GHz, offer a viable alternative by enabling non-intrusive monitoring without direct touch or line-of-sight. Wi-Fi signals, which range in frequency from 2.4 GHz to 7.2 GHz, have become an effective means of tracking activity and identifying vital signs. These signals find use in fall detection and real-time health monitoring. By integrating these Wi-Fi-based devices with IoT frameworks, healthcare providers can receive useful data, and smart-home features like automatically altering environmental parameters depending on vital signs can be improved. But compared to wearable and visual-based systems, Wi-Fi and other contactless technologies have some limitations in terms of precision and dependability, notwithstanding their potential.

It is anticipated that advancements in big data and machine learning, as well as the development of sophisticated semiconductor technologies like CMOS and Bi-CMOS, would lead to improvements in wireless sensing for medical applications. Technologies, including mHealth, 6LoWPAN, and Institute of Electrical and Electronics Engineers (IEEE 11073, Health informatics—Medical/health device communication), are being investigated for remote health monitoring in smart-city frameworks as IoT solutions in healthcare progress from basic designs to more elaborate smart systems [110]. Furthermore, new technologies like NFC, RF transmitters, and carbon nanotubes are opening new possibilities for data extrapolation in smart healthcare applications. With the help of multidisciplinary research, these technologies should continue to progress, opening new ways for contactless, affordable healthcare monitoring, especially when issues with accessibility and dependability are resolved [110–112]. Recently, wearable sensors for various purposes, including energy storage and health monitoring, have been transformed by recent developments in biodegradable and superhydrophobic materials-based and MXenes-based devices [24,86,113]. Furthermore, multimodal sensing e-skins fall into three categories, such as systems that combine chemical, electrophysiological, and physical sensors, and are investigated to simulate human skin for next-generation smart technology [5]. Recently, self-powered Triboelectric Nanogenerator (TENG) sensors monitoring ethanol during simulated exhalation and humidity using copper-oxide and polyethylene-oxide nanofibres have been investigated, which are suitable for non-invasive breath analysis applications. Key characteristics of the flexible and stretchable sensors created in this manner include their biocompatibility, scalability, lightweight design, flexibility, and wearability. Additionally, connection with a triboelectric source has made it possible to develop a self-powered, wearable, portable sensor. In conclusion, a wearable sensor for healthcare systems that runs on piezoelectric and triboelectric nanogenerators has been demonstrated. Smart sensors are useful because they can be created rapidly and affordably, worn conveniently, have a long lifespan, react swiftly, and may be used anywhere and at any time. They are also resistant to water and other factors [2,3].

However, SAR, which is regulated to prevent dangerous exposure, assesses the rate at which the human body absorbs energy from electromagnetic fields (EMF) in wearable devices. However, millimetre waves, which are now increasingly used in technologies like 5G, raise additional concerns [31]. Table 2 summarises various physiological signal-sensing devices.

Table 2. Various available wearable devices in literature.

References	Various Research Areas That Used Wearable Devices in the Literature
Healthcare/medical technologies for physiological signals monitoring systems	(1) Technologies (RF and microwave), (2) materials, and (3) applications and research directions
2017 to 2022 [113,114]	Wearable design and systems: healthcare, clinic, and hospital applications
2018 to 2022 [115,116]	Wearable design and systems: Energy harvesting, holistic systems, and self-healing composite
2021 to 2022 [117–119]	Next enabled device systems: Backscattering
2013 to 2022 [120–128]	Enabled device systems: RFID and RF sensing devices, far-field, and RF rectenna systems

Table 2. *Cont.*

References	Various Research Areas That Used Wearable Devices in the Literature
2022 to 2023 [129–131]	Soft materials: Enabled electronic for medicine and human interfaces, and soft, skin-integrated multifunctional microfluidic systems
2013 to 2023 [132–137]	Liquid metal-based electronic: Sensing and multi-site sensing capabilities
2022 to 2023 [138–144]	Deep learning-based: Blockchain and artificial intelligence technology

3. Fabrication Methods

Fabrication methods play an important role in the accuracy and precision of the measurements and impose various limitations on the cost and fabrication time. Various fabrication methods have been utilised for wearable devices, which are summarised in [145]. There is a growing need for affordable and readily accessible fabrication methods that enable the mass production of wearable and flexible electronic devices [11–13]. DW is an AM method that directly transfers onto a substrate material and that can be used in various industries, including microelectronics, optics, and biomedical engineering. DW is a flexible fabrication method that eliminates the need for moulds or masks and allows for exact material deposition on substrates. Because DW makes high-precision patterning possible, it can be applied to the production of microdevices, PCBs, and biosensing. With the help of the extrusion based DW assembly technique, a large variety of inks may be patterned with feature sizes as small as 250 nm in both planar and three-dimensional forms (3D). However, compressed air is utilised in this procedure to force inks with regulated rheological characteristics via a single nozzle (diameter: 1 to 500 μm). The DW assembly uses a regulated printing speed and pressure that are dependent on the rheology of the ink and the diameter of the nozzle to deposit inks at room temperature or in an appropriate coagulation reservoir. DW assembly has, nevertheless, been used to handle a broad variety of inks, including colloidal suspensions and gels, nanoparticle-filled inks, polymer melts, fugitive organic inks, hydrogels, sol-gel, and polyelectrolyte inks. While some inks, like polyelectrolyte inks, must be written into reservoir-induced coagulation in order to facilitate 3D printing, other inks, like sol-gel inks, can be printed directly in air and offer superior control over the deposition process, allowing the ink flow to be repeatedly started and stopped during assembly by layer-by-layer extrusion of viscoelastic inks.

DIW, a subset of DW, enables the creation of 3D structures with unique shapes. Applications of DIW include energy storage devices, soft robotics, and tissue scaffolding. DIW, called direct-write assembly, is an extrusion-based 3D AM printing method in which liquid or semisolid colloidal inks are dispensed via small nozzles under controlled pressure and flow rates. Several DIW methods are available that can make 3D patterns inside the materials. They can be divided into the following two categories: droplet-based techniques like ink-jet printing and hot-melt printing, and filamentary-based techniques like micro-pen writing, robocasting (or robotic deposition), and fused deposition.

Numerous ink patterns have been used, such as concentrated polyelectrolyte complexes, highly shear-thinning colloidal suspensions, colloidal gels, polymer melts, and diluted colloidal fluids. These inks solidify via solvent or temperature-induced phase changes or by liquid evaporation gelation. It is possible to create 3D structures with high aspect ratios (like parallel walls), spanning features, or continuous solids by carefully adjusting the ink composition, rheological behaviour, and printing conditions. Because the later structures require self-supporting elements to bridge gaps in the underlying layer or layers, they present the most difficulty for ink designers. To write 3D periodic architectures with filamentary features ranging in size from hundreds of micrometres to sub-micrometres, this feature article primarily focuses on our recent efforts to design concentrated colloidal, fugitive organic, and polyelectrolyte inks. These inks can be used as functional composites, microfluidic, and templates for photonic bandgap materials and inorganic–organic hybrid structures. Thanks to these recent developments, it is now possible to print metals and dielectrics simultaneously, overcoming earlier difficulties with print resolution and material compatibility. Examples of these uses include the construction of novel microelectronics, biocompatible medical scaffolds, and high-performance energy storage systems [27,28]. However, both DIW and DW are briefly explained, with some examples provided for clarity.

3.1. DIW Method

Writing plays a ubiquitous role in our daily lives, enabling us to document and communicate through various signs and symbols. As writing tools were invented and writing capabilities expanded, it significantly propelled human civilisation forward. In the context of designing electronic devices for wearable applications, Figure 4 shows four commonly utilised types of writing instruments: brush-type simple pen, pencil, fountain pen, and ballpoint pen [146]. Several writing materials, such as ink or graphite, are used to create signs and symbols on many substrates. Various pen-based techniques have been developed to fabricate electronics for wearable applications, including those utilising paper as a substrate. Each writing approach used in the creation of electronic devices has its own set of advantages and disadvantages. It is crucial to select an appropriate writing process based on the desired quality of electrical components and their specific application domains. Subsequently, the methods employed for device writing can be categorised into two groups: pen-based techniques, which comprise fountain pen, ballpoint pen, brush pen, and pencil methods; and pen-analogue approaches, as illustrated in Figure 5a–d. While the writing process for electronics using pen-based techniques is fast and portable, it is not the most convenient or cost-effective method due to the challenges associated with post-treatment procedures. In certain situations, extended durations, high temperatures, and vacuum settings are necessary. Moreover, despite the ability to write various components, pen-based direct writing methods are limited in their capability to fabricate complex components and devices, such as integrated circuits.

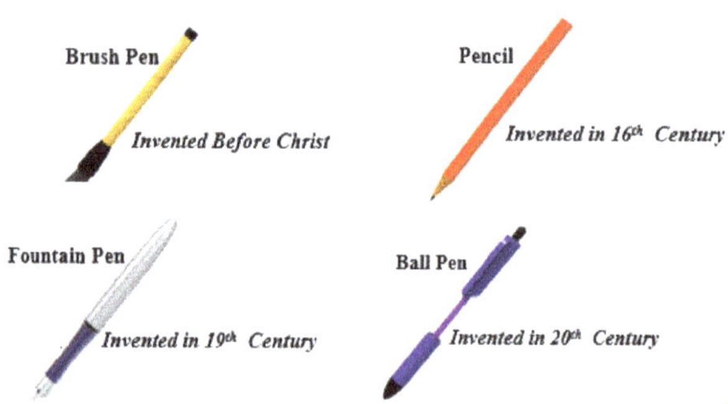

Figure 4. Development of direct-writing electronics for wearable and flexible applications [146].

Ideal electronics for diverse applications should possess qualities such as affordability, user-friendliness, high performance, and rapid prototyping capabilities. Recently, there has been great interest in the development of flexible electronics due to its many potential uses, including wearable functional gadgets [147], medical monitoring systems [148], and flexible energy storage and conversion [149]. For instance, a polyester-based integrated epidermal electronic system resembling a tattoo has been created, holding promise to enable doctors to timely monitor patient vital signs straightforwardly and effectively [150]. This system can sense and record real-time muscle movement and heart activities and detect brainwaves without expensive, bulky medical instruments. To create electronic devices that offer enhanced performance at a reduced cost, considerable attention has been devoted to the development of various elements closely linked to flexible electronics. These include substrates, conductive materials, and production processes. [151,152]. Numerous methods, including coating, sputtering, and printing methods such as gravure, flexography, inkjet, and screen printing, have been utilised to produce flexible electronics. Printing techniques allow for the rapid development of parts and gadgets with precise three-dimensional (3D) architecture, which makes it easier to create usable 3D electronics. As an example, the development of strain and pressure sensors in a stretchable silicon-based polymer substrate has shown a feasible solution in recent 3D-printed devices, showcasing a huge potential to produce sensing electronic systems with arbitrary shapes used in different wearable applications. Nevertheless, these advanced techniques may not fully address the increasing need for low-cost, easily accessible, and efficient prototyping of flexible electronics. This is due to coating processing times, high energy consumption associated with sputtering, and the expensive nature of required printing equipment [153,154]. These research activities represent a ground-breaking revolution in the advancement of circuit systems, sensors, energy devices, communication elements, and more. It encompasses a wide range of areas, including materials innovation and transformative manufacturing approaches [152,154].

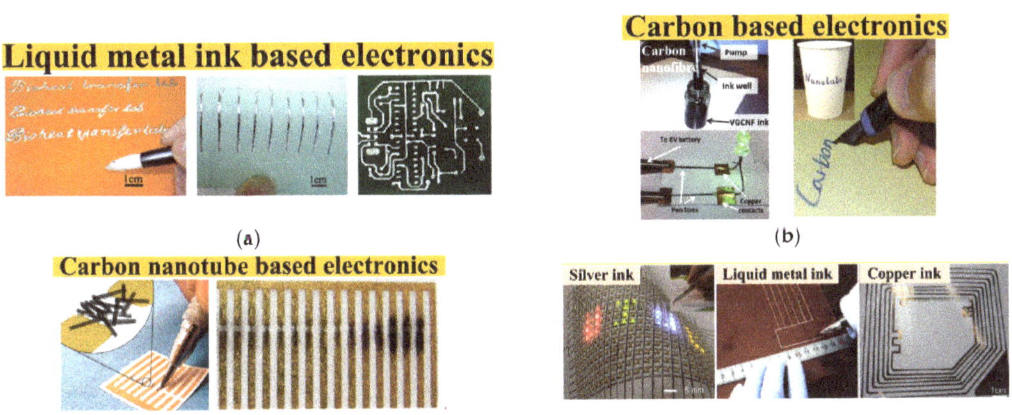

Figure 5. Writing electronics used in the literature: (**a**) brush pen, (**b**) fountain pen, (**c**) pencil, and (**d**) ball pen [146].

Recently, a novel approach has surfaced for creating electronic circuits using handwriting, where an array of writing instruments, including brush pens, pencils, fountain pens, and ballpoint pens, are employed to directly write on dielectric substrates and create electronic devices. This alternative method has gained attention as a promising way to integrate traditional writing tools with electronic functionalities. This technique of direct writing with a pen offers a convenient and rapid method for depositing electrically conductive materials on substrates, benefiting new biomedical and electrochemical sensing applications, electronic components, and energy storage devices. Pen-based writing enables end users to create and deploy sensor devices on-site as per their requirements. Additionally, when combined with printing methods, pen writing offers simplicity, cost-effectiveness, mass production capabilities, and high-resolution printing, known as the pen-analogue writing method. This method primarily encompasses applications like electrohydrodynamic and micro-plasma-based drop-on-demand (DoD) techniques. An example of a precise and versatile pen plotter, the AxiDraw machine, is demonstrated in Figure 6, which is used to design various items such as postcards, invitations, or drawings [155]. The researchers have used hand-writing electronics (HWE), which can provide the same benefits as printed electronics (PE) at a lower cost, making it simpler to prototype and create do-it-yourself (DIY) electronics. Figure 6 demonstrates mechanical printing processes, in contrast to conventional direct painting or writing methods, which enhance the accuracy of liquid metal-based flexible electronic circuits, a method that can be utilised for healthcare applications and wearable devices [156,157].

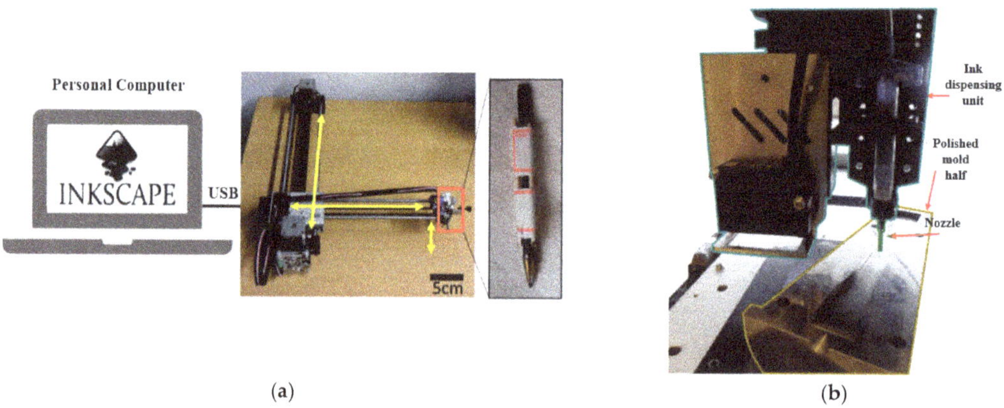

Figure 6. (**a**) Ultra-High Frequency (UHF) RFID tags [155], and (**b**) HWE system [157].

A variety of electronic devices use silver nanoparticle ink [158]. For example, to print on a mould surface, an automated micro-pen type dispensing device (such as Nordson Pro4 EFD) was utilised. After printing, the mould and printed traces were subjected to a 30-min sintering process in an ISOTEMP 282A vacuum oven [158]. The printed traces were positioned at a specific distance from the injection point on the mould surface to investigate the impact of polymer pressure on the interlocking and adhesion of ink/polymer. Each trace was printed in an inline form, measuring 20 mm in length and 50 μm in height. As discussed in detail in [158], a pen-based 3D printer provides a direct 3D structure, but using a pen in space and on a 2D surface is not convenient. Recently, researchers created a pen-based 4D printing (4Dp), which includes dipping the devices in a monomer solution to turn 2D prints into 3D shapes. One material used in 4Dp is MXene inks, which are produced from titanium carbide (Ti3C2) sediments that resemble clay. Ti3C2 inks can be utilised for stamping, printing, painting, and writing [159]. A pen-type setup was utilised to design glass nozzles using a micropipette puller (Sutter-based P-1000 instrument, as the company is located in Novato, CA, USA) that provides varied apertures, as shown in Figure 7 [160].

Direct laser writing (DLW) offers the capability to create precise patterns without the need for masks or chemicals. In a specific application, DLW was employed to fabricate a three-dimensional (3D) print of reduced graphene oxide (RGO). The researchers utilised a programmable approach that involved scanning a laser beam voxel-by-voxel, with each voxel measuring approximately 300 nm in diameter. By focusing the laser on concentrated flakes of graphene oxide (GO), the overlapping RGO flakes formed a 3D-printed structure. The treated area exhibited enhanced photoluminescence compared to the untreated region, enabling clear visualisation of the 3D pattern created by DLW [161]. Cellulose paper with microscale fibres can be used as a substrate for flexible-type organic light-emitting displays. Various facile pen-drawing can be used. Pens offer the versatility to utilise a variety of conductive and light-emitting materials for designing anode and hole-type injection layers. These materials can include different transfer layers, emission electron injection layers, and cathodes [162].

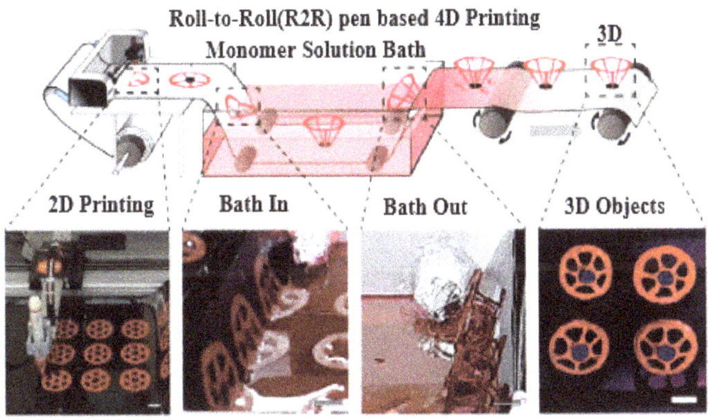

Figure 7. Flexible electronic: pen-based 4D printing [158].

3.2. DW Methods

Three-dimensional printed designs utilise printing layer-by-layer and additive manufacturing methods (AMs). Three-dimensional printing (3Dp) has various advantages as compared to other methods. It provides design flexibility and high resolution as well as low cost [163]. AM is an advanced technique that encompasses various methods such as fused deposition modelling (FDM), selective laser melting (SLM), electron-beam melting (EBM), aerosol jetting, inkjet printing, and more, which have become increasingly accessible in recent times. The new generation of AM techniques allows for the utilisation of a wide range of materials, including polymers, metallic nano inks, ceramics, nanocomposites, and alloys. These AM methods enable the production of flexible sensor devices. Notably, extrusion methods like FDM and FFF (fused filament fabrication) are commonly employed in AM for creating 3D designs. Some examples are depicted in Figure 8, including a rigid device shown in Figure 8a [164], a flexible device shown in Figure 8b [165], and a microsensor shown in Figure 8c [166,167].

For the successful printing of complex and 3D surfaces, precise control of the laser beam is crucial. To achieve this, a shutter is typically placed in front of the nozzle to modulate the beam. Additionally, maintaining an optimal distance between the nozzle and the substrate is essential, typically ranging from 1 mm to $\geq$10 mm. Deviating from these specified boundary conditions can lead to undesirable overspray defects in the final design [168,169]. Three-dimensional printing technologies have significantly impacted the biomedical field, emerging as an innovative method with diverse applications such as tissue engineering, organ fabrication, medicine, drug delivery, and more. In the biomedical domain, various 3Dp methods are utilised, including FDM, selective laser sintering (SLA), and DIW. These methods make use of a wide range of materials, which are summarised in Figure 9 [164].

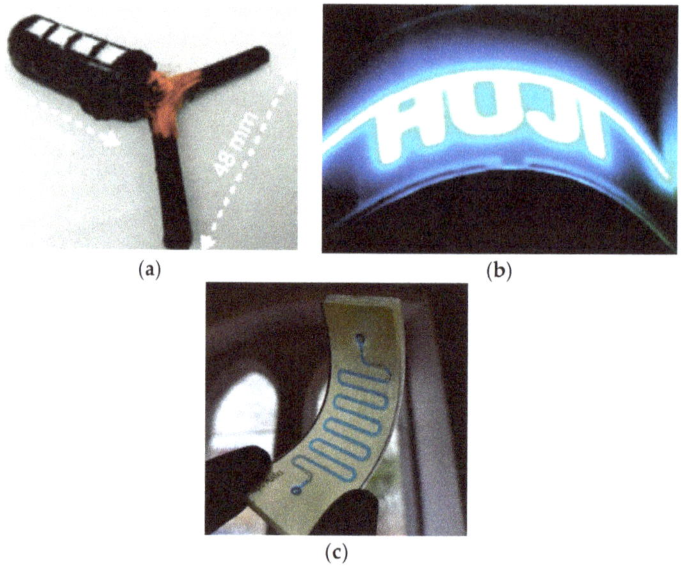

Figure 8. Flexible devices: (a) gradient echo device [164], (b) electroluminescence device [165], and (c) serpentine channel with integrated coverslips [167].

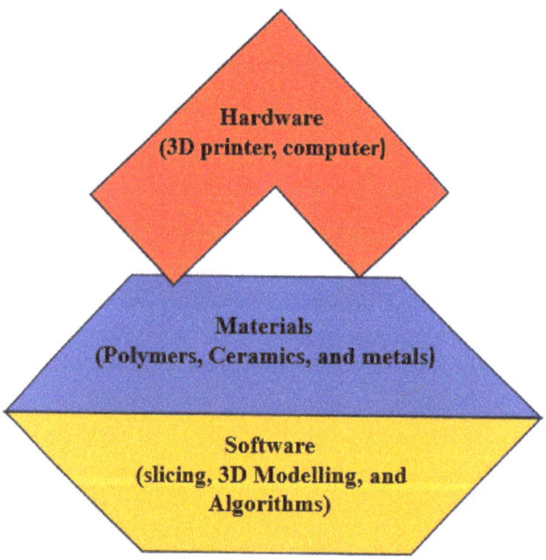

Figure 9. Fabricated devices using traditional 3D-AM methods.

Three-dimensional printing is utilised in the fabrication of various medical devices and structures, such as patient-specific orthoses, prostheses, craniofacial implants, and medical devices based on individual data. Three-dimensional printing utilises low-melting-point metal ink in an additive mode, offering versatile applications across various fields. In [170], liquids such as gallium, bismuth, and indium alloy were used to fabricate 3D devices. In the realm of 3D devices, certain inks such as Bi35In48.6Sn16Zn0.4 exhibit unique characteristics wherein they do not absorb or release heat while behaving like metals during phase changes. Leveraging these novel ink properties, along with silicone rubber (a nonmetal), hybrid 3D-printed structures for wearable applications have been successfully fabricated. These structures possess remarkable mechanical and electrical properties while also exhibiting thermal conductivity. The exceptional insulating properties of the non-metal ink contribute to the durability and functionality of the 3D-printed designs, ensuring their effective performance even in demanding environments. Three-dimensional printing offers the capability to create a diverse range of antennas, encompassing polymer, metallic, ceramic, composite, and integrated antennas with multiple materials. For instance, an e-band lens antenna was designed using an automatic synthesis algorithm and realised through polyjet 3D printing technology using VeroWhite materials [171]. Furthermore, the field of electronics, sensors, and electrochemical devices has seen the utilisation of several conductive polymers and nanocomposites, enabling advancements in their respective applications [172].

Various 3D-printed multilayer-phased array antenna systems have been designed for WBAN applications [173]. In [174], a common freeze-casting technique was used to fabricate liquid-based 3D antennas. Researchers tested micro components of the liquid metal alloy for 3Dp (at 30 °C). Additionally, liquid metal has emerged as a viable option for creating injectable 3D bio-electrodes used in ECG and EEG measurements. Advanced, high-resolution, 3D-reconfigurable liquid-metal structures with stretchable properties have been designed to integrate seamlessly into the human body, as depicted in Figure 10a [175]. This design enables the creation of reconfigurable antennas that can be tuned by altering their structures, as well as reversible, movable interconnections that function as mechanical switches. In the realm of 3D printing, the intricate housing structure of the antenna, which may include components like a microfluidic channel, can be fabricated directly onto the substrate using primarily the FDM technique, as illustrated in Figure 10b. Furthermore, researchers have developed a heat sink composed of silicon micro pin fins to efficiently cool a Stratix 10 GX-based field-programmable gate array (FPGA) that consists of multiple heterogeneous dice [176]. To enhance the thermal performance of a design, a piece of 3D-printed plastic with an embedded silicon micro pin-fin heat sink was used to seal the tops of the pin fins and connect the heat sink to the inlet and outlet tubing.

Various 3D-printing methods of liquid metal are available to design antenna and electronic devices, such as liquid-metal melt deposition printing, liquid-phase printing, suspension printing, microcontact printing, and in vivo 3D-printed moulding methods [177]. An ultra-compact implanted antenna that is integrated with a Nanostim leadless cardiac pacemakers (LCP) system is presented for use in biotelemetry [178]. A compact antenna that can work in the 2.4 GHz ISM band is created for integration with an LCP. Wearable antennas that are primarily functioning in *ultra-high frequency* (UHF) and microwave bands have been developed using AM [179]. In [180], it is reported that 3D polymer-conductive hybrid materials were designed using the 3D-FDM method that can be used in wearable applications; an example is shown in Figure 10c. RFID antennas may be designed utilising a 3D-DW method on a textile material, for example, see the design reported in [181]. In recent developments, the integration of FDM for the substrate and syringe dispensing for metallic layers has enabled the creation of a 5G antenna embedded within a medallion [182]. It is suggested to employ removable fingernail-mounted 3D mm-wave and microwave antennas for on-body communications [183]. For this purpose, EGaIn was utilised as the conductive part of the radiator, while Visijet M3 crystal resin was employed as a dielectric for 3D-printed planar and helical antennas [184]. A groundbreaking approach involves the utilisation of 3D printing to create new and innovative gradient refractive index (GRIN) lenses for an H-plane horn antenna [185]. These lenses are constructed using high-permittivity ceramic ZrO_2, resulting in enhanced performance and functionality. As reported in [186], a 3D-printed wire electrostatic discharge (WESD) structure comprising six upper capacitors was designed, enabling verification of the curved surface. This demonstrates the capability of 3D-printing technology to seamlessly integrate WESDs into various wearable devices, including sensors, energy harvesting systems, and light emitting diodes (LEDs), as shown in Figure 10d.

Utilising the DIW method and the PEDOT: PSS solution, researchers have successfully developed a high-resolution 3D-conducting polymer. This conducting polymer, along with its composites, has been extensively employed to construct flexible electrodes in diverse applications. The DIW method, in combination with the PEDOT: PSS solution, offers a versatile approach for fabricating flexible electrodes with exceptional precision. By precisely depositing the conducting polymer onto desired substrates, researchers have been able to create intricate electrode structures with enhanced conductivity and flexibility [187]. An innovative double-layered graphene conductor has been developed to enhance the optical and electrical properties of organic LEDs, offering a high surface area for improved performance [188]. A novel method called consecutive ink writing (CIW) has been developed to enable the printing of conducting polymer composites onto various flexible plastic substrates. This method involves an iterative extrusion process that produces a filament with a consistent diameter of 1.75 mm. By incorporating matrix polymers and carbon fillers, composite materials can be effectively used in 3D-printing applications. The CIW method provides a versatile approach for fabricating conducting polymer composites with precise control over the deposition process. The iterative extrusion ensures a uniform filament, which enhances the consistency and reliability of the printed structures. This method allows the integration of carbon fillers, enabling the customisation of electrical conductivity and mechanical properties of the printed composites.

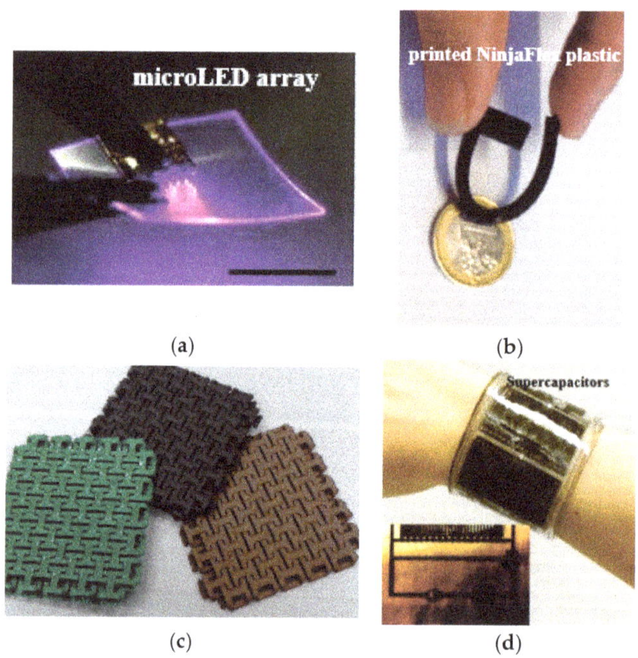

Figure 10. 3D designs: (**a**) micro-LED array [175], (**b**) fluidic antenna [184], (**c**) hybrid materials [180], and (**d**) six capacitors [186].

The ability to print conducting polymer composites on flexible plastic substrates opens a wide range of possibilities in the field of flexible electronics. These composites can be used to create functional components, such as sensors, electrodes, and interconnects, with enhanced electrical conductivity and mechanical flexibility. The versatility of the CIW method enables the fabrication of complex 3D structures, further expanding the potential applications of these composite materials in various industries, including electronics, robotics, and healthcare [189]. This novel CIW process makes it possible to quickly print composite materials with a lot of graphene filler onto different substrates in 2D and 3D shapes. The CIW method allows for 3D printing of PEDOT: PSS and graphene ink on both rigid glass and flexible polyimide substrates. This versatile technique offers precise control over the deposition process, enabling accurate and reliable printing of conducting polymers and graphene materials. The CIW method has significant potential in various fields, such as flexible electronics, sensors, and energy devices. Further research and development in this area will optimise the CIW technique, opening possibilities for creating complex and functional 3D-printed structures on different substrates; an example is shown in Figure 11a [190]. Recently, polymer 3D printing has been utilised in devices for healthcare applications; for example, tissue scaffolds have been designed using polyjet and stereolithography printing that can be used to achieve hierarchical forms that can mimic the bone. A conductive polymer can be used to create implants that do not degrade in the human body. Polymer 3D-printing methods such as extrusion, resin-based printing, and powder-based printing provide flexibility in material choices and support a wide range of design shapes, responses, layouts, etc. [191]. To create a textile-based sensing device to be used in healthcare applications, several designers have incorporated 3Dp technology, including FDM and SLS; an example is shown in Figure 11b [192].

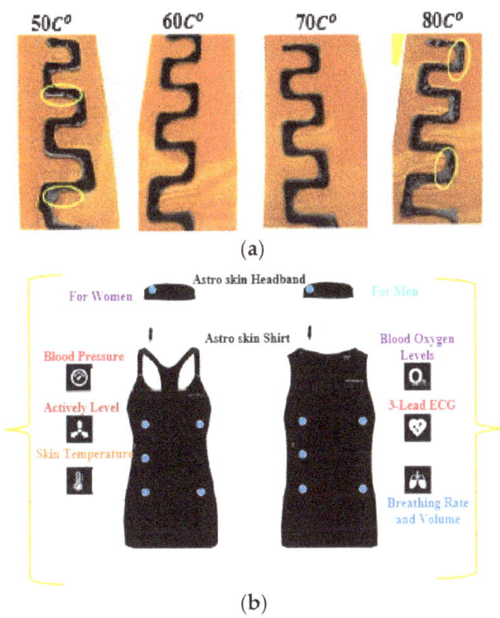

Figure 11. (**a**) EEG/PEDOT: PSS composite electrodes [190], and (**b**) flexible biosensor [192].

Spray coating is a DW method that can make a film deposition on a substrate, and which changes its dielectric values (ε) (Figure 12). This method has been broadly utilised because of its high speed (200 mm/s). Inks with a wide range of viscosities from 0.7 to 2500 mPa can be printed. Antennas, which can be used on IoT devices, have been easily fabricated using a simple spray-on method. As an example, researchers fabricated a 1-mm-thick MXene RFID tag with a reading range of 8 m at 860 MHz [193]. In [194], a liquid metal was sprayed to shape conductive paths on the elastomeric substrate. However, a textile duplex antenna utilising knit, and nonwoven fabrics was designed. The dielectric value (ε) of the fabric was enhanced due to a DW spray coating using nanoparticle ink.

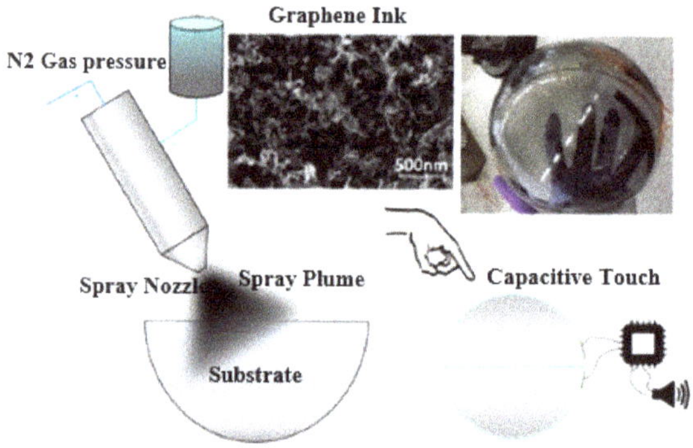

Figure 12. Spray coating process and semi-transparent capacitive-touch device [193].

3.3. Other Emerging Methods

Four-dimensional printing is a groundbreaking advancement in 3D-printing technologies that introduces the concept of self-transformation and functionality in printed materials. As illustrated in Figure 13a, 4D printing enables materials to change shape and perform specific functions in response to stimuli such as osmotic pressure, heat, current, ultraviolet light, and various energy sources. Unlike traditional 3D printing, which operates in three dimensions (x, y, and z axes), 4D printing adds a fourth dimension: time. This new dimension allows for dynamic and programmable behaviour in printed objects, revolutionising the design possibilities and applications of 3D-printing technology [195]. This fourth dimension allows for dynamic changes in the geometry of printed objects in response to external stimuli, thanks to the unique properties of smart materials. The design process of 4D printing enables the creation of objects that can undergo shape transformations and adapt to their environment, enhancing their functionality and versatility [196]. The development of 4D-printed devices involves the selection of various components, including printing technologies, materials, structures, stimuli, and applications. One of the key features of 4D designs is their ability to change shape, which can be achieved through three different approaches: using stimuli-responsive materials, employing a multi-material, or artificially inducing internal tensions [197].

The first approach, utilising stimuli-responsive materials, is widely popular and often considered the primary mechanism in 4D printing. In this technique, the 4D-printed structure initially possesses a temporary shape, and it transforms into its intended permanent shape when exposed to the appropriate stimulus. The second approach, known as multi-material, structures, leverages the unique characteristics of different materials, such as elasticity, swelling ratio, or thermal expansion coefficient, to achieve shape-changing capabilities. The multi-material structures combine many materials with varying properties, giving them special qualities, as discussed in 3D- to 6D-printing technologies. Different thermoplastics, polymers, and smart materials, including hydrogels and shape memory polymers (SMPs), can be combined with these structures. Combining these materials allows the production of objects with dynamic behaviours, such as the capacity to react to mechanical forces, temperature changes, and pressure, in addition to strength and durability. In biomedical applications, implants or devices must adjust to shifting biological conditions or mechanical stresses. The use of smart materials is the main factor driving the mechanism of form change in these multi-material structures. In 4D- and 6D-printing methods, these materials are designed to undergo controlled deformations in reaction to external stimuli. SMPs, for example, can be configured to hold a given shape and then revert to that form in response to a given stimulus, like heat or light. In contrast, hydrogels can expand or contract in response to variations in temperature or humidity. The capacity of these materials to recall and return to their original form when necessary is a crucial component of this shape-changing mechanism. This characteristic is essential for medical applications, like the creation of implants that may alter shape in response to changes in the surrounding tissue or the patient's growth. The integration of smart material science and additive manufacturing techniques is the foundation of these technologies' entire operation. The first step in the process is 3D printing, which constructs an object layer by layer. With the introduction of other dimensions like time and multi-directional production, this develops into 4D and 5D printing. The process of 6D printing combines the dynamic, stimulus-responsive properties of 4D printing with the structural benefits of 5D printing by employing five degrees of freedom to produce complex structures that can alter over time. Because of their intricate geometries, the printed items may now interact with their surroundings and have enhanced strength. This opens new possibilities in terms of self-healing, adaptation, and improved medical functions [139–141]. The third method involves intentionally inducing stress mismatching between layers, either during the design phase or during the printing process itself.

By carefully modifying the design and printing parameters, complex mechanical changes can be achieved when the stimulus is applied, effectively releasing the accumulated mechanical pressures [196]. Polyurethane was employed as a material for designing shape memory polymers that exhibit a time-dependent change, triggering cellular response and morphological transformations [198]. Another approach discussed in [199] involves the development of a thermally responsive theragripper using biocompatible and biodegradable poly materials. This theragripper is designed to deliver drugs through its layers and pores. Four-dimensional printing has been implemented using various DIW fabrication methods, making it accessible and compatible with a wide range of materials. The use of 4D printing has enabled the creation of smart, multi-material designs that are highly flexible, deformable, and efficient, with a self-enhancing capability. Numerous examples demonstrate the ability to achieve temporary and recovered morphologies in 4D-printed structures through diverse printing techniques, as illustrated in Figure 13b [200]. In [201–203], researchers developed more sophisticated structures through the combination of 4Dp materials. For example, these materials were made of nylon laminated with carbon fibre. They presented a broad range of frequency BW and radiation characteristics that can be utilised for 5G flexible and wearable devices. 4D printing is an innovation of a current 3D-printing method by incorporating time. However, 4D shapes can be deformed along with time upon the stimuli when they are exposed. Four-dimensional materials can respond to water, heat, magnetic and electric forces, light, and stress, as well as strain, pressure, etc.

The research presented in [204] introduces a novel hybrid engineering process called 5D printing (5Dp), as depicted in Figure 14a, which combines additive and subtractive manufacturing methods. This technique enables the printing of intricate structures with three movement axes (dimension: x, y, and z) and two rotational axes. Five-dimensional printing has found applications in creating artificial bones for surgical procedures and food packaging. The printer utilises a printing head and two rotational printing platforms to accommodate the curved surface of bones and provide enhanced strength to the implantable bone structures. Comparative analysis shows that 5D printing is four times stronger than traditional 3D printing and can withstand greater pressure. This capability makes 5D printing suitable for complex surgical components used in surgery, surgical planning, teaching, and learning. Additionally, 5D printing has potential applications in the food-packaging industry for creating intricate boxes, tableware, and other complex designs. Figure 14b showcases the ability of 5D printing to design sophisticated and curved shapes. The versatility of 5D printing extends to automotive and medical applications, as highlighted in [205]. The equipment utilised in medical research needs to be stronger and have more intricate shapes. Five-dimensional printing can produce any curved surface, which can be employed in automotive and medical devices. For instance, a concave type of cap cannot be 3Dp; it requires a complex design, numerous fillers, and support. Five-dimensional printing facilitates printing by having the ability to print curved layers. Soon, 5Dp will surpass all 3D and 4D printing methods. Five-dimensional printing is a new method to design several orthodontic components, crowns, implanted devices, and accessories used in healthcare and dental surgery. Furthermore, 5Dp devices will be more durable than 3D- and 4D-based printing devices. The emerging technology of 5Dp holds significant promise for creating curved shape designs in various fields, including dentistry. Applications such as dentures, artificial bones, stents, and complex dental tools can greatly benefit from 5Dp's ability to produce intricate and custom-shaped objects. Moreover, this innovative technology is expected to facilitate the development of wearable sensors for dental applications. With its unique capabilities, 5Dp is poised to enable the creation of complex wearable instruments that can enhance dental diagnostics and treatment. As a result, we can anticipate a wide range of successful applications for 5Dp in the field of dentistry and beyond.

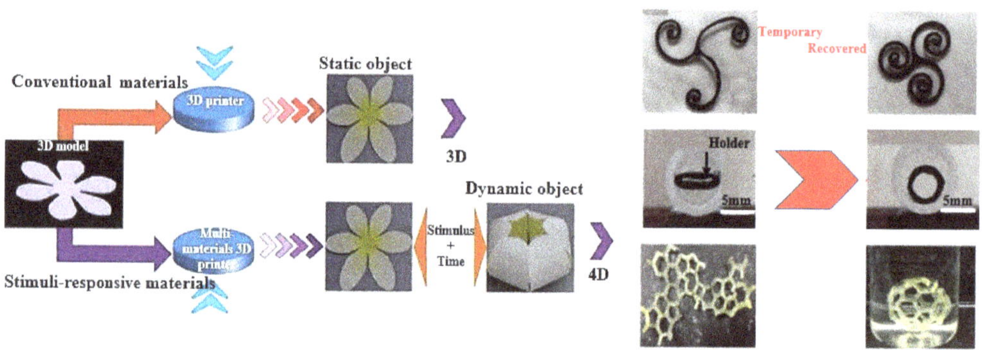

Figure 13. (**a**) 3Dp and 4Dp [197], and (**b**) temporary and recovered shapes of 4Dp shapes [200].

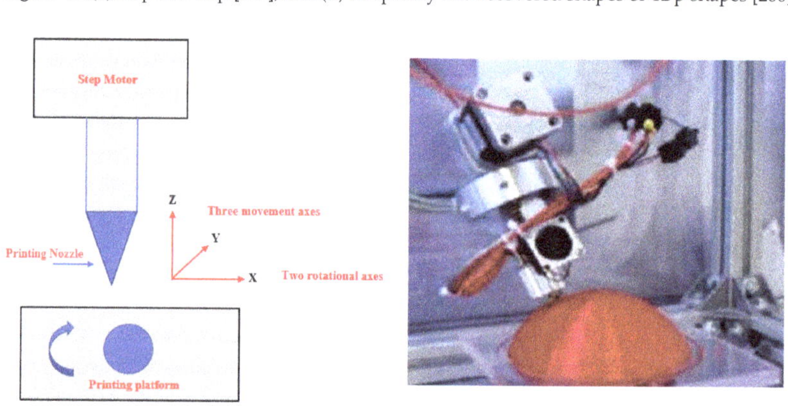

Figure 14. (**a**) Schematic diagram of 5D printing [204], and (**b**) printing of cap using 5Dp [204].

A recent study conducted in Greece introduced the concept of 6D printing (6Dp) in the literature. Figure 15 showcases the integration of 4Dp and 5Dp technologies within the 6Dp process. The study suggests that 6Dp will utilise five dimensions to generate designs that possess intelligent properties and the ability to change shape in response to environmental stimuli. By leveraging the advancements made in 5Dp, the development of smart materials has paved the way for the creation of 6Dp products with unique and intelligent behaviour [206]. Moreover, the utilisation of 6Dp in the food industry holds promise for the creation of robust and highly responsive materials. In orthopaedics, 6Dp technology can be employed to develop smart casts that offer innovative solutions for treating conditions like clubfoot and bone fractures, including congenital talipes equinovarus. The utilisation of smart materials (SMs) in 6Dp technology allows for the precise application of mechanical loads, facilitating optimal alignment of fractured forearms or malleoli in orthopaedic applications. The smart cast can adapt its shape based on the presence or absence of oedema in the affected limb. This approach offers significant advantages, including the potential to eliminate the need for serial casting and the ability to customise the cast to gently guide the foot into the correct position for clubfoot treatment. Furthermore, the inherent flexibility of the 6Dp process, along with efficient setup configurations, may lead to reduced material consumption and faster processing times [204].

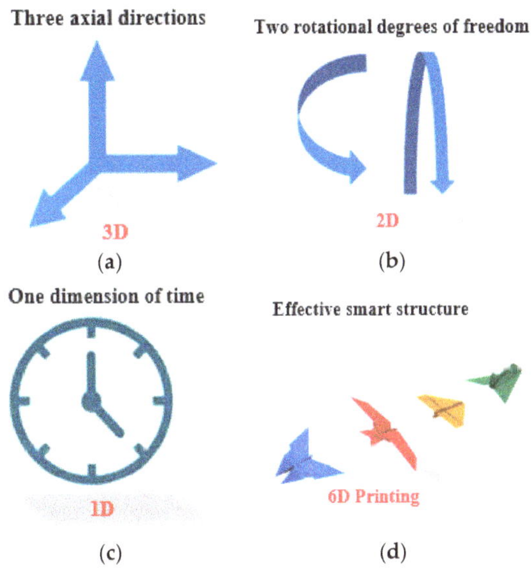

Figure 15. 1D to 6D printing method in the visual (**a**–**d**): dimensionality clarification [205,206].

The concept for the new printing method, which combines the 4Dp and 5Dp, is depicted in Figure 15 [205–207]. To summarise, the emerging technologies of 4D, 5D, and 6D printing are currently being explored by researchers to enhance the design and production of intelligent structures using AM techniques. The integration of multi-dimensional capabilities with 4D printing offers the potential to create highly advanced smart objects, leading to significant advancements in medical and orthopaedic applications. The continuous investigation and improvement of these techniques will contribute to the overall impact of AM methods in various domains, benefiting society. Six-dimensional printing is a relatively new area of AM; it is anticipated to expedite the research into finding complex materials. It produces lighter, stronger designs that are more sensitive to their environment. However, the setup and operation costs for 6Dp are anticipated to be significant. DIW techniques will continue to advance toward diversifying the materials, architectures, and fabrication techniques. However, 6D printing offers a promising solution in various fields, such as healthcare, the food industry, and so on. In general, these fields are not fully explored and have a high potential for research as well as commercial applications. Combining conventional fabrication with newly developed DW and DIW in 4Dp, 5Dp, and 6Dp would be beneficial, as shown in Figure 16. Table 3 summarises the printed devices in the literature.

Table 3. Summarised 3D to 6D techniques and their characteristics are available in the literature.

AM Methods/Types (References)	*Equipment*	*Summarised Review (1) Equipment (2) Features (3) Limitations*
2010–2021 3Dp FDM, CJP, SLA, SLS, DLP, MJM [208–210]	3D printer	(1) Design layer-by-layer along a vertical axis (2) Provide tautness on the layer of the product and no intelligence
2013–2021 4Dp [211–214] 3DP methods+ SMP, SMA, LCE, Hydrogels	3D printer, intelligence materials	(1) Design smart products (2) Limits of layer intelligence, restricted programmable and directivity
2016–2021 5Dp [215,216] 3D CAD file, 3D Scanner, and 3D printing material	5D printers or robotic arms	(1) Design in every direction Less material and process time, (2) No intelligence, higher cost
2021~ 6Dp [215,217,218] 5Dp with smart materials	5D printers, intelligence raw material	(1) Layer in all directions, with less material and process time to design products., and easily programmable. (2) Higher setup cost and additional calibration required

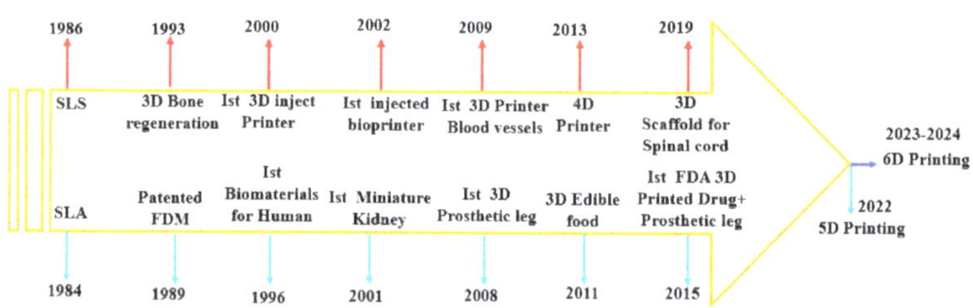

Figure 16. Previously designed 3D to 6D wearable and flexible devices in the literature review [219].

4. Human Body Models

In a body-centric device, it is necessary to assess the interactions between the human body and the device. In the initial research studies, the system was routinely evaluated with the help of tissue-equivalent phantom models. These phantoms could be numerical ones used in the simulation or physical ones used in experimental studies. These phantoms provided invaluable insight before moving to the animal or clinical study phases. This section provides insight into the state-of-the-art human body models utilised for the prediction of the body effects on the performance of wearable devices. Some examples in each area are provided for more clarification.

4.1. Type of the Human Body Models

The link between the sensor antenna and the human body is critical for wearable devices. The heterogeneous body tissues are lossy, significantly impacting the sensor antenna and its performance. The anatomical differences between patients increase the complexity of a suitable design. Many factors should be considered for designing wearable sensor antennas; for example, their performance in the presence of a human body, and their interaction with it. Many human numerical and physical phantom models exist. The first safety concern is that the level of heat created due to radiated power by the antenna may be absorbed and converted to heat, increasing the tissue temperature. Therefore, the level of the input power should be precisely controlled. Commonly, the specific absorption rate (SAR) is calculated to determine the input power level that is safe [220,221]. Several rules and regulations for heat generated by radiated waves have been implemented based on scientific evidence to prevent any adverse effects on the human body. The American National Standards Institute (ANSI), the Institute of Electrical and Electronics Engineers (IEEE), and the International Commission on Non-Ionizing Radiation Protection (ICNIRP) established standards for the maximum permissible SAR. Table 4 summarises the references which calculated the SAR values based on these standards. Standard limits for the SAR values concerning frequencies are summarised in Table 5 and are calculated as follows by [222]:

$$\mathrm{SAR} = \frac{|\mathrm{E}|^2 \sigma_e}{2\rho} \tag{1}$$

where E is the electric field strength (V/m), σ_e is the effective electrical conductivity (S/m), and ρ is the mass density of the tissue (kg/m^3).

4.1.1. Simulation Based Human Body Models

Near-field provides a strong coupling effect with the surrounding lossy media; as a result, the presence of a human body may affect the performance of a wearable device. An appropriate selection of the human body model is essential to accurately predict the interaction between the wearable device and the body. In the literature, various human body models have been shown and measured. In the numerical simulations, one may use a simple layered model of the body [223], like the one shown in Figure 17. These types of models, although they do not have the anatomical details of the body, are an effective tool for fast simulations and often are accurate enough for small devices. The advantage of a layered model is the fast simulation time that makes it possible to run optimisations for the best performance [224]. It is expected that after the simple simulations, a more realistic body model or an experimental study will follow to ensure the accuracy of the results. A good review of body phantoms used for the simulations and experimental validations in WBAN systems is presented in [225]. Multiple dielectric values of the biological tissues of the human body are briefly explained in [226].

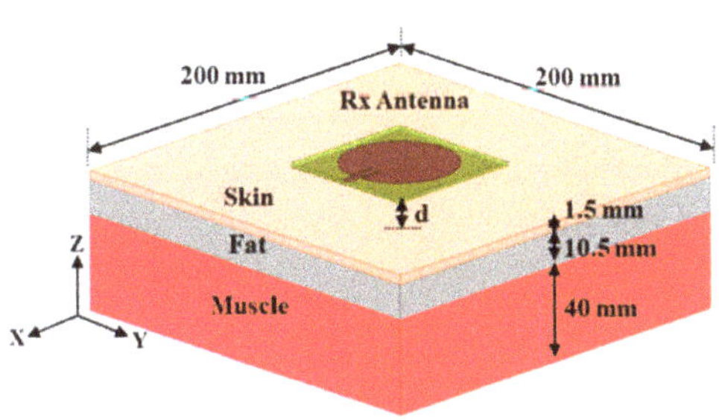

Figure 17. Layered model of the human body tissues used in the literature [227].

Table 4. A summary of SAR values calculations in the literature.

References	Models	Body Parts	Frequency (MHz)
[228,229]	Adult, adult	Head	900
[230]	Pregnant woman/fetus in the uterus	Torso/whole body	900, 2100, 2450, 2600
[231]	Adult	Whole body	98 to 2450
[232]	Adult	Head/torso	850, 900, 2100, 2600, 5100
[233]	Adult phantom	Brain/eye/skin	2300 to 2400
[234]	Adult/SAM	Head/brain/muscle	900, 2500, 3500
[235]	Adult	Brain/skin/fat/bone	900
[236]	Adult	Head/skin/bone	900, 1800
[237]	Adult, child	Whole body/head/torso	20 to 2400
[238]	Adult (male), child (Boy, girl)	Head/brain/eye	900, 1800
[239]	Adult (male, female), child	Head/brain	900, 1800, 2100, 2400
[240]	Adult, child	Head/fat/skin/bone/brain/eye	900, 800, 1800, 2100

Table 5. SAR limits for healthcare applications (W/Kg) [217,219].

Standard	Cube Mass (W/kg)	Whole-Body Average	Head and Trunk	Limb
ICNIRP	10 g	0.08	2	4
IEEE	1 g	2	1.6	4

Many factors, including application, computation time, and resource constraints, influence the choice of body models. It is usual to choose a theoretical or canonical phantom in the early stages of a design system for the following two reasons: to save time and computational resources. A more complex assessment phase will require realistic and voxel phantoms. Therefore, the simulation tools that use an accurate body phantom should not be underestimated. Another fact to consider is that fluctuations in the electrical properties of the body tissues may lead to changes in performance. This dependency on the body model and the type of phantom employed calls for a design that provides good isolation from the body and is not significantly affected by the body's dielectric properties [241]. Some phantom models and existing animal tissues employed in the literature are depicted in Table 6. In the literature, simple geometrical shapes are utilised to evaluate the antenna performance in human body models. For example, rectangular cuboids are commonly used phantom models because they are simple and planar [242]. Cylindrical [243] and spherical phantoms [244] are often used to model the arm [245], neck [246], head [247], or eyeball [248]. In [249], ellipsoid shapes are used to represent the trunk. In [250], semi-spherical phantoms represented the breast models and were used to model hyperthermia treatment devices. In [251], canonical forms were used to model the entire human body. In [252], different geometries were used to model different parts of a pregnant female body. Various important factors in the design of wearable sensor antennas include the shape of the phantom models, the distance between the phantom surface and the sensor, the orientation of the sensor to the body surface, the frequency bandwidth, temperature, and tissue.

4.1.2. Voxel Phantoms

Voxel phantoms represent more realistic shapes and arrangements of biological tissues. Various medical imaging systems have been used to provide such realistic human-body phantoms. One of these models is named "Hugo", which was based on the visible human project in 1994, and it was the first accessible voxel human model, as seen in Figure 18a [253]. Later, several human body models were created that included more high-resolution representations of the entire body. The Japanese male and female models [254] and the virtual family models [255] are illustrated in Figure 18b. Due to the complexity of the model that requires computational resources, whenever possible, it is common practice to restrict the body model to cut a specific section of the voxel model. This is especially possible when the device is small and is utilised outside of the body, such as mobile phones, microwave hyperthermia devices, or microwave imaging setups, as well as small wearable sensors [256]. These phantom models offer tissue data about the desired parts needed for applications. For the SAR computation of mobile phones, specific anthropomorphic mannequin (SAM) phantoms are used [257]. A volume shaped like a human head filled with material resembling the head tissues' average dielectric characteristics is used to model SAM phantoms. Realistic voxel breast models were employed in [258] for modelling microwave hyperthermia breast cancer treatment, as seen in Figure 18c. Various types of breasts, based on the American College of Radiology categorisation, were considered. The breast models may also be utilised for microwave breast cancer imaging [253]. Recently, there has been some interest in improving personalised treatment by creating patient-specific body phantom models. Furthermore, commercial electromagnetic solvers offer integrated modules that reconstruct 3D human models from magnetic resonance imaging (MRI) or computed tomography (CT) scan data. For instance, Sim4Life-based supports the integration of the medical image segmentation (MIS) tool Set (iSEG), as well as the ALBA 3D Seg-based module, as shown in Figure 18d [11–14,30,259–261], which was developed to integrate into the CST microwave studio software to design button antenna system [262]. The reconstruction of the 3D human models enhances the patient-specific simulations for real wearable applications. However, these computational human phantoms (CHPs) and their uses have achieved tremendous advancements in recent decades [262]. For example, CHPs characteristics enable the analysis of variations in the performance of wearable antennas and sensors due to the body's movement and help fabricate the wearable structures accurately [263–265].

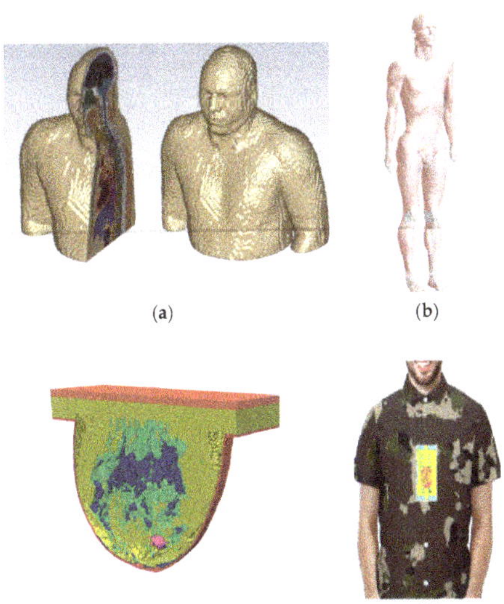

(a) (b) (c) (d)

Figure 18. Human phantoms: (**a**) the Hugo [253], (**b**) the duke (virtual family) [255], (**c**) the breast model [266], and (**d**) button antenna test on 3D human body model [260].

4.1.3. Experimental Phantoms

To validate the performance of fabricated wearable devices, experimental phantoms have been employed. These phantoms can be constructed using simple materials or derived from ex vivo or in vitro tissue. During measurements, the electrical and thermal properties of body models are influenced by the materials utilised, underscoring the significance of understanding these material properties. Various phantom shapes, including the human body head, abdomen, torso, and other anatomical regions, are utilised in measurements to replicate real-world scenarios [265].

4.1.4. Man-Made Phantoms

Man-made simple phantoms are created using chemical materials and may be liquid-based or gel-based. A simple liquid-based phantom may be created using saline water and salt [267,268]. Sometimes, a cavity is filled with a homogeneous lossy medium. This is a phantom used for various standard testing that may be utilised for wearable or implanted sensing devices, or a polyvinyl alcohol (PVA) type phantom was utilized, illustrated in Figure 19a [269]. Various recipes can be found in the literature. The composition of gel-based tissue phantoms can be tailored to replicate specific human tissues, such as muscle, adipose tissue, or breast tissue [270]. By adjusting the concentration of gel components and incorporating additives, the electrical conductivity and permittivity of the phantoms can be controlled to closely resemble those of real human tissues. In [271], a method to create various breast tissue-mimicking phantoms is presented. Schmid and Partner Engineering AG (SPEAG) designed a body phantom and a head phantom model for electromagnetic evaluations (POPEYE), which covers a frequency range of 300 MHz to 6 GHz, mimic a real situation, as shown in Figure 19b [272], and Figure 19c [273]. Creating a real body phantom with anatomical details is a complex process due to its complexity, which includes heterogeneous media with dispersive materials that exhibit changes in the dielectric values over a wide range of frequency bands.

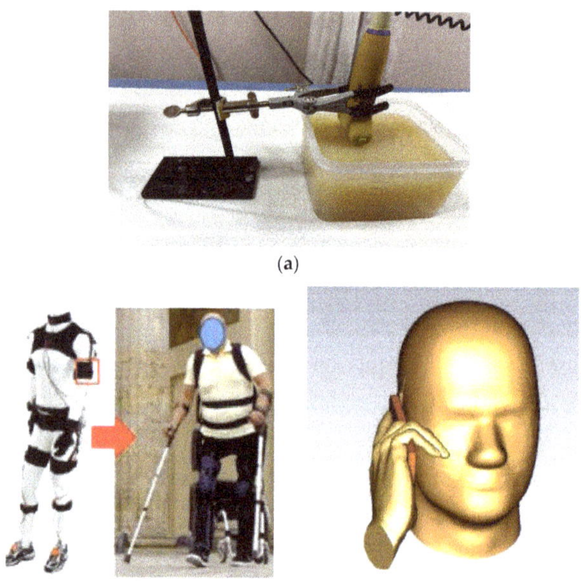

(a) (b) (c)

Figure 19. Human models: (**a**) multi-layers phantom [269], (**b**) device under test using human body with movable parts [272], and (**c**) SAM head [273].

4.1.5. Ex-Vivo Animal Phantoms

The ex vivo phantoms often utilise parts of animal bodies or minced meat. Several examples of the applications of such models can be found; for example, they have been used in validating microwave hyperthermia [274], microwave ablation [275], and implanted biotelemetry [276] devices. An example is shown in Figure 20, stress monitoring system (WSMS) that use multi-sensors system. Some studies have used biopsy tissue samples from human patients. One consideration is that after the tissue is excised from the body, the dielectric properties may change [274–276]. This variation should be considered in the assessment of the performance of the device. Preclinical testing for developing innovative treatments for diseases involves a combination of in vitro cell examinations and in vivo animal studies. However, it is important to note that outcomes observed in vitro often do not accurately reflect those observed in vivo. This might be brought on by the absence of the complex natural bone environment in vitro. A paradigm where cells are kept in their natural three-dimensional habitat is offered by ex vivo bone explant cultures.

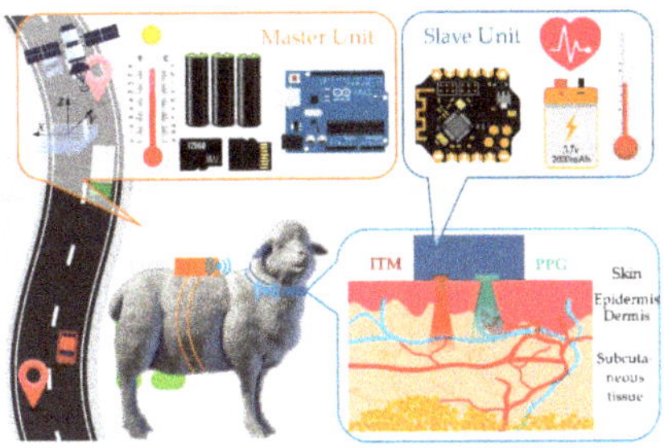

Figure 20. Examples of ex-vivo testing model [274].

Table 6. Different types of human body models.

Phantom Models References	This Table Summarises the Man-Made Models Used in the Literature
Realistic model [264]	Visible human project, Japanese male and female, virtual family, CST Microwave Studio voxel family, SAM phantoms, realistic breast phantom
Liquid phantom models [55,76,277]	Medium of the canonical shield, liquid, and gel, multilayers
Animal tissue [278,279]	Human body models, Minced meat, minced pork, piece of pork, porcine abdominal tissues, porcine eye, rat skin, chicken breast

5. Conclusions

In this manuscript, a comprehensive review of emerging technologies for designing wearable and flexible sensor devices utilised in WBAN is presented. We reviewed up-to-date design systems in WBAN systems, which can enable the next generation of wearable and flexible sensor devices. Advanced wearable sensors for many applications in healthcare, self-energy, and smart technologies have been made possible by recent advancements in biodegradable, flexible, and stretchable materials, including MXene ink and multimodal e-skin sensor devices. In the next step, various DW and DIW fabrication methods were reviewed, with examples provided. These fabrication methods included 3D- to 6D-based. These methods utilise a wide range of substrates and conductive materials to design complex structures across diverse domains. Next, to evaluate the effectiveness and performance of wearable and flexible sensor devices, an accurate body phantom model is essential at an earlier stage of the designing process. However, various voxel body and corresponding tissue models were discussed and presented, with examples provided for clarity. By incorporating these requirements, this manuscript offers a unique view that simultaneously considers three dynamic developments that have not been extensively studied together. Currently, wearable, and flexible sensor devices face key challenges related to design size, power supply, durability, and biocompatibility, as well as needed improvement in manufacturing processes for fully flexible and stretchable sensor devices. SAR, which is regulated to prevent dangerous exposure, assesses the rate at which the human body absorbs energy from electromagnetic fields (EMF) in wearable devices. Millimetre waves, which are now increasingly used in technologies like 5G, raise additional concerns. However, there is the possibility that the next-generation healthcare sensor devices will be small enough that they will connect with the terahertz (THz) frequency and will be used for sensing applications and cancer detection, etc. However, the researchers have shown serious concerned about the adverse health effects to use radiofrequency (RF) fields above 6 GHz, especially for the 5 G mobile phone network [280,281].

Author Contributions: S.M.A. prepared the final manuscript, and it was edited and reviewed by S.N. and Z.U.K., including S.A. (Saeed Alzahrani) and S.A. (Saad Alharbi), M.A. and R.A. framing the structure of the manuscript, conducting literature reviews and shared related materials, organizing, and interpreting relevant findings, drafting sections of the manuscript, reviewing, and providing feedback for revisions. All authors have read and agreed to the published version of the manuscript.

Funding: This research received no external funding.

Acknowledgments: The work of Zia Ullah Khan was supported by the 2021–2025 National Measurement System Programme of the UK government's Department for Science, Innovation and Technology (DSIT), under Science Theme Reference EMT24 of that Programme.

Conflicts of Interest: The authors declared no conflicts of interest.

References

1. Brophy, K.; Davies, S.; Olenik, S.; Çotur, Y.; Ming, D.; Van Zalk, N.; O'Hare, D.; Guder, F.; Yetisen, A.K. *The Future of Wearable Technologies*; Imperial College London: London, UK, 2021. [CrossRef]
2. Wang, Z.; Yi, N.; Zheng, Z.; Zhou, J.; Zhou, P.; Zheng, C.; Chen, H.; Shen, G.; Weng, M. Self-powered and degradable humidity sensors based on silk nanofibers and its wearable and human–machine interaction applications. *Chem. Eng. J.* **2024**, *497*, 154443. [CrossRef]
3. Mohamadbeigi, N.; Shooshtari, L.; Fardindoost, S.; Vafaiee, M.; Zad, A.I.; Mohammadpour, R. Self-powered triboelectric nanogenerator sensor for detecting humidity level and monitoring ethanol variation in a simulated exhalation environment. *Sci. Rep.* **2024**, *14*, 1562. [CrossRef] [PubMed]
4. Dai, B.; Gao, C.; Xie, Y. Flexible wearable devices for intelligent health monitoring. *View* **2022**, *3*, 20220027. [CrossRef]
5. Tu, J.; Wang, M.; Li, W.; Su, J.; Li, Y.; Lv, Z.; Li, H.; Feng, X.; Chen, X. Electronic skins with multimodal sensing and perception. *Soft Sci.* **2023**, *3*, 25. [CrossRef]
6. Zheng, Y.; He, Z.; Gao, Y.; Liu, J. Direct desktop printed-circuits-on-paper flexible electronics. *Sci. Rep.* **2013**, *3*, srep01786. [CrossRef]
7. Zhang, Y.; Shi, G.; Qin, J.; Lowe, S.E.; Zhang, S.; Zhao, H.; Zhong, Y.L. Recent Progress of Direct Ink Writing of Electronic Components for Advanced Wearable Devices. *ACS Appl. Electron. Mater.* **2019**, *1*, 1718–1734. [CrossRef]
8. Jiang, X.; Napier, C.; Hannigan, B.; Eng, J.J.; Menon, C. Estimating vertical ground reaction force during walking using a single inertial sensor. *Sensors* **2020**, *20*, 4345. [CrossRef]
9. Nate, K.; Tentzeris, M.M. A novel 3-D printed loop antenna using flexible NinjaFlex material for wearable and IoT applications. In Proceedings of the 2015 IEEE 24th Conference on Electrical Performance of Electronic Packaging and Systems, EPEPS 2015, San Jose, CA, USA, 25–28 October 2015; Institute of Electrical and Electronics Engineers Inc.: Piscataway, NJ, USA, 2015; pp. 171–174. [CrossRef]

10. Jun, S.; Sanz-Izquierdo, B.; Summerfield, M. UWB antenna on 3D printed flexible substrate and foot phantom. In Proceedings of the 2015 Loughborough Antennas & Propagation Conference (LAPC), Loughborough, UK, 2–3 November 2015; Institute of Electrical and Electronics Engineers Inc.: Piscataway, NJ, USA, 2015; pp. 1–5. [CrossRef]
11. Mohamadzade, B.; Hashmi, R.M.; Simorangkir, R.B.V.B.; Gharaei, R.; Rehman, S.U.; Abbasi, Q.H. Recent advances in fabrication methods for flexible antennas in wearable devices: State of the art. *Sensors* **2019**, *19*, 2312. [CrossRef]
12. Baeg, K.; Lee, J. Flexible Electronic Systems on Plastic Substrates and Textiles for Smart Wearable Technologies. *Adv. Mater. Technol.* **2020**, *5*, 2000071. [CrossRef]
13. Iqbal, S.M.A.; Mahgoub, I.; Du, E.; Leavitt, M.A.; Asghar, W. Advances in healthcare wearable devices. *NPJ Flex. Electron.* **2021**, *5*, 9. [CrossRef]
14. Wanekaya, A.K.; Chen, W.; Myung, N.V.; Mulchandani, A. Nanowire-Based Electrochemical Biosensors. *Electroanalysis* **2006**, *18*, 533–550. [CrossRef]
15. Park, J.; Hwang, J.C.; Kim, G.G.; Park, J. Flexible electronics based on one-dimensional and two-dimensional hybrid nanomaterials. *InfoMat* **2020**, *2*, 33–56. [CrossRef]
16. Shen, Z.; Yi, J.; Li, X.; Lo, M.H.P.; Chen, M.Z.Q.; Hu, Y.; Wang, Z. A soft stretchable bending sensor and data glove applications. *Robot. Biomimetics* **2016**, *3*, 22. [CrossRef]
17. Kim, S.; Amjadi, M.; Lee, T.-I.; Jeong, Y.; Kwon, D.; Kim, M.S.; Kim, K.; Kim, T.-S.; Oh, Y.S.; Park, I. Wearable, Ultrawide-Range, and Bending-Insensitive Pressure Sensor Based on Carbon Nanotube Network-Coated Porous Elastomer Sponges for Human Interface and Healthcare Devices. *ACS Appl. Mater. Interfaces* **2019**, *11*, 23639–23648. [CrossRef]
18. Su, X.; Borayek, R.; Li, X.; Herng, T.S.; Tian, D.; Lim, G.J.H.; Wang, Y.; Wu, J.; Ding, J. Integrated wearable sensors with bending/stretching selectivity and extremely enhanced sensitivity derived from agarose-based ionic conductor and its 3D-shaping. *Chem. Eng. J.* **2020**, *389*, 124503. [CrossRef]
19. Xia, Y.; Khamis, M.; Fernandez, F.A.; Heidari, H.; Butt, H.; Ahmed, Z.; Wilkinson, T.; Ghannam, R. State-of-the-Art in Smart Contact Lenses for Human–Machine Interaction. *IEEE Trans. Hum.-Mach. Syst.* **2023**, *53*, 187–200. [CrossRef]
20. Haque, R.; Islam, R.; Bassiri, Z.; Sazonov, E.; Martelli, D.; Shen, X. Smart Lacelock: A Novel Shoelace Tensioning Device for Human Motion Sensing. *IEEE Sens. J.* **2022**, *22*, 18349–18358. [CrossRef]
21. Carneiro, M.R.; Majidi, C.; Tavakoli, M. Multi-Electrode Printed Bioelectronic Patches for Long-Term Electrophysiological Monitoring. *Adv. Funct. Mater.* **2022**, *32*, 2205956. [CrossRef]
22. Atoche-Ensenat, R.; Perez, E.; Hernandez-Benitez, A.; Balam, A.; Estrada-Lopez, J.J.; Vazquez-Castillo, J.; Aviles, F.; Castillo-Atoche, A. A Smart Tactile Sensing System Based on Carbon Nanotube/Polypropylene Composites for Wearable Applications. *IEEE Sens. J.* **2023**, *23*, 2948–2955. [CrossRef]
23. Dai, Z.; Lei, M.; Ding, S.; Zhou, Q.; Ji, B.; Wang, M.; Zhou, B. Durable superhydrophobic surface in wearable sensors: From nature to application. *Exploration* **2023**, *4*, 20230046. [CrossRef]
24. Cai, X.; Xiao, Y.; Zhang, B.; Yang, Y.; Wang, J.; Chen, H.; Shen, G. Surface Control and Electrical Tuning of MXene Electrode for Flexible Self-Powered Human Machine Interaction. *Adv. Funct. Mater.* **2023**, *33*, 2304456. [CrossRef]
25. Wagih, M.; Balocchi, L.; Benassi, F.; Carvalho, N.B.; Chiao, J.-C.; Correia, R.; Costanzo, A.; Cui, Y.; Georgiadou, D.; Gouveia, C.; et al. Microwave-Enabled Wearables: Underpinning Technologies, Integration Platforms, and Next-Generation Roadmap. *IEEE J. Microw.* **2022**, *3*, 193–226. [CrossRef]
26. Nan, X.; Wang, X.; Kang, T.; Zhang, J.; Dong, L.; Dong, J.; Xia, P.; Wei, D. Review of Flexible Wearable Sensor Devices for Biomedical Application. *Micromachines* **2022**, *13*, 1395. [CrossRef]
27. Lehmhus, D.; Wuest, T.; Wellsandt, S.; Bosse, S.; Kaihara, T.; Thoben, K.-D.; Busse, M. Cloud-based automated design and additive manufacturing: A usage data-enabled paradigm shift. *Sensors* **2015**, *15*, 32079–32122. [CrossRef]
28. Khan, M.U.A.; Raad, R.; Tubbal, F.; Theoharis, P.I.; Liu, S.; Foroughi, J. Bending analysis of polymer-based flexible antennas for wearable, general IOT applications: A review. *Polymers* **2021**, *13*, 357. [CrossRef]
29. Zhao, Z.; Soni, S.; Lee, T.; Nijhuis, C.A.; Xiang, D. Smart Eutectic Gallium–Indium: From Properties to Applications. *Adv. Mater.* **2023**, *35*, e2203391. [CrossRef] [PubMed]
30. Ali, U.; Ullah, S.; Kamal, B.; Matekovits, L.; Altaf, A. Design, Analysis and Applications of Wearable Antennas: A Review. *IEEE Access* **2023**, *11*, 14458–14486. [CrossRef]
31. Kazanskiy, N.L.; Хонина, С.Н.; Ali Butt, M. Smart Contact Lenses—A Step towards Non-Invasive Continuous Eye Health Monitoring. *Biosensors* **2023**, *13*, 933. [CrossRef] [PubMed]
32. Simpson, R.; Teixeira, A.; Almonacid, S. Advances with intelligent on-line retort control and automation in thermal processing of canned foods. *Food Control* **2007**, *18*, 821–833. [CrossRef]
33. Centenaro, M.; Vangelista, L.; Zanella, A.; Zorzi, M. Long-range communications in unlicensed bands: The rising stars in the IoT and smart city scenarios. *IEEE Wirel. Commun.* **2016**, *23*, 60–67. [CrossRef]
34. Wang, Z.; Yang, Z.; Dong, T. A Review of wearable technologies for elderly care that can accurately track indoor position, recognize physical activities and monitor vital signs in real time. *Sensors* **2017**, *17*, 341. [CrossRef] [PubMed]

35. Patron, D.; Mongan, W.; Kurzweg, T.P.; Fontecchio, A.; Dion, G.; Anday, E.K.; Dandekar, K.R. On the Use of Knitted Antennas and Inductively Coupled RFID Tags for Wearable Applications. *IEEE Trans. Biomed. Circuits Syst.* **2016**, *10*, 1047–1057. [CrossRef]
36. Xu, L.D.; He, W.; Li, S. Internet of things in industries: A survey. *IEEE Trans. Ind. Inform.* **2014**, *10*, 2233–2243. [CrossRef]
37. Tseng, K.C.; Lin, B.-S.; Liao, L.-D.; Wang, Y.-L. Development of a wearable mobile electrocardiogram monitoring system by using novel dry foam electrodes. *IEEE Syst. J.* **2014**, *8*, 900–906. [CrossRef]
38. De, D.; Bharti, P.; Das, S.K.; Chellappan, S. Small Wearable Internet Multimodal Wearable Sensing for Fine-Grained Activity Recognition in Healthcare. 2015. Available online: https://www.computer.org/internet/ (accessed on 25 January 2024).
39. Pierleoni, P.; Belli, A.; Palma, L.; Pellegrini, M.; Pernini, L.; Valenti, S. A High Reliability Wearable Device for Elderly Fall Detection. *IEEE Sens. J.* **2015**, *15*, 4544–4553. [CrossRef]
40. Setz, C.; Arnrich, B.; Schumm, J.; La Marca, R.; Tröster, G.; Ehlert, U. Discriminating stress from cognitive load using a wearable eda device. *IEEE Trans. Inf. Technol. Biomed.* **2010**, *14*, 410–417. [CrossRef] [PubMed]
41. Garbarino, M.; Lai, M.; Tognetti, S.; Picard, R.; Bender, D. Empatica E3—A wearable wireless multi-sensor device for real-time computerized biofeedback and data acquisition. In Proceedings of the 2014 4th International Conference on Wireless Mobile Communication and Healthcare—Transforming Healthcare Through Innovations in Mobile and Wireless Technologies (MOBI-HEALTH), Athens, Greece, 3–5 November 2014; Institute for Computer Sciences, Social Informatics and Telecommunications Engineering (ICST): Portland, ME, USA, 2014. [CrossRef]
42. Fatihah, S.N.; Dewa, G.R.R.; Park, C.; Sohn, I. Self-Optimizing Bluetooth Low Energy Networks for Industrial IoT Applications. *IEEE Commun. Lett.* **2022**, *27*, 386–390. [CrossRef]
43. Institute of Electrical and Electronics Engineers; Massachusetts Institute of Technology; Media Laboratory; Lincoln Laboratory. *2013 IEEE International Conference on Body Sensor Networks, Cambridge, MA, USA, 6–9 May 2013*; MIT Media Lab: Cambridge, MA, USA, 2013.
44. Li, K.; Warren, S. A wireless reflectance pulse oximeter with digital baseline control for unfiltered photoplethysmograms. *IEEE Trans. Biomed. Circuits Syst.* **2012**, *6*, 269–278. [CrossRef] [PubMed]
45. Guo, D.; Tay, F.E.; Xu, L.; Yu, L.; Nyan, M.; Chong, F.; Yap, K.; Xu, B. A long-term wearable vital signs monitoring system using BSN. In Proceedings of the 11th EUROMICRO Conference on Digital System Design Architectures, Methods and Tools, DSD 2008, Parma, Italy, 3–5 September 2008; pp. 825–830. [CrossRef]
46. Mendelson, Y.; Duckworth, R.J.; Comtois, G. A Wearable Reflectance Pulse Oximeter for Remote Physiological Monitoring. In Proceedings of the 2006 International Conference of the IEEE Engineering in Medicine and Biology Society, New York, NY, USA, 30 August–3 September 2006.
47. Alsenwi, M.; Yaqoob, I.; Pandey, S.R.; Tun, Y.K.; Bairagi, A.K.; Kim, L.W.; Hong, C.S. Towards Coexistence of Cellular and WiFi Networks in Unlicensed Spectrum: A Neural Networks Based Approach. *IEEE Access* **2019**, *7*, 110023–110034. [CrossRef]
48. *International Conference on Mechatronic Sciences, Proceedings/2013 International Conference on Mechatronic Sciences, Electric Engineering and Computer (MEC), Shenyang, China, 20–22 December 2013*; Institute of Electrical and Electronics Engineers (IEEE): Piscataway, NJ, USA, 2014.
49. Mahmood, S.N.; Ishak, A.J.; Ismail, A.; Soh, A.C.; Zakaria, Z.; Alani, S. ON-OFF Body Ultra-Wideband (UWB) Antenna for Wireless Body Area Networks (WBAN): A Review. *IEEE Access* **2020**, *8*, 150844–150863. [CrossRef]
50. Lu, Y.; Zhao, G.; Su, F. Adaptive Ant-based Dynamic Routing Algorithm. In Proceedings of the Fifth World Congress on Intelligent Control and Automation, Hangzhou, China, 15–19 June 2004. [CrossRef]
51. Fang, Q.; Lee, S.-Y.; Permana, H.; Ghorbani, K.; Cosic, I. Developing a wireless implantable body sensor network in MICS band. *IEEE Trans. Inf. Technol. Biomed.* **2011**, *15*, 567–576. [CrossRef] [PubMed]
52. Mubashar, R.; Siddique, M.A.B.; Rehman, A.U.; Asad, A.; Rasool, A. Comparative performance analysis of short-range wireless protocols for wireless personal area network. *Iran J. Comput. Sci.* **2021**, *4*, 201–210. [CrossRef]
53. Shaikh, F.K.; Zeadally, S.; Exposito, E. Enabling technologies for green internet of things. *IEEE Syst. J.* **2017**, *11*, 983–994. [CrossRef]
54. Ramírez-Faz, J.; Fernández-Ahumada, L.M.; Fernández-Ahumada, E.; López-Luque, R. Monitoring of temperature in retail refrigerated cabinets applying iot over open-source hardware and software. *Sensors* **2020**, *20*, 846. [CrossRef]
55. Zhang, K.; Sun, H.; Li, X. Aging Population Spatial Distribution Discrepancy and Impacting Factor. *Sustainability* **2022**, *14*, 9528. [CrossRef]
56. Ku, M.; Hwang, J.C.; Oh, B.; Park, J.-U. Smart Sensing Systems Using Wearable Optoelectronics. *Adv. Intell. Syst.* **2020**, *2*, 1900144. [CrossRef]
57. Liu, Y.; Pharr, M.; Salvatore, G.A. Lab-on-Skin: A Review of Flexible and Stretchable Electronics for Wearable Health Monitoring. *ACS Nano* **2017**, *11*, 9614–9635. [CrossRef] [PubMed]
58. Nyein, H.Y.Y.; Bariya, M.; Tran, B.; Ahn, C.H.; Brown, B.J.; Ji, W.; Davis, N.; Javey, A. A wearable patch for continuous analysis of thermoregulatory sweat at rest. *Nat. Commun.* **2021**, *12*, 1823. [CrossRef]
59. Gao, F.; Liu, C.; Zhang, L.; Liu, T.; Wang, Z.; Song, Z.; Cai, H.; Fang, Z.; Chen, J.; Wang, J.; et al. Wearable and flexible electrochemical sensors for sweat analysis: A review. *Microsyst. Nanoeng.* **2023**, *9*, 1. [CrossRef]

60. Kennedy, D.A.; Lee, T.; Seely, D. A Comparative review of thermography as a breast cancer screening technique. *Integr. Cancer Ther.* **2009**, *8*, 9–16. [CrossRef]
61. Alexeev, V.L.; Das, S.; Finegold, D.N.; Asher, S.A. Photonic crystal glucose-sensing material for noninvasive monitoring of glucose in tear fluid. *Clin. Chem.* **2004**, *50*, 2353–2360. [CrossRef] [PubMed]
62. Kim, J.; Imani, S.; de Araujo, W.R.; Warchall, J.; Valdés-Ramírez, G.; Paixao, T.R.L.C.; Mercier, P.P.; Wang, J. Wearable salivary uric acid mouthguard biosensor with integrated wireless electronics. *Biosens. Bioelectron.* **2015**, *74*, 1061–1068. [CrossRef]
63. Arakawa, T.; Tomoto, K.; Nitta, H.; Toma, K.; Takeuchi, S.; Sekita, T.; Minakuchi, S.; Mitsubayashi, K. A Wearable Cellulose Acetate-Coated Mouthguard Biosensor for In Vivo Salivary Glucose Measurement. *Anal. Chem.* **2020**, *92*, 12201–12207. [CrossRef]
64. Guk, K.; Han, G.; Lim, J.; Jeong, K.; Kang, T.; Lim, E.-K.; Jung, J. Evolution of wearable devices with real-time disease monitoring for personalized healthcare. *Nanomaterials* **2019**, *9*, 813. [CrossRef]
65. Dias, D.; Paulo Silva Cunha, J. Wearable health devices—Vital sign monitoring, systems and technologies. *Sensors* **2018**, *18*, 2414. [CrossRef]
66. Kang, H.S.; Exworthy, M. Wearing the Future—Wearables to Empower Users to Take Greater Responsibility for Their Health and Care: Scoping Review. *JMIR mHealth uHealth* **2022**, *10*, e35684. [CrossRef] [PubMed]
67. Mukhopadhyay, S.C.; Suryadevara, N.K.; Nag, A. Wearable Sensors for Healthcare: Fabrication to Application. *Sensors* **2022**, *22*, 5137. [CrossRef]
68. Gambhir, S.S.; Ge, T.J.; Vermesh, O.; Spitler, R.; Gold, G.E. Continuous health monitoring: An opportunity for precision health. *Sci. Transl. Med.* **2021**, *13*, eabe5383. [CrossRef] [PubMed]
69. Heikenfeld, J.; Jajack, A.; Rogers, J.; Gutruf, P.; Tian, L.; Pan, T.; Li, R.; Khine, M.; Kim, J.; Wang, J.; et al. Wearable sensors: Modalities, challenges, and prospects. *Lab Chip* **2018**, *18*, 217–248. [CrossRef]
70. Quer, G.; Radin, J.M.; Gadaleta, M.; Baca-Motes, K.; Ariniello, L.; Ramos, E.; Kheterpal, V.; Topol, E.J.; Steinhubl, S.R. Wearable sensor data and self-reported symptoms for COVID-19 detection. *Nat. Med.* **2021**, *27*, 73–77. [CrossRef] [PubMed]
71. Feng, J.; Ao, H.; Cao, P.; Yang, T.; Xing, B. Flexible tactile sensors with interlocking serrated structures based on stretchable multiwalled carbon nanotube/silver nanowire/silicone rubber composites. *RSC Adv.* **2024**, *14*, 13934–13943. [CrossRef] [PubMed]
72. Jiang, W.; Majumder, S.; Kumar, S.; Subramaniam, S.; Li, X.; Khedri, R.; Mondal, T.; Abolghasemian, M.; Satia, I.; Deen, M.J. A Wearable Tele-Health System towards Monitoring COVID-19 and Chronic Diseases. *IEEE Rev. Biomed. Eng.* **2022**, *15*, 61–84. [CrossRef]
73. Guo, S.; Wu, K.; Li, C.; Wang, H.; Sun, Z.; Xi, D.; Zhang, S.; Ding, W.; Zaghloul, M.E.; Wang, C.; et al. Integrated contact lens sensor system based on multifunctional ultrathin MoS2 transistors. *Matter* **2021**, *4*, 969–985. [CrossRef] [PubMed]
74. Jiang, N.; Montelongo, Y.; Butt, H.; Yetisen, A.K. Microfluidic Contact Lenses. *Small* **2018**, *14*, 1704363. [CrossRef]
75. Qiao, Y.; Luo, J.; Cui, T.; Liu, H.; Tang, H.; Zeng, Y.; Liu, C.; Li, Y.; Jian, J.; Wu, J.; et al. Soft Electronics for Health Monitoring Assisted by Machine Learning. *Nano-Micro Lett.* **2023**, *15*, 66. [CrossRef]
76. Kim, D.; Choi, Y. Applications of Smart Glasses in Applied Sciences: A Systematic Review. *Appl. Sci.* **2021**, *11*, 4956. [CrossRef]
77. Xiang, L.; Zeng, X.; Xia, F.; Jin, W.; Liu, Y.; Hu, Y. Recent Advances in Flexible and Stretchable Sensing Systems: From the Perspective of System Integration. *ACS Nano* **2020**, *14*, 6449–6469. [CrossRef] [PubMed]
78. Wong, W.F.; Ang, K.P.; Sethi, G.; Looi, C.Y. Recent Advancement of Medical Patch for Transdermal Drug Delivery. *Medicina* **2023**, *59*, 778. [CrossRef]
79. Mondal, S.; Zehra, N.; Choudhury, A.; Iyer, P.K. Wearable Sensing Devices for Point of Care Diagnostics. *ACS Appl. Bio Mater.* **2021**, *4*, 47–70. [CrossRef]
80. Wu, Z.; Cheng, T.; Wang, Z.L. Self-powered sensors and systems based on nanogenerators. *Sensors* **2020**, *20*, 2925. [CrossRef]
81. Seshadri, D.R.; Davies, E.V.; Harlow, E.R.; Hsu, J.J.; Knighton, S.C.; Walker, T.A.; Voos, J.E.; Drummond, C.K. Wearable Sensors for COVID-19: A Call to Action to Harness Our Digital Infrastructure for Remote Patient Monitoring and Virtual Assessments. *Front. Digit. Health* **2020**, *2*, 8. [CrossRef]
82. Gao, J.; Fan, Y.; Zhang, Q.; Luo, L.; Hu, X.; Li, Y.; Song, J.; Jiang, H.; Gao, X.; Zheng, L.; et al. Ultra-Robust and Extensible Fibrous Mechanical Sensors for Wearable Smart Healthcare. *Adv. Mater.* **2022**, *34*, e2107511. [CrossRef] [PubMed]
83. Ahmadzadeh, S.; Luo, J.; Wiffen, R. Review on Biomedical Sensors, Technologies and Algorithms for Diagnosis of Sleep Disordered Breathing: Comprehensive Survey. *IEEE Rev. Biomed. Eng.* **2022**, *15*, 4–22. [CrossRef] [PubMed]
84. Verma, D.; Singh, K.R.; Yadav, A.K.; Nayak, V.; Singh, J.; Solanki, P.R.; Singh, R.P. Internet of things (IoT) in nano-integrated wearable biosensor devices for healthcare applications. *Biosens. Bioelectron. X* **2022**, *11*, 100153. [CrossRef]
85. Berryhill, S.; Morton, C.J.; Dean, A.; Berryhill, A.; Provencio-Dean, N.; Patel, S.I.; Estep, L.; Combs, D.; Mashaqi, S.; Gerald, L.B.; et al. Effect of wearables on sleep in healthy individuals: A randomized crossover trial and validation study. *J. Clin. Sleep Med.* **2020**, *16*, 775–783. [CrossRef]
86. Hajian, S.; Maddipatla, D.; Narakathu, B.B.; Atashbar, M.Z. MXene-based flexible sensors: A review. *Front. Sens.* **2022**, *3*, 1006749. [CrossRef]

87. Xue, N.; Wang, C.; Liu, C.; Sun, J. Highly integrated MEMS-ASIC sensing system for intracorporeal physiological condition monitoring. *Sensors* **2018**, *18*, 107. [CrossRef] [PubMed]
88. Mucchi, L.; Jayousi, S.; Martinelli, A.; Caputo, S.; Intrieri, E.; Gigli, G.; Gracchi, T.; Mugnai, F.; Favalli, M.; Fornaciai, A.; et al. A flexible wireless sensor network based on ultra-wide band technology for ground instability monitoring. *Sensors* **2018**, *18*, 2948. [CrossRef]
89. The many faces of wearables. *Nat. Electron.* **2022**, *5*, 709. [CrossRef]
90. Powers, R.; Etezadi-Amoli, M.; Arnold, E.M.; Kianian, S.; Mance, I.; Gibiansky, M.; Trietsch, D.; Alvarado, A.S.; Kretlow, J.D.; Herrington, T.M.; et al. Smartwatch inertial sensors continuously monitor real-world motor fluctuations in Parkinson's disease. *Sci. Transl. Med.* **2021**, *13*, eabd7865. [CrossRef]
91. Sreejith, S.; Joseph, L.L.; Kollem, S.; Vijumon, V.; Ajayan, J. Biodegradable sensors: A comprehensive review. *Measurement* **2023**, *219*, 113261. [CrossRef]
92. Trafton, A. MIT engineers develop sensors for face masks that help gauge fit. *MIT News | Massachusetts Institute of Technology*, 20 October 2022. Available online: https://news.mit.edu/2022/sensors-face-masks-fit-1020?utm_source=chatgpt.com (accessed on 1 March 2024).
93. Raja, J.M.; Elsakr, C.; Roman, S.; Cave, B.; Pour-Ghaz, I.; Nanda, A.; Maturana, M.; Khouzam, R.N. Apple Watch, Wearables, and Heart Rhythm: Where do we stand? *Ann. Transl. Med.* **2019**, *7*, 417. [CrossRef]
94. Ullah, H.; Wahab, A.; Will, G.; Karim, M.R.; Pan, T.; Gao, M.; Lai, D.; Lin, Y.; Miraz, M.H. Recent Advances in Stretchable and Wearable Capacitive Electrophysiological Sensors for Long-Term Health Monitoring. *Biosensors* **2022**, *12*, 630. [CrossRef]
95. Yang, J.; Liu, Y.; Morgan, P.L. Human–machine interaction towards Industry 5.0: Human-centric smart manufacturing. *Digit. Eng.* **2024**, *2*, 100013. [CrossRef]
96. Sjoding, M.W.; Dickson, R.P.; Iwashyna, T.J.; Gay, S.E.; Valley, T.S. Racial Bias in Pulse Oximetry Measurement. *N. Engl. J. Med.* **2020**, *383*, 2477–2478. [CrossRef] [PubMed]
97. Tang, X.; Zhao, L.; Chong, J.; You, Z.; Zhu, L.; Ren, H.; Shang, Y.; Han, Y.; Li, G. 5G-based smart healthcare system designing and field trial in hospitals. *IET Commun.* **2021**, *16*, 1–13. [CrossRef]
98. Mishra, L.; Vikash; Varma, S. Seamless Health Monitoring Using 5G NR for Internet of Medical Things. *Wirel. Pers. Commun.* **2021**, *120*, 2259–2289. [CrossRef]
99. Lloret, J.; Parra, L.; Taha, M.; Tomás, J. An architecture and protocol for smart continuous eHealth monitoring using 5G. *Comput. Netw.* **2017**, *129*, 340–351. [CrossRef]
100. Mishra, A.; Agrawal, D.P. Continuous health condition monitoring by 24 × 7 sensing and transmission of physiological data over 5-G cellular channels. In Proceedings of the 2015 International Conference on Computing, Networking and Communications, ICNC 2015, Garden Grove, CA, USA, 16–19 February 2015; Institute of Electrical and Electronics Engineers Inc.: Piscataway, NJ, USA, 2015; pp. 584–590. [CrossRef]
101. Zheng, J.; Wang, Y.; Zhang, J.; Guo, W.; Yang, X.; Luo, L.; Jiao, W.; Hu, X.; Yu, Z.; Wang, C.; et al. 5G ultra-remote robot-assisted laparoscopic surgery in China. *Surg. Endosc.* **2020**, *34*, 5172–5180. [CrossRef]
102. Veletić, M.; Apu, E.H.; Simić, M.; Bergsland, J.; Balasingham, I.; Contag, C.H.; Ashammakhi, N. Implants with Sensing Capabilities. *Chem. Rev.* **2022**, *122*, 16329–16363. [CrossRef] [PubMed]
103. Bai, Y.; Meng, H.; Li, Z.; Wang, Z.L. Degradable piezoelectric biomaterials for medical applications. *MedMat* **2024**, *1*, 40–49. [CrossRef]
104. Muntoni, G.; Casula, G.A.; Montisci, G. The 3D-Printed Non-Radiating Edge Gap-Coupled Curved Patch Antenna. *IEEE Open J. Antennas Propag.* **2023**, *4*, 229–236. [CrossRef]
105. Hashemi, M.R.M.; Fikes, A.C.; Gal-Katziri, M.; Abiri, B.; Bohn, F.; Safaripour, A.; Kelzenberg, M.D.; Warmann, E.L.; Espinet, P.; Vaidya, N.; et al. A flexible phased array system with low areal mass density. *Nat. Electron.* **2019**, *2*, 195–205. [CrossRef]
106. Liu, C.; Kim, J.; Yang, D.S.; Cho, D.; Yoo, S.; Madhvapathy, S.R.; Jeong, H.; Yang, T.; Luan, H.; Avila, R.; et al. Multifunctional Materials Strategies for Enhanced Safety of Wireless, Skin-Interfaced Bioelectronic Devices. *Adv. Funct. Mater.* **2023**, *33*, 2302256. [CrossRef]
107. Kim, C.H.; Lee, D.H.; Youn, J.; Lee, H.; Jeong, J. Simple and cost-effective microfabrication of flexible and stretchable electronics for wearable multi-functional electrophysiological monitoring. *Sci. Rep.* **2021**, *11*, 14823. [CrossRef] [PubMed]
108. Jeerapan, I.; Khumngern, S. Printed Devices for Wearable Biosensors: Laboratory to Emerging Markets. *IEEE J. Flex. Electron.* **2023**, *2*, 358–365. [CrossRef]
109. Liu, S.; Xue, B.; Yan, W.; Rwei, A.Y.; Wu, C. Recent Advances and Design Strategies Towards Wearable Near-Infrared Spectroscopy. *IEEE Open J. Nanotechnol.* **2023**, *4*, 25–35. [CrossRef]
110. Ge, Y.; Taha, A.; Shah, S.A.; Dashtipour, K.; Zhu, S.; Cooper, J.M.; Abbasi, Q.H.; Imran, M.A. Contactless WiFi Sensing and Monitoring for Future Healthcare—Emerging Trends, Challenges, and Opportunities. *IEEE Rev. Biomed. Eng.* **2023**, *16*, 171–191. [CrossRef] [PubMed]

111. Li, C.; Un, K.-F.; Mak, P.-I.; Chen, Y.; Munoz-Ferreras, J.-M.; Yang, Z.; Gomez-Garcia, R. Overview of Recent Development on Wireless Sensing Circuits and Systems for Healthcare and Biomedical Applications. *IEEE J. Emerg. Sel. Top. Circuits Syst.* **2018**, *8*, 165–177. [CrossRef]
112. Alshamrani, M. IoT and artificial intelligence implementations for remote healthcare monitoring systems: A survey. *J. King Saud Univ.—Comput. Inf. Sci.* **2021**, *34*, 4687–4701. [CrossRef]
113. Zhan, Y.; Li, W.; Amirfazli, A.; Yu, S. Recent advances in shape memory superhydrophobic surfaces: Concepts, mechanism, classification, applications and challenges. *Polymer* **2022**, *256*, 125193. [CrossRef]
114. Kim, H.; Kwon, Y.; Lim, H.; Kim, J.; Kim, Y.; Yeo, W. Recent Advances in Wearable Sensors and Integrated Functional Devices for Virtual and Augmented Reality Applications. *Adv. Funct. Mater.* **2020**, *31*, 2005692. [CrossRef]
115. Lemey, S.; Agneessens, S.; Rogier, H. Wearable Smart Objects: Microwaves Propelling Smart Textiles: A Review of Holistic Designs for Wireless Textile Nodes. *IEEE Microw. Mag.* **2018**, *19*, 83–100. [CrossRef]
116. Costanzo, A.; Benassi, F.; Monti, G. Wearable, Energy-Autonomous RF Microwave Systems. *IEEE Microw. Mag.* **2022**, *23*, 24–38. [CrossRef]
117. Liu, W.; Huang, K.; Zhou, X.; Durrani, S. Next generation backscatter communication: Systems, techniques, and applications. *EURASIP J. Wirel. Commun. Netw.* **2019**, *2019*, 69. [CrossRef]
118. Torres, R.; Correia, R.; Carvalho, N.B.; Daskalakis, S.N.; Goussetis, G.; Ding, Y.; Georgiadis, A.; Eid, A.; Hester, J.; Tentzeris, M.M. Backscatter Communications. *IEEE J. Microw.* **2021**, *1*, 864–878. [CrossRef]
119. Song, C.; Ding, Y.; Eid, A.; Hester, J.G.D.; He, X.; Bahr, R.; Georgiadis, A.; Goussetis, G.; Tentzeris, M.M. Advances in Wirelessly Powered Backscatter Communications: From Antenna/RF Circuitry Design to Printed Flexible Electronics. *Proc. IEEE* **2022**, *110*, 171–192. [CrossRef]
120. Mezzanotte, P.; Palazzi, V.; Alimenti, F.; Roselli, L. Innovative RFID Sensors for Internet of Things Applications. *IEEE J. Microw.* **2021**, *1*, 55–65. [CrossRef]
121. Tedjini, S.; Karmakar, N.; Perret, E.; Vena, A.; Koswatta, R.; E-Azim, R. Hold the chips: Chipless technology, an alternative technique for RFID. *IEEE Microw. Mag.* **2013**, *14*, 56–65. [CrossRef]
122. Grosinger, J.; Pachler, W.; Bosch, W. Tag Size Matters: Miniaturized RFID Tags to Connect Smart Objects to the Internet. *IEEE Microw. Mag.* **2018**, *19*, 101–111. [CrossRef]
123. Bianco, G.M.; Occhiuzzi, C.; Panunzio, N.; Marrocco, G. A Survey on Radio Frequency Identification as a Scalable Technology to Face Pandemics. *IEEE J. Radio Freq. Identif.* **2022**, *6*, 77–96. [CrossRef]
124. Grosinger, J.; Michalowska-Forsyth, A. Space Tag. *IEEE Microw. Mag.* **2022**, *23*, 55–71. [CrossRef]
125. Wagih, M.; Weddell, A.S.; Beeby, S. Rectennas for radio-frequency energy harvesting and wireless power transfer: A review of antenna design [Antenna Applications Corner]. *IEEE Antennas Propag. Mag.* **2020**, *62*, 95–107. [CrossRef]
126. Valenta, C.R.; Durgin, G.D. Harvesting wireless power: Survey of Energy-harvester conversion efficiency in far-field, wireless power transfer systems. *IEEE Microw. Mag.* **2014**, *15*, 108–120. [CrossRef]
127. Wagih, M.; Weddell, A.S.; Beeby, S. Millimeter-Wave Power Harvesting: A Review. *IEEE Open J. Antennas Propag.* **2020**, *1*, 560–578. [CrossRef]
128. Lu, X.; Wang, P.; Niyato, D.; Kim, D.I.; Han, Z. Wireless networks with rf energy harvesting: A contemporary survey. *IEEE Commun. Surv. Tutor.* **2015**, *17*, 757–789. [CrossRef]
129. Choi, J.; Bandodkar, A.J.; Reeder, J.T.; Ray, T.R.; Turnquist, A.; Kim, S.B.; Nyberg, N.; Hourlier-Fargette, A.; Model, J.B.; Aranyosi, A.J.; et al. Soft, Skin-integrated multifunctional microfluidic systems for accurate colorimetric analysis of sweat biomarkers and temperature. *ACS Sens.* **2019**, *4*, 379–388. [CrossRef]
130. Chugh, V.; Basu, A.; Kaushik, A.; Basu, A.K. E-skin—Based advanced wearable technology for Health Management. *Curr. Res. Biotechnol.* **2023**, *5*, 100129. [CrossRef]
131. Herbert, R.; Jeong, J.-W.; Yeo, W.-H. Soft material-enabled electronics for medicine, healthcare, and human-machine interfaces. *Materials* **2020**, *13*, 517. [CrossRef] [PubMed]
132. Kim, K.; Ahn, J.; Jeong, Y.; Choi, J.; Gul, O.; Park, I. All-soft multiaxial force sensor based on liquid metal for electronic skin. *Micro Nano Syst. Lett.* **2021**, *9*, 2. [CrossRef]
133. Ping, B.; Zhou, G.; Zhang, Z.; Guo, R. Liquid metal enabled conformal electronics. *Front. Bioeng. Biotechnol.* **2023**, *11*, 1118812. [CrossRef]
134. Yu, Y.; Zhang, J.; Liu, J. Biomedical Implementation of Liquid Metal Ink as Drawable ECG Electrode and Skin Circuit. *PLoS ONE* **2013**, *8*, e58771. [CrossRef]
135. Won, P.; Jeong, S.; Majidi, C.; Ko, S.H. Recent advances in liquid-metal-based wearable electronics and materials. *iScience* **2020**, *24*, 102698. [CrossRef]
136. Jeong, Y.R.; Kim, J.; Xie, Z.; Xue, Y.; Won, S.M.; Lee, G.; Jin, S.W.; Hong, S.Y.; Feng, X.; Huang, Y.; et al. A skin-attachable, stretchable integrated system based on liquid GaInSn for wireless human motion monitoring with multi-site sensing capabilities. *NPG Asia Mater.* **2017**, *9*, e443. [CrossRef]

137. Guo, R.; Wang, X.; Yu, W.; Tang, J.; Liu, J. A highly conductive and stretchable wearable liquid metal electronic skin for long-term conformable health monitoring. *Sci. China Technol. Sci.* **2018**, *61*, 1031–1037. [CrossRef]
138. Bi, H.; Liu, J.; Kato, N. Deep Learning-Based Privacy Preservation and Data Analytics for IoT Enabled Healthcare. *IEEE Trans. Ind. Inform.* **2022**, *18*, 4798–4807. [CrossRef]
139. Guo, Z.; Shen, Y.; Wan, S.; Shang, W.-L.; Yu, K. Hybrid Intelligence-Driven Medical Image Recognition for Remote Patient Diagnosis in Internet of Medical Things. *IEEE J. Biomed. Health Inform.* **2021**, *26*, 5817–5828. [CrossRef]
140. Qiu, X.; Liu, L.; Chen, W.; Hong, Z.; Zheng, Z. Online deep reinforcement learning for computation offloading in blockchain-empowered mobile edge computing. *IEEE Trans. Veh. Technol.* **2019**, *68*, 8050–8062. [CrossRef]
141. Abdel-Basset, M.; Moustafa, N.; Hawash, H.; Razzak, I.; Sallam, K.M.; Elkomy, O.M. Federated Intrusion Detection in Blockchain-Based Smart Transportation Systems. *IEEE Trans. Intell. Transp. Syst.* **2022**, *23*, 2523–2537. [CrossRef]
142. Zhang, W.; Fan, W.; Zhang, G.; Mao, S. Learning-based joint service caching and load balancing for MEC blockchain networks. *China Commun.* **2023**, *20*, 125–139. [CrossRef]
143. Nguyen, D.C.; Ding, M.; Pathirana, P.N.; Seneviratne, A.; Li, J.; Poor, H.V. Cooperative Task Offloading and Block Mining in Blockchain-based Edge Computing with Multi-agent Deep Reinforcement Learning. *IEEE Trans. Mob. Comput.* **2021**, *22*, 2021–2037. [CrossRef]
144. Guo, Z.; Yu, K.; Kumar, N.; Wei, W.; Mumtaz, S.; Guizani, M. Deep-Distributed-Learning-Based POI Recommendation Under Mobile-Edge Networks. *IEEE Internet Things J.* **2023**, *10*, 303–317. [CrossRef]
145. Khan, S.; Lorenzelli, L.; Dahiya, R.S. Technologies for printing sensors and electronics over large flexible substrates: A review. *IEEE Sens. J.* **2015**, *15*, 3164–3185. [CrossRef]
146. Ali, S.; Sovuthy, C.; Imran, M.; Socheatra, S.; Abbasi, Q.; Abidin, Z. Recent Advances of Wearable Antennas in Materials, Fabrication Methods, Designs, and Their Applications: State-of-the-Art. *Micromachines* **2020**, *11*, 888. [CrossRef]
147. Liao, X.; Liao, Q.; Yan, X.; Liang, Q.; Si, H.; Li, M.; Wu, H.; Cao, S.; Zhang, Y. Flexible and highly sensitive strain sensors fabricated by pencil drawn for wearable monitor. *Adv. Funct. Mater.* **2015**, *25*, 2395–2401. [CrossRef]
148. Campana, A.; Cramer, T.; Simon, D.T.; Berggren, M.; Biscarini, F. Electrocardiographic recording with conformable organic electrochemical transistor fabricated on resorbable bioscaffold. *Adv. Mater.* **2014**, *26*, 3874–3878. [CrossRef] [PubMed]
149. Jung, S.; Lee, J.; Hyeon, T.; Lee, M.; Kim, D. Fabric-based integrated energy devices for wearable activity monitors. *Adv. Mater.* **2014**, *26*, 6329–6334. [CrossRef] [PubMed]
150. Kim, D.-H.; Lu, N.; Ma, R.; Kim, Y.-S.; Kim, R.-H.; Wang, S.; Wu, J.; Won, S.M.; Tao, H.; Islam, A.; et al. Epidermal Electronics. *Science* **2011**, *333*, 838–843. [CrossRef] [PubMed]
151. Sun, Y.; Rogers, J.A. Inorganic semiconductors for flexible electronics. *Adv. Mater.* **2007**, *19*, 1897–1916. [CrossRef]
152. Han, Y.L.; Liu, H.; Ouyang, C.; Lu, T.J.; Xu, F. Liquid on paper: Rapid prototyping of soft functional components for paper electronics. *Sci. Rep.* **2015**, *5*, 11488. [CrossRef]
153. Gao, Y.; Li, H.; Liu, J. Direct Writing of Flexible Electronics through Room Temperature Liquid Metal Ink. *PLoS ONE* **2012**, *7*, e45485. [CrossRef] [PubMed]
154. Tai, Y.-L.; Yang, Z.-G. Fabrication of paper-based conductive patterns for flexible electronics by direct-writing. *J. Mater. Chem.* **2011**, *21*, 5938–5943. [CrossRef]
155. Loghin, F.C.; Salmerón, J.F.; Lugli, P.; Becherer, M.; Falco, A.; Rivadeneyra, A. Optimization of a handwriting method by an automated ink pen for cost-effective and sustainable sensors. *Chemosensors* **2021**, *9*, 264. [CrossRef]
156. Hou, Z.; Lu, H.; Li, Y.; Yang, L.; Gao, Y. Direct Ink Writing of Materials for Electronics-Related Applications: A Mini Review. *Front. Mater.* **2021**, *8*, 647229. [CrossRef]
157. Loaldi, D.; Piccolo, L.; Brown, E.; Tosello, G.; Shemelya, C.; Masato, D. Hybrid process chain for the integration of direct ink writing and polymer injection molding. *Micromachines* **2020**, *11*, 509. [CrossRef]
158. Song, S.W.; Lee, S.; Choe, J.K.; Kim, N.-H.; Kang, J.; Lee, A.C.; Choi, Y.; Choi, A.; Jeong, Y.; Lee, W.; et al. Direct 2D-to-3D transformation of pen drawings. *Sci. Adv.* **2021**, *7*, eabf3804. [CrossRef]
159. Quain, E.; Mathis, T.S.; Kurra, N.; Maleski, K.; Van Aken, K.L.; Alhabeb, M.; Alshareef, H.N.; Gogotsi, Y. Direct Writing of Additive-Free MXene-in-Water Ink for Electronics and Energy Storage. *Adv. Mater. Technol.* **2019**, *4*, 1800256. [CrossRef]
160. Aabith, S.; Caulfield, R.; Akhlaghi, O.; Papadopoulou, A.; Homer-Vanniasinkam, S.; Tiwari, M.K. 3D direct-write printing of water soluble micromoulds for high-resolution rapid prototyping. *Addit. Manuf.* **2022**, *58*, 103019. [CrossRef]
161. You, R.; Liu, Y.; Hao, Y.; Han, D.; Zhang, Y.; You, Z. Laser Fabrication of Graphene-Based Flexible Electronics. *Adv. Mater.* **2020**, *32*, e1901981. [CrossRef] [PubMed]
162. Xu, Y.; Fei, Q.; Page, M.; Zhao, G.; Ling, Y.; Stoll, S.B.; Yan, Z. Paper-based wearable electronics. *iScience* **2020**, *24*, 102736. [CrossRef] [PubMed]
163. Jandyal, A.; Chaturvedi, I.; Wazir, I.; Raina, A.; Haq, M.I.U. 3D printing—A review of processes, materials and applications in industry 4.0. *Sustain. Oper. Comput.* **2022**, *3*, 33–42. [CrossRef]

164. Smart pill from MIT monitors and medicates via Bluetooth—The Engineer. The Engineer, 14 December 2018. Available online: https://www.theengineer.co.uk/content/news/smart-pill-from-mit-monitors-and-medicates-via-bluetooth (accessed on 27 March 2024).
165. Tan, H.W.; Choong, Y.Y.C.; Kuo, C.N.; Low, H.Y.; Chua, C.K. 3D printed electronics: Processes, materials, and future trends. *Prog. Mater. Sci.* **2022**, *127*, 100945. [CrossRef]
166. Lin, Y.; Chen, J.; Tavakoli, M.M.; Gao, Y.; Zhu, Y.; Zhang, D.; Kam, M.; He, Z.; Fan, Z. Printable Fabrication of a Fully Integrated and Self-Powered Sensor System on Plastic Substrates. *Adv. Mater.* **2019**, *31*, e1804285. [CrossRef] [PubMed]
167. Capabilities | Rapid Fluidics Ltd. 2016. Available online: https://www.rapidfluidics.com/capabilities (accessed on 27 March 2024).
168. Wiklund, J.; Karakoç, A.; Palko, T.; Yiğitler, H.; Ruttik, K.; Jäntti, R.; Paltakari, J. A Review on printed electronics: Fabrication methods, inks, substrates, applications, and environmental impacts. *J. Manuf. Mater. Process.* **2021**, *5*, 89. [CrossRef]
169. Nyabadza, A.; Vázquez, M.; Coyle, S.; Fitzpatrick, B.; Brabazon, D. Review of materials and fabrication methods for flexible nano and micro-scale physical and chemical property sensors. *Appl. Sci.* **2021**, *11*, 8563. [CrossRef]
170. Wang, X.; Liu, J. Recent advancements in liquid metal flexible printed electronics: Properties, technologies, and applications. *Micromachines* **2016**, *7*, 206. [CrossRef] [PubMed]
171. Ligon, S.C.; Liska, R.; Stampfl, J.; Gurr, M.; Mülhaupt, R. Polymers for 3D Printing and Customized Additive Manufacturing. *Chem. Rev.* **2017**, *117*, 10212–10290. [CrossRef]
172. Ryan, K.R.; Down, M.P.; Hurst, N.J.; Keefe, E.M.; Banks, C.E. Additive manufacturing (3D printing) of electrically conductive polymers and polymer nanocomposites and their applications. *eScience* **2022**, *2*, 365–381. [CrossRef]
173. Zhao, Y.-Q.; Peng, Z. Three-dimensional phased array antenna analysis and simulation. In Proceedings of the 2009 3rd IEEE International Symposium on Microwave, Antenna, Propagation and EMC Technologies for Wireless Communications, Beijing, China, 27–29 October 2009. [CrossRef]
174. Gannarapu, A.; Gozen, B.A. Freeze-Printing of Liquid Metal Alloys for Manufacturing of 3D, Conductive, and Flexible Networks. *Adv. Mater. Technol.* **2016**, *1*, 1600047. [CrossRef]
175. Park, Y.G.; An, H.S.; Kim, J.Y.; Park, J.U. High-resolution, reconfigurable printing of liquid metals with three-dimensional structures. *Sci. Adv.* **2019**, *5*, eaaw2844. [CrossRef] [PubMed]
176. Sarvey, T.E.; Kaul, A.; Rajan, S.K.; Dasu, A.; Gutala, R.; Bakir, M.S. Microfluidic Cooling of a 14-nm 2.5-D FPGA with 3-D Printed Manifolds for High-Density Computing: Design Considerations, Fabrication, and Electrical Characterization. *IEEE Trans. Compon. Packag. Manuf. Technol.* **2019**, *9*, 2393–2403. [CrossRef]
177. Dickey, M.D. Emerging Applications of Liquid Metals Featuring Surface Oxides. *ACS Appl. Mater. Interfaces* **2014**, *6*, 18369–18379. [CrossRef] [PubMed]
178. Zada, M.; Shah, I.A.; Basir, A.; Yoo, H. Ultra-Compact Implantable Antenna with Enhanced Performance for Leadless Cardiac Pacemaker System. *IEEE Trans. Antennas Propag.* **2021**, *69*, 1152–1157. [CrossRef]
179. Kiourti, A. RFID Antennas for Body-Area Applications: From Wearables to Implants. *IEEE Antennas Propag. Mag.* **2018**, *60*, 14–25. [CrossRef]
180. Fafenrot, S.; Grimmelsmann, N.; Wortmann, M.; Ehrmann, A. Three-dimensional (3D) printing of polymer-metal hybrid materials by fused deposition modeling. *Materials* **2017**, *10*, 1199. [CrossRef] [PubMed]
181. Jun, S.Y.; Elibiary, A.; Sanz-Izquierdo, B.; Winchester, L.; Bird, D.; McCleland, A. 3-D printing of conformal antennas for diversity wrist worn applications. *IEEE Trans. Compon. Packag. Manuf. Technol.* **2018**, *8*, 2227–2235. [CrossRef]
182. Fawaz, M.; Jun, S.; Oakey, W.; Mao, C.; Elibiary, A.; Sanz-Izquierdo, B.; Bird, D.; McClelland, A. 3D printed patch Antenna for millimeter wave 5G wearable applications. In Proceedings of the 12th European Conference on Antennas and Propagation (EuCAP 2018), London, UK, 9–13 April 2018.
183. Njogu, P.; Sanz-Izquierdo, B.; Elibiary, A.; Jun, S.Y.; Chen, Z.; Bird, D. 3D Printed Fingernail Antennas for 5G Applications. *IEEE Access* **2020**, *8*, 228711–228719. [CrossRef]
184. Paracha, K.N.; Butt, A.D.; Alghamdi, A.S.; Babale, S.A.; Soh, P.J. Liquid Metal Antennas: Materials, Fabrication and Applications. *Sensors* **2019**, *20*, 177. [CrossRef]
185. Oh, Y.; Kordsmeier, N.; Askari, H.; Adams, J.J. Low Profile GRIN Lenses with Integrated Matching Using 3-D Printed Ceramic. *IEEE Open J. Antennas Propag.* **2023**, *4*, 12–22. [CrossRef]
186. Zheng, F.; Jiang, H.-Y.; Yang, X.-T.; Guo, J.-H.; Sun, L.; Guo, Y.-Y.; Xu, H.; Yao, M.-S. Reviews of wearable healthcare systems based on flexible gas sensors. *Chem. Eng. J.* **2024**, *490*, 151874. [CrossRef]
187. Yuk, H.; Lu, B.; Lin, S.; Qu, K.; Xu, J.; Luo, J.; Zhao, X. 3D printing of conducting polymers. *Nat. Commun.* **2020**, *11*, 1604. [CrossRef] [PubMed]
188. Wu, X.; Li, F.; Wu, W.; Guo, T. Flexible organic light emitting diodes based on double-layered graphene/PEDOT: PSS conductive film formed by spray-coating. *Vacuum* **2014**, *101*, 53–56. [CrossRef]

189. Jeon, H.; Kim, Y.; Yu, W.-R.; Lee, J.U. Exfoliated graphene/thermoplastic elastomer nanocomposites with improved wear properties for 3D printing. *Compos. Part B Eng.* **2020**, *189*, 107912. [CrossRef]
190. Lee, H.; Kim, Y.; Kim, J.; Moon, S.Y.; Lee, J.U. Consecutive Ink Writing of Conducting Polymer and Graphene Composite Electrodes for Foldable Electronics-Related Applications. *Polymers* **2022**, *14*, 5294. [CrossRef]
191. Oderinde, O.; Liu, S.; Li, K.; Kang, M.; Imtiaz, H.; Yao, F.; Fu, G. Multifaceted polymeric materials in three-dimensional processing (3DP) technologies: Current progress and prospects. *Polym. Adv. Technol.* **2018**, *29*, 1586–1602. [CrossRef]
192. Wu, S.; Zeng, T.; Liu, Z.; Ma, G.; Xiong, Z.; Zuo, L.; Zhou, Z. 3D Printing Technology for Smart Clothing: A Topic Review. *Materials* **2022**, *15*, 7391. [CrossRef] [PubMed]
193. Carey, T.; Jones, C.; Le Moal, F.; Deganello, D.; Torrisi, F. Spray-Coating Thin Films on Three-Dimensional Surfaces for a Semitransparent Capacitive-Touch Device. *ACS Appl. Mater. Interfaces* **2018**, *10*, 19948–19956. [CrossRef]
194. Zhou, Y.; Soltani, S.; Li, B.M.; Wu, Y.; Kim, I.; Soewardiman, H.; Werner, D.H.; Jur, J.S. Direct-write spray coating of a full-duplex antenna for e-textile applications. *Micromachines* **2020**, *11*, 1056. [CrossRef]
195. 4D Printing: A Technology Coming from the Future. Sculpteo. Available online: https://www.sculpteo.com/en/3d-learning-hub/best-articles-about-3d-printing/4d-printing-technology/ (accessed on 5 March 2024).
196. Kuang, X.; Roach, D.J.; Wu, J.; Hamel, C.M.; Ding, Z.; Wang, T.; Dunn, M.L.; Qi, H.J. Advances in 4D Printing: Materials and Applications. *Adv. Funct. Mater.* **2018**, *29*, 1805290. [CrossRef]
197. Huang, Y.; Tao, Y.; Wang, Y. Shape-Memory Effect of 4D-Printed Gamma-Irradiated Low-Density Polyethylene. *Crystals* **2024**, *14*, 717. [CrossRef]
198. Hendrikson, W.J.; Rouwkema, J.; Clementi, F.; A van Blitterswijk, C.; Farè, S.; Moroni, L. Towards 4D printed scaffolds for tissue engineering: Exploiting 3D shape memory polymers to deliver time-controlled stimulus on cultured cells. *Biofabrication* **2017**, *9*, 031001. [CrossRef] [PubMed]
199. Malachowski, K.; Breger, J.; Kwag, H.R.; Wang, M.O.; Fisher, J.P.; Selaru, F.M.; Gracias, D.H. Stimuli-responsive theragrippers for chemomechanical controlled release. *Angew. Chem. Int. Ed.* **2014**, *53*, 8045–8049. [CrossRef]
200. Zhang, F.; Wang, L.; Zheng, Z.; Liu, Y.; Leng, J. Magnetic programming of 4D printed shape memory composite structures. *Compos. Part A Appl. Sci. Manuf.* **2019**, *125*, 105571. [CrossRef]
201. Ahmed, A.; Arya, S.; Gupta, V.; Furukawa, H.; Khosla, A. 4D printing: Fundamentals, materials, applications, and challenges. *Polymer* **2021**, *228*, 123926. [CrossRef]
202. Choong, Y.Y.C.; Maleksaeedi, S.; Eng, H.; Wei, J.; Su, P.-C. 4D printing of high-*performance* shape memory polymer using stereolithography. *Mater. Des.* **2017**, *126*, 219–225. [CrossRef]
203. Wei, H.; Zhang, Q.; Yao, Y.; Liu, L.; Liu, Y.; Leng, J. Direct-write fabrication of 4D active shape-changing structures based on a shape memory polymer and its nanocomposite. *ACS Appl. Mater. Interfaces* **2017**, *9*, 876–883. [CrossRef] [PubMed]
204. Nida, S.; Moses, J.A.; Anandharamakrishnan, C. Emerging applications of 5D and 6D printing in the food industry. *J. Agric. Food Res.* **2022**, *10*, 100392. [CrossRef]
205. Anas, S.; Khan, M.Y.; Rafey, M.; Faheem, K. Concept of 5D printing technology and its applicability in the healthcare industry. *Mater. Today Proc.* **2022**, *56*, 1726–1732. [CrossRef]
206. Vasiliadis, A.V.; Koukoulias, N.; Katakalos, K. From Three-Dimensional (3D)- to 6D-Printing Technology in Orthopedics: Science Fiction or Scientific Reality. *J. Funct. Biomater.* **2022**, *13*, 101. [CrossRef] [PubMed]
207. Mondal, K.; Tripathy, P.K. Preparation of smart materials by additive manufacturing technologies: A review. *Materials* **2021**, *14*, 6442. [CrossRef]
208. Lee, J. A 3D food printing process for the new normal Era: A review. *Processes* **2021**, *9*, 1495. [CrossRef]
209. Idà, E.; Nanetti, F.; Mottola, G. An Alternative Parallel Mechanism for Horizontal Positioning of a Nozzle in an FDM 3D Printer. *Machines* **2022**, *10*, 542. [CrossRef]
210. Shah, J.; Snider, B.; Clarke, T.; Kozutsky, S.; Lacki, M.; Hosseini, A. Large-scale 3D printers for additive manufacturing: Design considerations and challenges. *Int. J. Adv. Manuf. Technol.* **2019**, *104*, 3679–3693. [CrossRef]
211. Reddy, S. Smart Materials for 4D Printing: A Review on Developments, Challenges and Applications. In *Lecture Notes in Mechanical Engineering*; Springer Science and Business Media Deutschland GmbH: Berlin, Germany, 2022; pp. 3–10. [CrossRef]
212. Joshi, S.; Rawat, K.; Karunakaran, C.; Rajamohan, V.; Mathew, A.T.; Koziol, K.; Thakur, V.K.; Balan, A.S.S. 4D printing of materials for the future: Opportunities and challenges. *Appl. Mater. Today* **2020**, *18*, 100490. [CrossRef]
213. Sahafnejad-Mohammadi, I.; Karamimoghadam, M.; Zolfagharian, A.; Akrami, M.; Bodaghi, M. 4D printing technology in medical engineering: A narrative review. *J. Braz. Soc. Mech. Sci. Eng.* **2022**, *44*, 233. [CrossRef]
214. Pugliese, R.; Regondi, S. Artificial Intelligence-Empowered 3D and 4D Printing Technologies toward Smarter Biomedical Materials and Approaches. *Polymers* **2022**, *14*, 2794. [CrossRef]
215. Georgantzinos, S.K.; Giannopoulos, G.I.; Bakalis, P.A. Additive manufacturing for effective smart structures: The idea of 6D printing. *J. Compos. Sci.* **2021**, *5*, 119. [CrossRef]

216. Improved Printing: 5D Printing—Omega Consulting. *Omega Consulting*, 16 June 2023. Available online: https://omegaconsulting.online/insight/5d-printing/ (accessed on 21 October 2024).
217. Dorenbos, P. The electronic structure of lanthanide doped compounds with 3d, 4d, 5d, or 6d conduction band states. *J. Lumin.* **2014**, *151*, 224–228. [CrossRef]
218. O'Neal, B. 6D Printing Now Available, As is 3D Moon Dust Printing, Invisible Filament & 3D Printed Wine—3DPrint.com | The Voice of 3D Printing/Additive Manufacturing. *3DPrint.com | The Voice of 3D Printing/Additive Manufacturing*, 1 April 2015. Available online: https://3dprint.com/55256/april-fools-2015/ (accessed on 21 October 2024).
219. Ates, H.C.; Nguyen, P.Q.; Gonzalez-Macia, L.; Morales-Narváez, E.; Güder, F.; Collins, J.J.; Dincer, C. End-to-end design of wearable sensors. *Nat. Rev. Mater.* **2022**, *7*, 887–907. [CrossRef] [PubMed]
220. Rubin, G.J.; Nieto-Hernandez, R.; Wessely, S. Idiopathic environmental intolerance attributed to electromagnetic fields (formerly 'electromagnetic hypersensitivity'): An updated systematic review of provocation studies. *Bioelectromagnetics* **2010**, *31*, 1–11. [CrossRef] [PubMed]
221. *IEEE Std C95.1-1991*; IEEE Standard for Safety Levels with Respect to Human Exposure to Radio Frequency Electromagnetic Fields, 3 kHz to 300 GHz. IEEE: Piscataway, NJ, USA, 2006; Volume 2005, pp. 1–238.
222. Ahlbom, A.; Bergqvist, U.; Bernhardt, J.H.; Cesarini, J.P.; Grandolfo, M.; Hietanen, M.; Mckinlay, A.F.; Repacholi, M.H.; Sliney, D.H.; Stolwijk, J.A.; et al. Guidelines for limiting exposure to time-varying electric, magnetic, and electromagnetic fields (up to 300 GHz). *Health Phys.* **1998**, *74*, 494–521.
223. Neethirajan, S. Recent advances in wearable sensors for animal health management. *Sens. Bio-Sens. Res.* **2017**, *12*, 15–29. [CrossRef]
224. Skrivervik, A.K.; Bosiljevac, M.; Sipus, Z. Propagation Considerations for Implantable Antennas. In Proceedings of the 11th European Conference on Antennas and Propagation (EuCAP 2017), Paris, France, 19–24 March 2017; pp. 1573–1577.
225. Kiourti, A.; Nikita, K.S. A review of implantable patch antennas for biomedical telemetry: Challenges and solutions [Wireless Corner]. *IEEE Antennas Propag. Mag.* **2012**, *54*, 210–228. [CrossRef]
226. Gabriel, C.; Gabriel, S.; Corthout, E. The dielectric properties of biological tissues: I. Literature survey. *Phys. Med. Biol.* **1996**, *41*, 2231–2249. [CrossRef] [PubMed]
227. Shaw, T.; Samanta, G.; Mitra, D.; Mandal, B.; Augustine, R. Design of Metamaterial Based Efficient Wireless Power Transfer System Utilizing Antenna Topology for Wearable Devices. *Sensors* **2021**, *21*, 3448. [CrossRef]
228. Institute of Electrical and Electronics Engineers; Universitatea Politehnică București; Facultatea de Inginerie Electrica. *10th International Symposium on Advanced Topics in Electrical Engineering (ATEE 2017), Bucharest, Romania, 23–25 March 2017*; Institute of Electrical and Electronics Engineers (IEEE): Piscataway, NJ, USA, 2017.
229. Stanković, V.; Jovanović, D.; Krstić, D.; Marković, V.; Cvetković, N. Temperature distribution and Specific Absorption Rate inside a child's head. *Int. J. Heat Mass Transf.* **2017**, *104*, 559–565. [CrossRef]
230. Hockicko, P.; Dubovan, J. *ELEKTRO 2020: 13th International Conference: Conference Proceedings (IEEE Xplore Compliant), Taormina, Italy*; IEEE: Piscataway, NJ, USA, 2020.
231. Gryz, K.; Zradziński, P.; Karpowicz, J. The role of the location of personal exposimeters on the human body in their use for assessing exposure to the electromagnetic field in the radiofrequency range 98–2450 MHz and compliance analysis: Evaluation by virtual measurements. *BioMed Res. Int.* **2015**, *2015*, 272460. [CrossRef] [PubMed]
232. Rashid, T.B.; Song, H.H. Analysis of biological effects of cell phone radiation on human body using specific absorption rate and thermoregulatory response. *Microw. Opt. Technol. Lett.* **2019**, *61*, 1482–1490. [CrossRef]
233. Christopher, B.; Khandaker, M.U.; Jojo, P.J. Empirical study on specific absorption rate of head tissues due to induced heating of 4G cell phone radiation. *Radiat. Phys. Chem.* **2021**, *178*, 108910. [CrossRef]
234. Zhang, H.H.; Gong, L.F.; Wang, J.; Sha, W.E.I.; Li, L.; Wang, J.X.; Wang, C. Specific absorption rate assessment of fifth generation mobile phones with specific anthropomorphic mannequin model and high-resolution anatomical head model. *Int. J. RF Microw. Comput. Eng.* **2022**, *32*, e23158. [CrossRef]
235. Seif, F.; Omidi, M.; Bayatiani, M.R.; Ghorbani, M. Measurement and Calculation of SAR Due to 900 MHz Electromagnetic Waves. *Front. Biomed. Technol.* **2022**, *9*, 38–44. [CrossRef]
236. Seetharaman, R.; Tharun, M.; Gayathri, S.; Mole, S.; Anandan, K. Analysis of specific absorption rate and heat transfer in human head due to mobile phones. *Mater. Today Proc.* **2022**, *51*, 2365–2374. [CrossRef]
237. Conil, E.; Hadjem, A.; Lacroux, F.; Wong, M.F.; Wiart, J. Variability analysis of SAR from 20 MHz to 2.4 GHz for different adult and child models using finite-difference time-domain. *Phys. Med. Biol.* **2008**, *53*, 1511–1525. [CrossRef] [PubMed]
238. Christ, A.; Gosselin, M.-C.; Christopoulou, M.; Kühn, S.; Kuster, N. Age-dependent tissue-specific exposure of cell phone users. *Phys. Med. Biol.* **2010**, *55*, 1767–1783. [CrossRef] [PubMed]
239. Wiart, J.; Hadjem, A.; Wong, M.F.; Bloch, I. Analysis of RF exposure in the head tissues of children and adults. *Phys. Med. Biol.* **2008**, *53*, 3681–3695. [CrossRef]

240. Bhargava, D.; Leeprechanon, N.; Rattanadecho, P.; Wessapan, T. Specific absorption rate and temperature elevation in the human head due to overexposure to mobile phone radiation with different usage patterns. *Int. J. Heat Mass Transf.* **2019**, *130*, 1178–1188. [CrossRef]
241. Celik, A.; Salam, K.N.; Eltawil, A. The Internet of Bodies: A Systematic Survey on Propagation the Internet of Bodies: A Systematic Survey on Propagation Characterization and Channel Modeling Characterization and Channel Modeling. *IEEE Internet Things J.* **2021**, *9*, 321–345. [CrossRef]
242. Kim, J.; Rahmat-Samii, Y. Implanted antennas inside a human body: Simulations, designs, and characterizations. *IEEE Trans. Microw. Theory Tech.* **2004**, *52*, 1934–1943. [CrossRef]
243. Occhiuzzi, C.; Contri, G.; Marrocco, G. Design of implanted RFID tags for passive sensing of human body: The STENTag. *IEEE Trans. Antennas Propag.* **2012**, *60*, 3146–3154. [CrossRef]
244. Schmidt, C.; Casado, F.; Arriola, A.; Ortego, I.; Bradley, P.D.; Valderas, D. Broadband uhf implanted 3-D conformal antenna design and characterization for in-off body wireless links. *IEEE Trans. Antennas Propag.* **2014**, *62*, 1433–1444. [CrossRef]
245. Alemaryeen, A.; Noghanian, S. On-Body Low-Profile Textile Antenna with Artificial Magnetic Conductor. *IEEE Trans. Antennas Propag.* **2019**, *67*, 3649–3656. [CrossRef]
246. Hasson, A.M.; Patel, R.D.; Sissoko, C.A.; Brattain, L.; Dion, G.R. 3D-printed neck phantoms with detailed anatomy for ultrasound-guided procedure and device testing. *Laryngoscope Investig. Otolaryngol.* **2024**, *9*, e1309. [CrossRef] [PubMed]
247. Kiourti, A.; Nikita, K.S. Miniature scalp-implantable antennas for telemetry in the MICS and ISM bands: Design, safety considerations and link budget analysis. *IEEE Trans. Antennas Propag.* **2012**, *60*, 3568–3575. [CrossRef]
248. Permana, H.; Fang, Q.; Rowe, W.S.T. Hermetic implantable antenna inside vitreous humor simulating fluid. *Prog. Electromagn. Res.* **2012**, *133*, 571–590. [CrossRef]
249. Kiourti, A.; Nikita, K.S. Detuning issues and performance of a novel implantable antenna for telemetry applications. In Proceedings of the 6th European Conference on Antennas and Propagation, EuCAP 2012, Prague, Czech Republic, 26–30 March 2012; pp. 746–749. [CrossRef]
250. Iero, D.A.M.; Crocco, L.; Isernia, T.; Korkmaz, E. Optimal focused electromagnetic hyperthermia treatment of breast cancer. In Proceedings of the 2016 10th European Conference on Antennas and Propagation, EuCAP 2016, Davos, Switzerland, 10–15 April 2016. [CrossRef]
251. Miry, C.; Gillard, R.; Loison, R. An application of the multi-level DG-FDTD to the analysis of the transmission between a dipole in free-space and an implanted antenna in a simplified body model with various positions. In Proceedings of the European Conference on Antennas and Propagation, EuCAP 2009, Berlin, Germany, 23–27 March 2009; pp. 67–70.
252. Zivkovic, Z.; Vujevic, E.; Poljak, D.; Drissi, K.E.K.; Sarolic, A. Computation of SAR in the simplified model of a pregnant woman exposed to RF radiation from 10 MHz to 1800 MHz. In Proceedings of the 2013 21st International Conference on Software, Telecommunications and Computer Networks, SoftCOM 2013, Split, Croatia, 18–20 September 2013. [CrossRef]
253. Canicatti, E.; Giampietri, E.; Brizi, D.; Fontana, N.; Monorchio, A. A Numerical Exposure Assessment of Portable Self-Protection, High-Range, and Broadband Electromagnetic Devices. *IEEE Open J. Antennas Propag.* **2021**, *2*, 555–563. [CrossRef]
254. Nagaoka, T.; Watanabe, S.; Sakurai, K.; Kunieda, E.; Watanabe, S.; Taki, M.; Yamanaka, Y. Development of realistic high-resolution whole-body voxel models of Japanese adult males and females of average height and weight, and application of models to radio-frequency electromagnetic-field dosimetry. *Phys Med Biol.* **2004**, *49*, 1. [CrossRef] [PubMed]
255. DUKE»IT'IS Foundation. Itis.swiss. 2024. Available online: https://itis.swiss/virtual-population/virtual-population/vip3/duke/ (accessed on 22 April 2024).
256. Alemaryeen, A.; Noghanian, S. Crumpling effects and specific absorption rates of flexible AMC integrated antennas. *IET Microw. Antennas Propag.* **2018**, *12*, 627–635. [CrossRef]
257. Panagamuwa, C.J.; Whittow, W.G.; Edwards, R.M.; Vardaxoglou, J.C. Experimental verification of a modified Specific Anthropomorphic Mannequin (SAM) head used for SAR measurements. In Proceedings of the 2007 Loughborough Antennas and Propagation Conference, LAPC 2007 Conference Proceedings, Loughborough, UK, 2–3 April 2007; pp. 261–264. [CrossRef]
258. Zastrow, E.; Davis, S.K.; Lazebnik, M.; Kelcz, F.; Van Veen, B.D.; Hagness, S.C. Development of Anatomically Realistic Numerical Breast Phantoms with Accurate Dielectric Properties for Modeling Microwave Interactions With the Human Breast. *IEEE Trans. Biomed. Eng.* **2008**, *55*, 2792–2800. [CrossRef]
259. Bhargava, D.; Rattanadecho, P. Microwave imaging of breast cancer: Simulation analysis of SAR and temperature in tumors for different age and type. *Case Stud. Therm. Eng.* **2022**, *31*, 101843. [CrossRef]
260. Saeidi, T.; Al-Gburi, A.J.A.; Karamzadeh, S. A Miniaturized Full-Ground Dual-Band MIMO Spiral Button Wearable Antenna for 5G and Sub-6 GHz Communications. *Sensors* **2023**, *23*, 1997. [CrossRef] [PubMed]
261. Getting Started with Seg3D. FieldTrip Toolbox. 2024. Available online: https://www.fieldtriptoolbox.org/getting_started/seg3d/ (accessed on 21 March 2024).

262. Carter, L.M.; Bellamy, M.B.; Choi, C.; Kim, C.H.; Bolch, W.E.; Jokisch, D.; Kesner, A.L. Influence of Body Posture on Internal Organ Dosimetry: Radiocesium Exposure Modeling Using Novel Posture-dependent Mesh Computational Phantoms. *Health Phys.* **2023**, *125*, 137–146. [CrossRef]
263. Li, C.-H.; Douglas, M.; Ofli, E.; Derat, B.; Gabriel, S.; Chavannes, N.; Kuster, N. Influence of the hand on the specific absorption rate in the head. *IEEE Trans. Antennas Propag.* **2012**, *60*, 1066–1074. [CrossRef]
264. Ding, S.; Koulouridis, S.; Pichon, L. Implantable wireless transmission rectenna system for biomedical wireless applications. *IEEE Access* **2020**, *8*, 195551–195558. [CrossRef]
265. Ito, K. Numerical and experimental human body phantoms. In Proceedings of the 2007 IET Seminar on Antennas and Propagation for Body-Centric Wireless Communications, London, UK, 24 April 2007.
266. Miaskowski, A.; Gas, P. Numerical Estimation of SAR and Temperature Distributions inside Differently Shaped Female Breast Tumors during Radio-Frequency Ablation. *Materials* **2022**, *16*, 223. [CrossRef]
267. Nguyen, C.D.; Edwards, S.A.; Iorizzo, T.W.; Longo, B.N.; Yaroslavsky, A.N.; Kaplan, D.L.; Mallidi, S. Investigation of silk as a phantom material for ultrasound and photoacoustic imaging. *Photoacoustics* **2022**, *28*, 100416. [CrossRef] [PubMed]
268. McGarry, C.K.; Grattan, L.J.; Ivory, A.M.; Leek, F.; Liney, G.P.; Liu, Y.; Miloro, P.; Rai, R.; Robinson, A.; Shih, A.J.; et al. Tissue mimicking materials for imaging and therapy phantoms: A review. *Phys. Med. Biol.* **2020**, *65*, 23TR01. [CrossRef]
269. Jawli, A.; Nabi, G.; Huang, Z. A Polyvinyl Alcohol (PVA)-Based Phantom for Prostate Cancer Detection Using Multiparametric Ultrasound: A Validation Study. *Bioengineering* **2024**, *11*, 1052. [CrossRef]
270. Hahn, C.; Noghanian, S. Heterogeneous breast phantom development for microwave imaging using regression models. *Int. J. Biomed. Imaging* **2012**, *2012*, 803607. [CrossRef]
271. Ustbas, B.; Kilic, D.; Bozkurt, A.; Aribal, M.E.; Akbulut, O.; Ustbas, B.; Kilic, D.; Bozkurt, A.; Aribal, M.E.; Akbulut, O. Silicone-based composite materials simulate breast tissue to be used as ultrasonography training phantoms. *Ultrasonics* **2018**, *88*, 9–15. [CrossRef] [PubMed]
272. Lee, J.K.; Han, S.J.; Kim, K.; Kim, Y.H.; Lee, S. Wireless Epidermal Six-Axis Inertial Measurement Units for Real-Time Joint Angle Estimation. *Appl. Sci.* **2020**, *10*, 2240. [CrossRef]
273. Montasel, A.M.; Mahmoud, K.R.; Elmikatf, H.A. An interaction study between PIFAS handset antenna and a human hand-head in personal communications. *Prog. Electromagn. Res. B* **2012**, *37*, 21–42. [CrossRef]
274. Cui, Y.; Zhang, M.; Li, J.; Luo, H.; Zhang, X.; Fu, Z. WSMS: Wearable Stress Monitoring System based on IoT Multi-Sensor Platform for Living Sheep Transportation. *Electronics* **2019**, *8*, 441. [CrossRef]
275. Sherman, J.L.; Darbinyan, A.; Magid, M.S.; Ong, P.; Weissman, B.; Benkov, K.; Lipskar, A.M. Pediatric colonic inflammatory myofibroblastic tumor presenting as colo-colonic intussusception: A case report and review of the literature. *J. Pediatr. Surg. Case Rep.* **2015**, *3*, 392–396. [CrossRef]
276. Curto, S.; Taj-Eldin, M.; Fairchild, D.; Prakash, P. Microwave ablation at 915 MHz vs. 2.45 GHz: A theoretical and experimental investigation. *Med. Phys.* **2015**, *42*, 6152–6161. [CrossRef]
277. Huang, F.-J.; Lee, C.-M.; Chang, C.-L.; Chen, L.-K.; Yo, T.-C.; Luo, C. H. Rectenna application of miniaturized implantable antenna design for triple-band biotelemetry communication. *IEEE Trans. Antennas Propag.* **2011**, *59*, 2646–2653. [CrossRef]
278. Warty, R.; Tofighi, M.-R.; Kawoos, U.; Rosen, A. Characterization of implantable antennas for intracranial pressure monitoring: Reflection by and transmission through a scalp phantom. *IEEE Trans. Microw. Theory Tech.* **2008**, *56*, 2366–2376. [CrossRef]
279. Joachimowicz, N.; Conessa, C.; Henriksson, T.; Duchene, B. Breast Phantoms for Microwave Imaging. *IEEE Antennas Wirel. Propag. Lett.* **2014**, *13*, 1333–1336. [CrossRef]
280. Karipidis, K.; Mate, R.; Urban, D.; Tinker, R.; Wood, A. 5G mobile networks and health—A state-of-the-science review of the research into low-level RF fields above 6 GHz. *J. Expo. Sci. Environ. Epidemiol.* **2021**, *31*, 585–605. [CrossRef] [PubMed]
281. Gezimati, M.; Singh, G. Advances in terahertz technology for cancer detection applications. *Opt. Quantum Electron.* **2023**, *55*, 151. [CrossRef] [PubMed]

sensors

MDPI

Review

Flexible Strain Sensors Based on Thermoplastic Polyurethane Fabricated by Electrospinning: A Review

Zhiyuan Zhou †, Weirui Tang †, Teer Xu, Wuyang Zhao, Jingjing Zhang * and Chuanwu Bai *

School of Tropical Agriculture and Forestry, Hainan University, Danzhou 571799, China; 20213006647@hainanu.edu.cn (Z.Z.); 20213005555@hainanu.edu.cn (W.T.); 20213005572@hainanu.edu.cn (T.X.); 20213005554@hainanu.edu.cn (W.Z.)

* Correspondence: 993958@hainanu.edu.cn (J.Z.); 993681@hainanu.edu.cn (C.B.)

† These authors contributed equally to this work.

Abstract: Over recent years, thermoplastic polyurethane (TPU) has been widely used as a substrate material for flexible strain sensors due to its remarkable mechanical flexibility and the ease of combining various conductive materials by electrospinning. Many research advances have been made in the preparation of flexible strain sensors with better ductility, higher sensitivity, and wider sensing range by using TPU in combination with various conductive materials through electrospinning. However, there is a lack of reviews that provide a systematic and comprehensive summary and outlook of recent research advances in this area. In this review paper, the working principles of strain sensors and electrospinning technology are initially described. Subsequently, recent advances in strain sensors based on electrospun TPU are tracked and discussed, with a focus on the incorporation of various conductive fillers such as carbonaceous materials, MXene, metallic materials, and conductive polymers. Moreover, the wide range of applications of electrospun TPU flexible strain sensors is thoroughly discussed. Finally, the future prospects and challenges of electrospun TPU flexible strain sensors in various fields are pointed out.

Keywords: flexible strain sensors; electrospinning; thermoplastic polyurethane; conductive materials; applications

1. Introduction

With the rapid advances in the Internet of Things (IoT), Industry 4.0, big data, artificial intelligence (AI), robotics, and digital health, there is an increasing demand for high-precision sensing at all places and times [1]. Strain sensors can convert various types of deformations into changes in electrical signals to detect the dynamic characteristics of physical states, which have garnered widespread attention from many scholars [2]. However, traditional strain sensors made from rigid materials such as metals, ceramics, and semiconductors face inherent performance limitations such as lack of flexibility and ductility, making it difficult to meet growing demands [3]. As shown in Figure 1, flexible strain sensors have been studied and applied in many fields of intelligent electronic devices such as intelligent electronic skin [4], human–computer interaction [5], sports performance monitoring [6], and flexible robotic [7] due to their excellent properties of flexibility, stretchability, breathability, high sensitivity, and strong stability. Flexible strain sensors can be trimmed to different sizes and even folded into different shapes for a variety of complex and changing applications, which greatly broadens their application areas [8].

Figure 1. Electrospun flexible strain sensors and potential applications in intelligent electronic skin, human–computer interaction, sports performance monitoring, movement detection, and flexible robotic. Reprinted with permission from Ref. [9]. Copyright 2023, Wiley.

The performance of flexible strain sensors mainly depends on the mechanical properties of the flexible polymer matrix and the electrical properties of conductive fillers [10]. Common materials used as flexible polymer matrices include thermoplastic polyurethane (TPU) [11], polyimide (PI) [12], polyetheretherketone (PEEK) [13], polyethersulfone (PES) [14], polycarbonate (PC) [15], polyethylene naphthalate (PEN) [16], hydrogel [17], and polyethylene terephthalate (PET) [18]. Among these, TPU exhibits superior overall performance. It not only has excellent abrasion resistance, exceeding that of natural rubber by a factor of five (far superior to PI) but also boasts high transparency—in contrast to PEEK, which is usually opaque or translucent [19,20]. TPU's strong flexibility and mechanical elasticity surpass those of PES, and its economic benefits over hydrogel are realized at a lower cost [21,22]. In terms of stability, TPU's chemical resistance is better than PET's, and it has excellent thermal stability and adaptability to low-temperature environments compared to PC [23,24]. Additionally, TPU is easy to process and can be recycled, making it more environmentally friendly than PEN [16]. These combined advantages make TPU one of the most widely used flexible polymer matrices for flexible strain sensors in practical applications. However, pure TPU is essentially a non-conductive material and cannot be directly used to fabricate sensing elements [25]. Electrospinning technology can further process TPU nanofibers in combination with various conductive fillers to endow the composite TPU nanofibers with high electrical conductivity, enabling the preparation of high-performance flexible strain sensors.

Currently, researchers have explored the use of carbon materials (e.g., carbon black, carbon nanotubes, and graphene) [26–28], metallic nanomaterials (e.g., nanoparticles, nanorods, and nanowires) [29–31], and conductive polymers (e.g., PPy and PANI) [32,33]. They have prepared numerous flexible strain sensors with excellent performance by combining these materials with electrospun TPU nanofiber substrates. However, there is a lack of review papers that provide a systematic and comprehensive summary and outlook of these recent research results. For example, Backes et al. explored TPU synthesis, fabrication techniques, blends, composites, and applications, but did not focus on the most advanced TPU prepared by electrospinning techniques and their application in flexible sensors [34]. Chen et al. introduced the research progress of flexible strain sensors based on conductive fillers and TPU but did not analyze the most fruitful electrospun TPU flexible strain sensors, and their review was published three years ago, which affects its novelty and timeliness [11]. In this review, Section 1 provides a general introduction to the working principles of strain sensors and electrospinning technology, highlighting the salient features of electrospun TPU strain sensors. Section 2 summarizes recent research advances in various conductive fillers for electrospun TPU flexible strain sensors, with a focus on carbonaceous materials, MXene, metallic materials, and conductive polymers. Section 3 offers a comprehensive discussion of various sensing applications of electrospun TPU flexible strain sensors, including motion detection, health monitoring, recognition of facial expressions and speech, electronic skin, and multifunctional electrospun TPU sensors. Lastly, the future research trends and challenges of electrospun TPU flexible strain sensors are proposed. This review comprehensively and systematically summarizes the recent advances in electrospun TPU-based flexible strain sensors, bridges the gap in the field, and serves as a unique guideline for subsequent research on electrospun TPU-based flexible strain sensors.

2. Electrospinning TPU Strain Sensors

2.1. Strain Sensors

Strain sensors can convert various external physical stimuli (e.g., stretching, compression, twisting, etc.) into recognizable electrical signal outputs. According to the transduction mechanism, strain sensors can be classified into resistive [35], capacitive [36], voltage (piezoelectric and triboelectric) [37,38], and inductance/magnetism types [39,40]. Resistive strain sensors are widely used due to their significant advantages, such as simple preparation processes, easy signal reception, and stable signal output [41]. Their mode of operation involves using the resistance strain gauge to quantify the response to the direction and degree of strain applied to the sensor through changes in the piezoresistance ratio or increases in the inherent resistance of the material when the conductive layer is deformed [42]. Initially, resistive strain sensors were used to detect the fatigue level of materials. However, with in-depth research on composite nanomaterials, flexible resistive strain sensors based on conductive polymer composites (CPCs) are now increasingly applied in high-end fields, such as intelligent wearable devices and electronic skin [43].

The working principle of CPC strain sensors is mainly based on the tunneling effect, percolation theory, and crack propagation. According to the electron tunneling effect, the conductive network is generated by the high-speed movement of electrons in the conductive pathway based on conductive filler loading [44,45]. There is a potential energy barrier between adjacent but unconnected conductive fillers [46]. Under applied stress, the spacing between adjacent conductive fillers changes, and according to quantum mechanics, electrons may pass through the potential barrier between adjacent conductive fillers, resulting in the rearrangement of the conductive network, which is converted into a macroscopic electrical signal output. Percolation theory suggests that when the conductive filler in the composite system is increased to a certain critical amount, the electrical resistance decreases sharply [47]. Meanwhile, the resistivity-conductive filler dosage curve shows a narrow region of abrupt change, where a small change in the amount of conductive filler leads to a significant change in resistivity [48]. The conductivity of CPC strain sensors is described using percolation theory based on the tunneling effect, with the following equation:

$$\sigma \propto (\varnothing - \varnothing_c)^t, \varnothing > \varnothing_c \quad (1)$$

where σ is the electrical conductivity of the composites, $\varnothing$ is the mass fraction of conductive filler, $\varnothing_c$ represents the percolation threshold, and t is a critical exponent in relation to the dimensionality of the conductive network. Crack propagation means that when the conductive filler is stretched to a certain degree, cracks will appear on the surface or inside the conductive filler, which will lead to a change in the contact area and thus cause a change in resistance [49]. This will be specifically analyzed in the next section.

GF is introduced to evaluate the response sensitivity of the strain sensor, defined as the ratio of the relative change in resistance to the mechanical strain. GF is analyzed using the slope of the change curve, which can be calculated from the following equation:

$$GF = \frac{\Delta R/R_0}{\varepsilon} = \frac{(R_t - R_0)/R_0}{\varepsilon}, \varepsilon = \Delta L/L_0 \quad (2)$$

where R_0 is the initial resistance, R_t is the test resistance, and ε is the applied strain.

2.2. Electrospinning Techniques

Electrospinning technology is a process in which a polymer solution undergoes jet stretching under the action of high-voltage electrostatic field forces, and through solvent evaporation, continuous, ultrafine fibers, as well as micron-sized, ultra-thin fiber membranes, are prepared [50–52]. The basic device consists of four main components: a high-voltage power supply that provides high-voltage electrostatic field force, a syringe with a jet needle, a syringe pump, and a grounded receiving device [53]. In the process of electrospinning, the liquid extruded from the nozzle forms hanging droplets under the action of surface tension, while the high-voltage electrostatic field produces a transient potential difference between the needle and the receiver, causing the extruded droplets to take on a charge, with the repulsive force between the surface charges of the droplets counteracting the surface tension [54]. As the voltage increases, the repulsive surface charge force dominates and the hemispherical liquid is elongated, deforming the droplet into a stable Taylor cone [55]. The charged fluid is ejected from the Taylor cone under the action of an electrostatic field and stretched within the electrostatic field [56]. Initially, the charged fluid undergoes linear stretching. Under the action of the electrostatic field, electrical bending instability (whipping instability) grows [57]. The ejected fluid thins under stretching and solidifies rapidly until solid fibers are deposited on the collector [58]. By controlling the parameters of electrostatic spinning, nanofibers with different morphologies can be prepared, such as uniaxially and biaxially arranged nanofiber meshes, ribbons, porous fibers, necklace-like fibers, nanowebs, hollow fibers, nanowire-in-microtube, and multichannel tubules [51] (Figure 2).

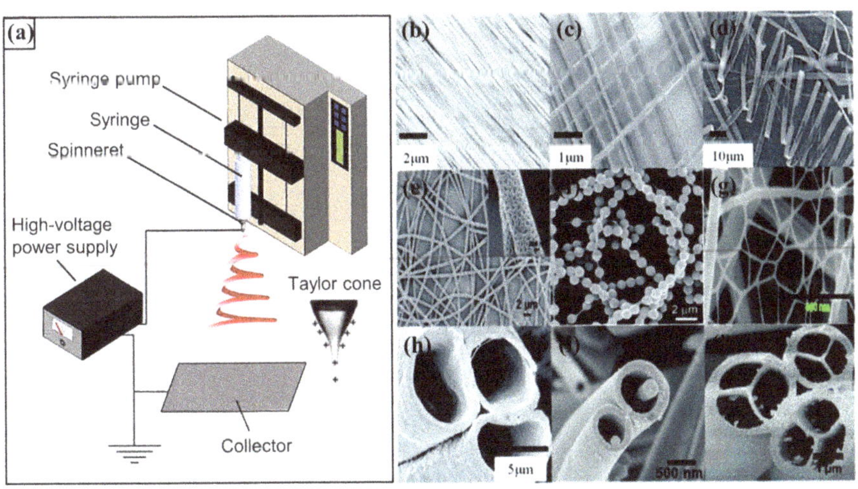

Figure 2. (**a**) Basic setup for electrospinning. Reprinted with permission from Ref. [42]. Copyright 2019, American Chemical Society. Different morphologies of electrospun polymer fibers: (**b**) uniaxially aligned, (**c**) biaxially oriented, (**d**) ribbon, (**e**) porous fibers, (**f**) Necklace-like, (**g**) nanowebs, (**h**) hollow, (**i**) nanowire-in-microtube, and (**j**) multichannel tubular. Reprinted with permission from Ref. [51]. Copyright 2014, the Royal Society Chemistry.

Electrospinning technology provides a unique and effective way to prepare TPU nanofiber materials. It has advantages in microstructure control and can produce nanofibers with a high specific surface area and fine pore structure [50]. Compared to other preparation processes (e.g., melt mixing, solution mixing, solvent casting, particle leaching, thermally induced phase separation (TIPS), 3D printing, and injection molding), electrospinning exhibits unique benefits: it offers precise control over microstructure, unlike melt blending; it reduces the use of hazardous organic solvents, showing better environmental compatibility than solution blending; and it simplifies the production process and shortens preparation time compared to TIPS [59]. Although 3D printing technology can produce complex geometries, electrospinning excels in microstructural consistency and orientation [60]. Injection molding, while suitable for rapid production, is less precise than electrospinning in microstructure control [61].

2.3. *Characterized Electrospun TPU Strain Sensors*

Table 1 demonstrates the effect of different electrospinning process parameters on the morphological properties of TPU [61]. Electrospinning technology endows TPU nanofibers with high porosity, high uniformity, and high-quality microstructure. The prepared electrospun TPU nanofibers have a large specific surface area, which is conducive to the bonding and loading of more conductive fillers [62]. Therefore, the most common and effective method to prepare electrospun TPU strain sensors is to use electrospun TPU fibers as a substrate and introduce conductive fillers. This method combines the high tensile properties of TPU nanofibers with the high electrical conductivity of conductive fillers. Highly conductive nanoparticles form continuous conductive pathways, which are carried and protected by a flexible matrix made of TPU nanofibers with high stretch-recovery properties [63].

Table 1. Effect of electrospinning process parameters on morphological properties of TPU.

Process Parameter	Description	Effect on Morphological Properties
Nozzle Axis Type	Single Axis Nozzle	Uniform diameters, limited control over orientation and porosity
	Coaxial Nozzle	Increases nanofiber diameter, more porous structure, enhanced control over morphology
Collector Type	Rotating Cylinder Collector	Larger diameters, less effective in fiber alignment
	Rotational Wire Collector	Thinner, more oriented fibers, better alignment due to smaller contact area and higher rotation speeds
Collector Speed	Low Speed	Random fiber orientations, limited impact on diameter
	High Speed (2000 rpm)	Improved orientation, increased fiber diameter due to relaxation post-spinning, optimal for aligned fiber applications
Filler Types	Inorganic Fillers (e.g., Salt Particles)	Thinnest, highly oriented fibers, enhanced alignment
	Organic Fillers (e.g., Polystyrene)	Modified morphology, impact on mechanical properties, and surface characteristics
Fiber Diameter	Increases with Collector Speed	Larger diameters at higher speeds due to relaxation
	Increases with Coaxial Nozzle	Thicker fibers compared to single-axis nozzles
Fiber Orientation	Enhanced with Higher Rotational Speeds	Speeds up to 2000 rpm needed for significant orientation, critical for specific applications
	Better with Wire Collector	Superior alignment compared to cylindrical collectors
Surface Porous Structure	Coaxial Technique	Highly porous surfaces with low boiling point solvents for core, formation of hollow and porous fibers
	Influenced by Solvent Properties	Control porosity and surface characteristics based on solvent boiling points

3. The Use of Conductive Fillers in Electrospun TPU Strain Sensors

Among all types of conductive fillers applied in electrospun TPU sensors, carbonaceous materials, MXene, metallic materials, and conductive polymers are the most common.

In this section, recent research advances in various conductive fillers for electrospun TPU flexible strain sensors are discussed, especially on carbonaceous materials, MXene, metallic materials, and conductive polymers. Details of some typical electrospun TPU strain sensors are summarized in Table 2.

Table 2. Main features of various flexible strain sensors.

Conductive Material	Materials	Structure	Sensitivity (at Strain %)	Sensing Range (%)	Response/Recovery Time (ms)	Durability (Cycles at Strain %)	Ref
0D CB	CB/TPU	membrane	8962.7 (160)	0–160	60	10,000 (10)	[64]
	CB/TPU/Ecoflex	sandwich	3186.4 (225)	0.5–225	70	5000 (40)	[65]
1D CNTs	DATPU/CNTs	membrane	1200 (710)	0–710	300	15,000 (100)	[66]
	CNTs/DA/TPU	membrane	10,528.53 (200)	0–200	188/211	300 (10)	[67]
	CNTs/F-TPU	mat	2.86 (150)	0–550	110	5000 (30)	[68]
	CNTs/TPU@SBS	double-percolation	32,411 (100)	0.2–100	214	500 (10)	[69]
	CNTs/TPU	membrane	1571 (400)	0–400	-	10,000 (10)	[70]
	TPU/CNTs	tube	57 (760)	2–760	45	10,000 (200)	[71]
2D Gr	PDA/rGO/TPU	mat	6583 (150)	0–150	100	9000 (50)	[72]
	GNPs/TPU/PDMS	crack-wrinkle	750 (24)	0.1–24	90	20,000 (5)	[73]
	TPU/GNPs	porous-convoluted	4047.5 (350)	0–350	-	10,000 (50)	[74]
multidimensionality	PDA/CB/CNF/TPU	membrane	312.4 (285)	0.12–285	287/457	1000 (50)	[75]
	TPU/CB-CNTs	fiber	6.0 (3.0)	0–280.5	248	2000 (5)	[76]
	PDMS/(CB+CNTs)/TPU	mat	49,863.5 (437.9)	0–437.9	-	1300 (50)	[77]
	CB/MWCNTs/TPU	membrane	5705.53 (150)	0–150	220	6000 (7)	[78]
	TPU/SCB@CNTs/F-SiO_2	film	60.42 (100)	0–100	75/100	1000 (70)	[79]
	TPU/CNTs-Gr	membrane	217 (172)	0–172	-	10,000 (30)	[80]
	CB/CNTs/GR/TPU	yarn	17.74 (100)	0–227	220	5000 (50)	[81]
MXene	MXene/PDA/TPU	membrane	2600 (200)	0.1–200	77/186	1000 (50)	[82]
	TPU/MXene/BNNS	sandwich	2080.9 (100)	0–100	-	2500 (50)	[83]
	MXene/CNTs/TPU	bilayer-conductive	2911 (330)	0–330	80	2600 (50)	[84]
	MXene-CNTs/TPU/Ecoflex	sandwich	1719.2 (251)	0.1–251	50	10,000 (50)	[85]
Ag	TPU/Ag/CNTs	membrane	6834 (604)	0.1–604	122/164	1000 (50)	[86]
	AgNWs/MXene/TPU	yarn	1.7 (1000)	0–1000	100	3000 (100)	[87]
LM	LM/TPU	membrane	0.2 (200)	0–200	290/360	9000 (75)	[88]
	TPU/LM/NFs	yarn	6 (200)	0–548	-	1000 (50)	[89]
conductive polymer	PDA/PPy/TPU-TPE	fiber	58.9 (184)	0–184	58.82/117.64	2000 (20)	[90]
	MWCNT/PEDOT:PSS/TPU	membrane	6008.3 (680)	0.5–680	200	6000 (50)	[91]
	rGO/PANI/TPU	mat	3000.2 (300)	0.1–300	90	10,000 (40)	[92]

3.1. *Carbonaceous Materials*

3.1.1. Zero-Dimensional Carbonaceous Material—CB

Zero-dimensional carbon black (CB) possesses outstanding advantages, including good dispersion in solvents, high conductivity, and small size, which make it an ideal conductive nanofiller [93]. The high specific surface area and low dimensionality of the conductive CB particles facilitate proper dispersion in the electrospun TPU substrate and establish stable interactions with the thermoplastic polyurethane chains, endowing the composites with excellent electromechanical properties [94]. However, the point-to-point mobile conductive network constructed by conductive CB particles in thermoplastic polyurethanes is easily deformed, which makes the composites exhibit high conductive sensitivity under tensile strain [95]. Conductive CB particles are widely used as conductive fillers in electrospun TPU strain sensors due to the ease of processing and economic practicality for large-scale industrial production [96].

A common strategy is to prepare CB/TPU composite nanofiber membranes by embedding CB particles in TPU nanofiber membranes using the ultrasonication method to construct the conductive network [64] (Figure 3a). The CB/TPU flexible strain sensor prepared from this composite nanofiber membrane exhibits a combination of high sensitivity under tensile strain (GF of 8962.7 at 155% strain), fast response time (60 ms), excellent stability and durability (greater than 10,000 cycles), and a wide workable tensile range (0–160%). On this basis, some researchers have prepared CB/TPU/Ecoflex strain sensors using Ecoflex to encapsulate ultrasonically modified CB/TPU composite nanofibers to form a sandwich structure [65] (Figure 3b). This sandwich structure provides effective protection for the conductive CB/TPU fiber network, giving the strain sensor excellent sensing performance, including a low detection limit (0.5% strain), a wide response range (up to 225% strain), ultra-high sensitivity (maximum GF of 3186.4 at 225% strain), fast response time (70 ms), and good repeatability (over 5000 tension/release cycles). In addition, the strain sensor has excellent immunity to external humidity and temperature.

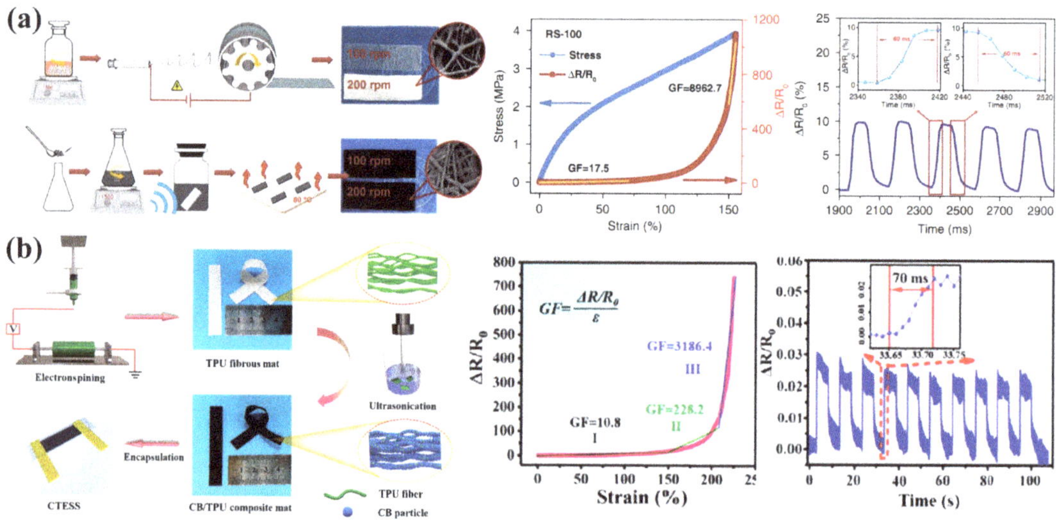

Figure 3. (**a**) Schematic illustration of the preparation process of CB/TPU strain sensor. $\Delta R/R_0$ of CB/TPU strain sensor toward the applied strain; the response time of CB/TPU strain sensor under 1% strain. Reprinted with permission from Ref. [64]. Copyright 2021, Springer. (**b**) Schematic illustration of the preparation process of CB/TPU/Ecoflex strain sensor (CTESS). $\Delta R/R_0$ of CTESS toward the applied strain; the response time of CTESS under 1% strain. Reprinted with permission from Ref. [65]. Copyright 2020, Elsevier.

Nevertheless, the application of conductive CB particles in electrospun TPU strain sensors faces certain limitations. Due to their zero-dimensional shape, CB particles are prone to aggregation in the electrospun TPU matrix, which inevitably results in an increase in conductivity percolation and affects the overall conductivity of the strain sensor [97]. Additionally, the interactions between CB particles are extremely weak, which makes the constructed point-to-point mobile conductive network easily susceptible to destruction under strain, affecting the overall tensile performance of the strain sensor [98].

The section discusses the use of zero-dimensional CB as a conductive nanofiller in electrospun thermoplastic polyurethane (TPU) strain sensors. CB is valued for its dispersion in solvents, high conductivity, and small size. It forms a point-to-point conductive network in TPU, enhancing electromechanical properties and sensitivity to tensile strain. However, CB's tendency to aggregate can increase conductivity percolation and affect sensor conductivity. The section also highlights the use of CB/TPU composite nanofiber membranes and the creation of sandwich structures for improved sensor performance, including sensitivity, response time, and durability.

3.1.2. One-Dimensional Carbonaceous Material—CNTs

Among carbonaceous materials, carbon nanotubes (CNTs) are considered the most suitable conductive fillers for application in electrospun TPU flexible sensors. Compared to CB, CNTs have a larger aspect ratio, which results in extremely low conductivity percolation in CNT/TPU conductive composites [99]. Additionally, CNTs possess a typical hexagonal and linear entangled structure that is perfectly connected through chemical bonds. This allows them to form conductive pathways and construct conductive networks in electrospun TPU substrates even at low concentrations [100]. CNTs are categorized into single-walled carbon nanotubes (SWCNTs) and multi-walled carbon nanotubes (MWCNTs) based on the number of graphene layers within the carbon nanotube structure. Ultra-high aspect ratio CNTs are widely used in electrospun TPU strain sensors because they create an end-to-end mobile conductive network in the electrospun TPU substrate, providing the sensors with excellent electrical conductivity [101].

The conductivity of electrospun TPU strain sensors depends on the content and dispersion state of the conductive filler within the electrospun TPU matrix [102]. When the CNT content in the electrospun TPU substrate is appropriately increased, the distance between nanotubes decreases, forming a more stable end-to-end mobile conductive network, thus improving the sensor's overall conductivity [103]. Therefore, one way to prepare highly conductive electrospun TPU strain sensors is to appropriately increase the CNT content within the electrospun TPU matrix. Some researchers have modified TPU nanofiber membranes by rapidly synthesizing dopamine (DA) to form a polydopamine (PDA) coating on the surface, significantly improving the membranes' ability to load CNTs [66,104]. This is primarily due to DA undergoing oxidative self-polymerization with oxygen in humid air in a weakly alkaline environment, forming oligomers that continuously cross-link into polymers with higher molecular weights. DA, its oxidation products, and its polymers self-assemble in solution through various covalent and non-covalent bonds, forming the PDA layer. PDA has extremely strong adhesion ability and can enhance the deposition fastness of CNTs on the TPU nanofiber membrane surface. Meanwhile, the abundant hydroxyl and amino groups in PDA molecules increase the hydrophilicity of the TPU nanofiber membrane. This facilitates more CNTs in the suspension to penetrate the pores of the TPU nanofiber membrane, thus improving deposition efficiency. Flexible strain sensors prepared based on the CNT/DA/TPU composite nanofiber membrane exhibit excellent sensitivity (GF of 10,528.53 at 200% strain), fast response/recovery time (188/211 ms), wide sensing range (up to 200%), and excellent stability and durability [67] (Figure 4a). On this basis, some researchers explored constructing hydrophobic layers consisting of polyhedral oligomeric silsesquioxane (POSS), and 1H,1H,2H,2H-perfluorooctyltrimethoxysilane (FAS) on CNT/DA/TPU composite nanofiber mats. POSS as a protective layer increased the sensing range of the flexible sensor to 550%, while the lower surface energy of the -CF2- and -CF3 groups in FAS made the sensor superhydrophobic [68] (Figure 4b).

However, the tensile properties of the composites are negatively affected by increasing the content of CNTs in the TPU matrix [105]. To improve the electrical conductivity of TPU composites while maintaining their tensile properties, methods such as controlling the content of carbon nanotubes and lowering the percolation threshold have been explored. Several researchers have fabricated a high-performance CNT/TPU@SBS flexible strain sensor by constructing a double-percolation structure consisting of a continuous CNT/TPU phase and a styrene-butadiene-styrene (SBS) phase [106,107]. The introduction of the double-percolation structure, consisting of SBS phases incompatible with TPU, lowered the percolation threshold of the CNT/TPU@SBS strain sensor to as low as 0.38 wt.%, while the conductivity of 1%-CNT/TPU@SBS (4.12×10^{-3} S/m) was two orders of magnitude higher compared to that of 1%-CNT/TPU (3.17×10^{-5} S/m) [69]. This double-percolation structure also provides the 1%-CNT/TPU@SBS strain sensor with a wide strain detection range (0.2–100%), ultra-high sensitivity (maximum GF of 32,411 at 100% strain), fast response time (214 ms), and stability (500 loading/unloading cycles).

Since CNTs usually cluster together in the TPU matrix due to van der Waals forces, this affects the overall conductivity of the flexible strain sensors. Therefore, improving the dispersion state of CNTs in the TPU matrix is beneficial for preparing flexible strain sensors with higher electrical conductivity [108]. The most economical and common strategy to improve the dispersion of CNTs is the ultrasonic treatment method. This involves ultrasonically dispersing CNTs into a suspension by adding CNTs to deionized water and then immersing TPU nanofiber membranes into it for further ultrasonic treatment to prepare CNT/TPU composite nanofiber membranes [70]. CNT/TPU strain sensors based on the composite nanofiber membranes offer an excellent combination of high sensitivity (maximum GF of 1571), superior tensile strength and toughness (stresses greater than 24 MPa and strains greater than 400%), outstanding durability (10,000 cycles at 10% strain), and a wide operating tensile range (0–400%). Some researchers prepared a CNT/TPU composite nanofiber yarn with an elongation at a break of up to 476% by ultrasonically mixing TPU dissolved in a CNT dispersion and applying a multi-needle liquid bath electrospinning technique to the mixture [25]. Ultrasonic mixing followed by electrospinning significantly improved the dispersion of CNTs in the composite nanofiber yarn. A highly conductive dip-coated CNT/TPU composite nanofiber yarn (1.02 kΩ/cm) was developed by simply dip-coating the CNT ink. The yarn-based flexible strain sensor exhibits a high relative resistance change (440%) with satisfactory linearity and repeatability (1250 cycles) at 140% strain. Building on this, some researchers developed a TPU/CNT composite nanofibrous tube consisting of wavy and oriented fibers coated with carboxylated CNTs using spraying and ultrasound-assisted deposition [71] (Figure 4c). The combination of spraying and ultrasound-assisted deposition resulted in 12 wt% CNT deposition on the composite nanofiber tube, forming an excellent conductive network with electrical conductivity up to 0.01 S/cm. The as-prepared strain sensors exhibit a wide strain sensing range of up to 760%, high sensitivity (GF of 57), low detection limit (2%), and fast response time (45 ms) while maintaining reliable long-term stability and durability. Moreover, TPU/MWCNT strain sensors were prepared by filtering an aqueous dispersion of MWCNTs through electrospun TPU nanofiber membranes, drying them, welding them directly onto T-shirts, and encapsulating them with a thin silica gel layer [109]. Flexible strain sensors prepared with KMnO4-oxidized MWCNTs have a GF of up to 46 at an applied strain of 10%. In addition, there are other ways to achieve better conductivity in flexible strain sensors, such as increasing the level of dispersion by incorporating other conductive materials that aid in the dispersion of CNTs. This approach will be explained later when introducing other conductive materials for electrospun TPU strain sensors.

This section examines the application of one-dimensional carbonaceous material, specifically CNTs, in electrospun TPU flexible sensors. CNTs, including single-walled and multi-walled varieties, are praised for their aspect ratio and low percolation threshold, which contribute to the formation of stable conductive pathways in TPU substrates. The section details methods to improve CNT dispersion and loading, such as using polydopamine coatings and ultrasonic treatment, resulting in enhanced sensor sensitivity, response/recovery times, and durability.

3.1.3. Two-Dimensional Carbonaceous Material—Gr

Graphene (Gr) is a two-dimensional carbonaceous material consisting of a single layer of sp^2-hybridized carbon atoms with intrinsic conductivity typically between 10^3 and 10^6 S/cm [110]. The laminated Gr nanosheets have a high surface area, large specific surface area, and two-dimensional structure, making it easier to form a layer-connected planar-to-planar conductive network in the TPU substrate [111]. Simultaneously, the excellent electrical conductivity of Gr lowers the percolation threshold, reducing the impact on the tensile properties of TPU composite nanomaterials [112]. Moreover, the two-dimensional layered structure of Gr enables it to be tightly bonded to the TPU substrate, responding promptly to external stimuli and exhibiting excellent electrical sensitivity [113]. Therefore, Gr and its derivatives (e.g., reduced graphene oxide (rGO) and graphene oxide (GO)) are widely used as conductive fillers in electrospun TPU strain sensors due to their excellent electrical, thermal, mechanical, and sensing properties.

A common strategy is to ultrasonicate the electrospun TPU nanofiber membrane in a mixed solution of rGO and sodium dodecyl sulfate (SDS), and then polymerize the treated composite nanofiber membrane in a DA solution to obtain a PDA/rGO/TPU flexible strain sensor [72]. Here, PDA containing hydrophilic groups (-OH and $-NH_2$) improves the hydrophilicity and breathability of the composite nanofiber mats and prevents rGO from falling off the mats. PDA/rGO/TPU flexible strain sensors exhibit an excellent combination of properties, including high sensitivity (gauge factor of 6583 at strains above 140%), wide sensing range (0–150%), fast response time (100 ms), excellent durability (9000 cycles at 50% strain), and excellent gas permeability (water vapor evaporation rate of 38%). Building on this, some researchers explored 1H,1H,2H,2H-perfluorodecanethiol (PFDT) fluorination of PDA/rGO/TPU composite nanofibrous membranes by water vapor gasification [114]. PFDT fluoride treatment imparts superhydrophobicity so that the sensor, under any strain condition, repels water, acid, and alkali and exhibits corrosion resistance, with the contact angle increasing from 117° to 150°. This makes the sensor applicable to various harsh and complex working environments.

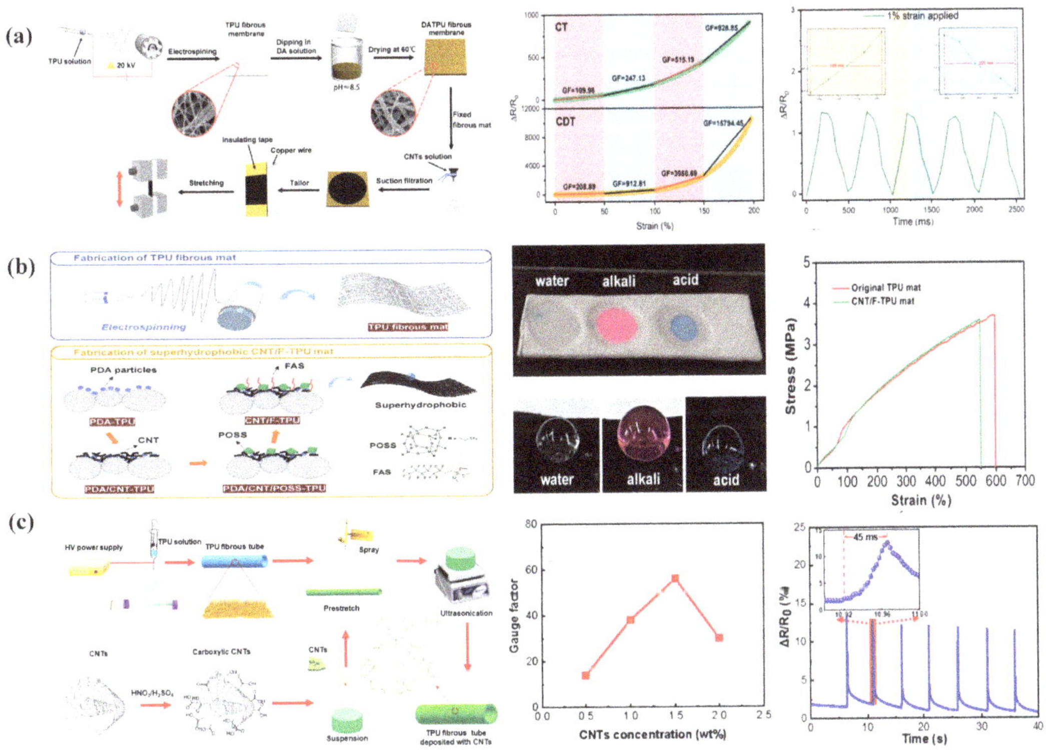

Figure 4. (**a**) Schematic illustration of the preparation process of a CNT/DA/TPU strain sensor; $\Delta R/R_0$ of CNT/TPU and CNT/DA/TPU strain sensor toward the applied strain; the response/recovery time of CNT/DA/TPU strain sensor under 1% strain. Reprinted with permission from Ref. [67]. Copyright 2023, Wiley. (**b**) Schematic illustration of the preparation process of CNT/F-TPU strain sensor; photographs of deionized water, 1 M alkaline, and 1 M acidic droplets on a raw TPU surface (top) and the CNT/F-TPU surface (bottom); comparison of the stress–strain curves of TPU and CNT/F-TPU. Reprinted with permission from Ref. [68]. Copyright 2023, American Chemical Society. (**c**) Schematic illustration of the preparation process of TPU/CNT strain sensor; the relation between GF of TPU/CNTs tube and CNT concentration; the response time of TPU/CNT strain sensor under 1% strain. Reprinted with permission from Ref. [71]. Copyright 2022, MDPI.

In addition, some researchers have explored the construction of various novel structures to further enhance Gr's conductivity in TPU flexible strain sensors. Researchers innovatively prepared thermoplastic polyurethane-based cellulose acetate composite nanomembranes (CA/TPU) with abundant mesopores by electrostatic spinning and then firmly anchored rGO and GO successively to them by ultrasonic impregnation to obtain an rGO/GO@CA/TPU flexible strain sensor [115] (Figure 5a). The porous structure, obtained by phase separation utilizing differences in solvent boiling point and solubility, has a high specific surface area and suitable pore size, greatly improving the adsorption degree and fastness of rGO and GO. The rGO on the surface of the CA/TPU composite nanofiber membrane interconnects to form a good conductive network, while GO prevents the overlapping of rGO during the stretching process, thereby improving the sensitivity and stability of the sensing electrical signal. This enables the rGO/GO@CA/TPU flexible strain sensor to achieve excellent sensitivity (gauge factor of 3.006) at extremely low strain (0.5%). However, increases in conductivity, sensitivity, and response time are often accompanied by decreases in ductility and sensing range. The simultaneous introduction of rGO and GO reduces the tensile resilience of the composite nanofiber membrane, and the rGO/GO@CA/TPU flexible strain sensor shows good testing results only in the low strain range (0.5–10%). Nevertheless, a team of researchers explored pressing and dragging layers of graphite nanoplates (GNPs) on a TPU substrate and then encapsulating them with polydimethylsiloxane (PDMS) as a protective layer to create a new crack-wrinkle structure (CWS) ultrasensitive strain sensor [73] (Figure 5b). The area proportions of cracked and wrinkled structures in this CWS flexible strain sensor are 31.8% and 9.5%, respectively. According to crack extension theory, when the CWS flexible strain sensor is subjected to a small stretch, the cracks in the GNP layer break, leading to a change in the contact area, which rapidly causes a significant change in resistance. This allows the CWS strain sensor to exhibit ultra-high sensitivity at low strains (gauge factor of 750 at 24% strain), ultra-low detection limits (a strain of 0.1%), and ultra-short response times (90 ms). Even though the introduction of the wrinkle structure reduces the irrecoverable damage of cracks and improves the recoverability and stability of the CWS flexible strain sensors to some extent, the CWS flexible strain sensors can still only work properly at a lower strain range (0.1–24%). To balance sensitivity and strain range, researchers prepared porous TPU–PEO composite nanofiber membranes by water-washed electrospinning and then stimulated them with anhydrous ethanol to obtain a porous curled network structure [74] (Figure 5c). The effective combination of porous and convoluted structures ensures high sensitivity (gauge factor of 4047.5 at 350% strain) of the composite nanofiber membrane after ultrasonic treatment with GNPs while achieving a wide strain operating range (0–350%). The TPU/GNP flexible strain sensor maintains excellent stability and durability after long-term operation (10,000 cycles).

Gr, a two-dimensional carbonaceous material, is explored in this section for its use in TPU strain sensors. Gr's layered structure and conductivity enable the creation of a planar conductive network in TPU, offering high sensitivity and a low percolation threshold. The section describes strategies to integrate Gr with TPU, such as ultrasonication and polymerization, to produce sensors with high sensitivity, fast response, and durability. It also discusses the construction of novel structures to enhance Gr's conductivity and the balance between sensitivity and strain range in sensor design.

3.1.4. Multi-Dimensional Hybrid Carbonaceous Materials

The conductive carbonaceous materials of different dimensions dispersed in the TPU substrate have different aggregation degrees and construct conductive networks with different structures. The various structures of the conductive network further affect the conductivity, percolation threshold, and tensile properties of the TPU composite nanomaterials, thus reflecting the differences in the sensing properties of the electrospun TPU flexible strain sensors in terms of tensile properties, sensitivity, sensing range, and other properties [75,116]. Generally speaking, the conductive network constructed by carbonaceous conductive materials with lower dimensionality is fragile in structure, easily fractured when stretched, and has higher electrical sensitivity, but can only work in a narrower strain range [117]. On the other hand, carbonaceous conductive materials with high dimensionality have high aspect ratios, which are easy to overlap and entangle with each other in the

TPU substrate, and the constructed conductive network is more stable and solid, showing a lower percolation threshold and a wider sensing range [76]. Hence, researchers have explored a variety of new ways to combine and tune conductive carbonaceous materials of different dimensions to combine their strengths, leading to the development of electrospun TPU flexible strain sensors with simultaneous high conductivity, high sensitivity, and wide sensing range.

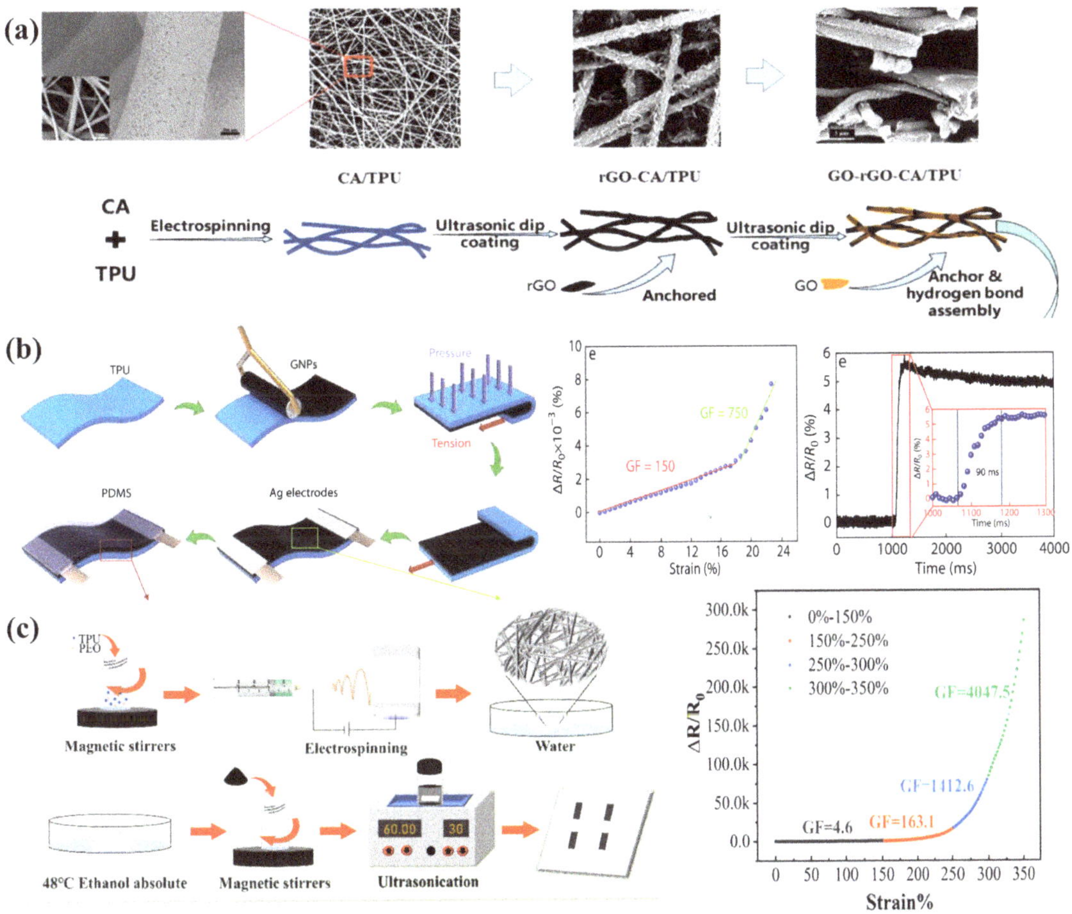

Figure 5. (**a**) Schematic illustration of the whole preparation process of the rGO/GO@CA/TPU strain sensor. Reprinted with permission from Ref. [115]. Copyright 2022, MDPI. (**b**) Schematic illustration of the preparation process of CWS strain sensor; $\Delta R/R_0$ of CWS strain sensor toward the applied strain; the response time of CWS strain sensor under 0.1% strain. Reprinted with permission from Ref. [73]. Copyright 2020, Springer Nature. (**c**) Schematic illustration of the preparation process of TPU/GNP strain sensor; relative change in resistance and tensile strain curve of TPU/GNP strain sensor. Reprinted with permission from Ref. [74]. Copyright 2024, American Chemical Society.

One research group utilized ultrasonic treatment to adsorb both CB and CNTs onto electrospun TPU nanofiber membranes, subsequently modifying them with PDMS to achieve low surface energy and create a superhydrophobic PDMS/(CB+CNTs)/TPU flexible strain sensor [77] (Figure 6a). CB, as the main filler, aggregates into clusters, while CNTs, serving as auxiliary fillers, intersperse between CB clusters to bridge the gaps, forming a more stable hybrid conductive network with "point-line" contact [78,79]. Meanwhile, PDMS acts as a binder, bonding the CB, CNTs, and TPU together, significantly improving the overall ductility of the composite nanofiber interface. This enables the PDMS/(CB+CNTs)/TPU flexible strain sensors to exhibit excellent sensitivity (GF up to 49,863.5 in the 431.5–437.9% strain range) and a wide strain operating range (0–437.9%). Another research group prepared a TPU/5CNT/5Gr flexible strain sensor by simultaneously depositing carboxylated CNTs and Gr on electrospun TPU nanofiber membranes in a 5:5 ratio via vacuum-assisted deposition and ultrasonication [80] (Figure 6b). Due to the π-π interaction, Gr nanosheets tend to stack, affecting the tensile properties of TPU composite nanofiber membranes [118]. However, the synergistic effect between Gr and CNTs mitigates this issue. Gr acts as a spacer to separate the intertwined CNTs, while the long and zigzag CNTs inhibit the overlapping of Gr nanosheets and fill the gaps between them, forming a three-dimensional electrically conductive network with "line-plane" contacts [119]. This design allows the TPU/5CNT/5Gr flexible strain sensor to simultaneously exhibit a wide operating range of 172% and a high strain measurement factor of approximately 217.

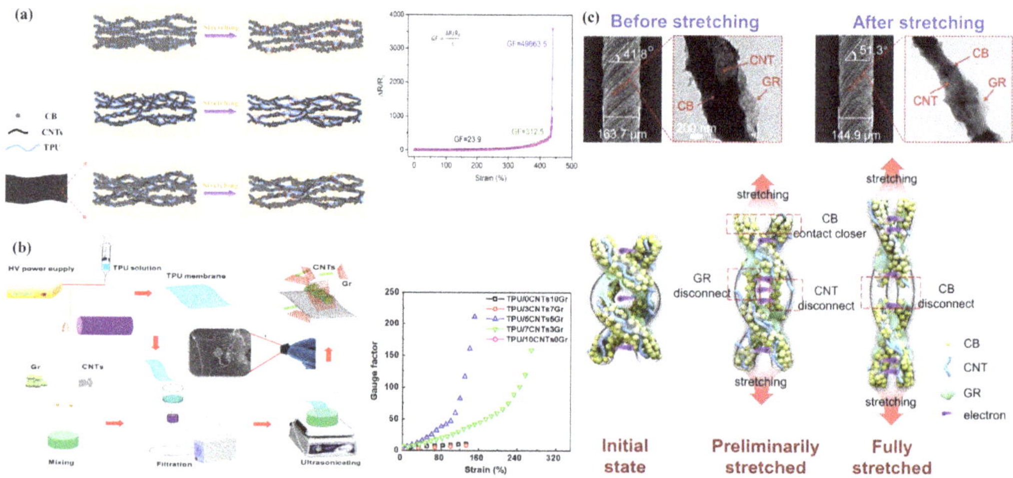

Figure 6. (**a**) Schematic diagram of the strain sensing mechanisms of the CB-TPU, CNT-TPU, (CB+CNTs)-TPU composites with different conductive networks; GF of PDMS/(CB+CNTs)/TPU strain sensor toward the applied strain. Reprinted with permission from Ref. [77]. Copyright 2022, Elsevier. (**b**) Schematic illustration of the preparation process of TPU/CNTs-Gr strain sensor; GF of TPU/CNTs-Gr strain sensor with different composition content toward the applied strain. Reprinted with permission from Ref. [80]. Copyright 2022, Wiley. (**c**) Structural evolution of the ternary conductive network SEM images of 24CB/3CNT/3GR-Y and TEM images of inside nanofiber before stretching and after stretching; schematic illustration of structural evolution of ternary conductive network under stretching. Reprinted with permission from Ref. [81]. Copyright 2022, Elsevier.

In addition, some researchers developed a highly stretchable and conductive MCN/TPU composite nanofiber yarn (MCN/TPU-NY) by combining CB, CNTs, and Gr ternary carbon nanomaterials with electrospun TPU nanofiber yarns via multi-needle liquid bath electrospinning and twisting techniques [81] (Figure 6c). By comparing the effects of different MCN compositions on sensor performance, it was found that CB-based fillers could achieve high electrical conductivity and maintain satisfactory mechanical properties under high loading (24 wt.%) conditions. When the concentration reaches the networking threshold (3 wt.%), 1D CNTs as auxiliary fillers can improve mechanical properties and strain operating range but weaken electrical conductivity. On the other hand, 2D Gr as auxiliary fillers can enhance electrical conductivity and sensitivity but reduce mechanical properties. The 24CB/3CNT/3GR flexible strain sensor, prepared by combining these three carbon nanofillers to construct a ternary conductive network, exhibits excellent sensing performance with high sensitivity (GF of 17.74 at 100% strain), a wide detection strain range (0–227%), fast response time (220 ms), and strong durability (5000 cycles at 50% and 100% strain).

This subsection explores the synergistic effects of combining carbonaceous materials of different dimensions in TPU strain sensors. Hybrid networks of zero-dimensional CB, one-dimensional CNTs, and two-dimensional Gr are designed to enhance conductivity, sensitivity, and strain range. Techniques like ultrasonic treatment and vacuum-assisted deposition are employed to achieve stable and efficient conductive networks. These hybrid structures address the limitations of individual fillers, offering improved sensor performance across various metrics.

3.2. MXene

Since its first synthesis in 2011, MXene (two-dimensional transition metal carbides and carbonitrides) has been widely used in applications such as wearable devices, energy storage, catalysis, and electromagnetic interference shielding [120]. Although more than 30 MXene materials have been synthesized in laboratory settings, the main form of MXene used in electrospun TPU flexible strain sensors is $Ti_3C_2T_x$. Here, T_x denotes various surface terminal groups (e.g., O, OH, F, and Cl) obtained by different synthesis methods [121]. By tuning the functional groups on the surface of MXene ($Ti_3C_2T_x$), it can be effectively combined with various conductive materials on the TPU substrate to form a synergistic effect, leading to the preparation of composite nanofibers with higher conductivity [122]. Furthermore, MXene ($Ti_3C_2T_x$) itself has excellent electrical conductivity, hydrophilicity, and mechanical properties, which are favorable for its dispersion in the electrospun TPU matrix, thus enabling the preparation of electrospun TPU flexible strain sensors with both excellent conductivity and ductility [123].

A basic strategy is to deposit the MXene sensing layer on the electrospun TPU nanofiber membrane by electrostatic spraying to prepare a sandwich-structured resistive MXene/TPU flexible strain sensor [124]. During the electrostatic spray deposition process, the agglomeration of MXene is reduced due to the Coulomb repulsion between conductive particles with the same charge. Meanwhile, the electrostatic spray deposition results in better adhesion of the prepared MXene sensing layer to the electrospun TPU nanofiber membrane compared with other deposition processes, leading to more uniform stress distribution during stretching and a wider strain operating range (extended to 70%) of the prepared MXene/TPU flexible strain sensor. On this basis, some researchers synthesized dopamine to form a polydopamine coating on electrospun TPU nanofiber membranes, then ultrasonically loaded MXene to prepare a high-performance MXene/PDA/TPU flexible strain sensor [82]. The PDA coating not only improves the adhesion between the MXene and the TPU but also combines with the MXene to enhance the EMI shielding performance of the prepared MXene/PDA/TPU flexible strain sensors (absolute effective shielding value of 8383.33 $dB \cdot cm^2 \cdot g^{-1}$). The MXene/PDA/TPU flexible strain sensor also exhibits a low detection threshold (0.1%), high sensitivity (GF up to 2600 at 200% strain), a wide strain operating range (as wide as 200%), and fast response/recovery times (77/186 ms). Furthermore, other researchers prepared a TPU/MXene/TPU-BNNS (TMTB) flexible sandwich strain sensor with advanced thermal regulation by introducing a TPU/boron nitride nanosheet (BNNS) thermally conductive layer assembled with the electrospun TPU substrate and the MXene sensing layer [83] (Figure 7a). The high orientation and strong van der Waals interactions between BNNS and TPU in the TPU/BNNS thermal conductivity layer facilitate the construction of effective thermal conductivity channels and reduce phonon scattering between the interacting units, greatly improving the thermal conductivity at a low filler permeability threshold (thermal conductivity up to 1.5 $W \cdot m^{-1} \cdot K^{-1}$). This unique structure simultaneously endows the prepared sensor with a wide strain operating range (0–100%), ultra-high GF (2080.9), and reliable biocompatibility and antimicrobial properties.

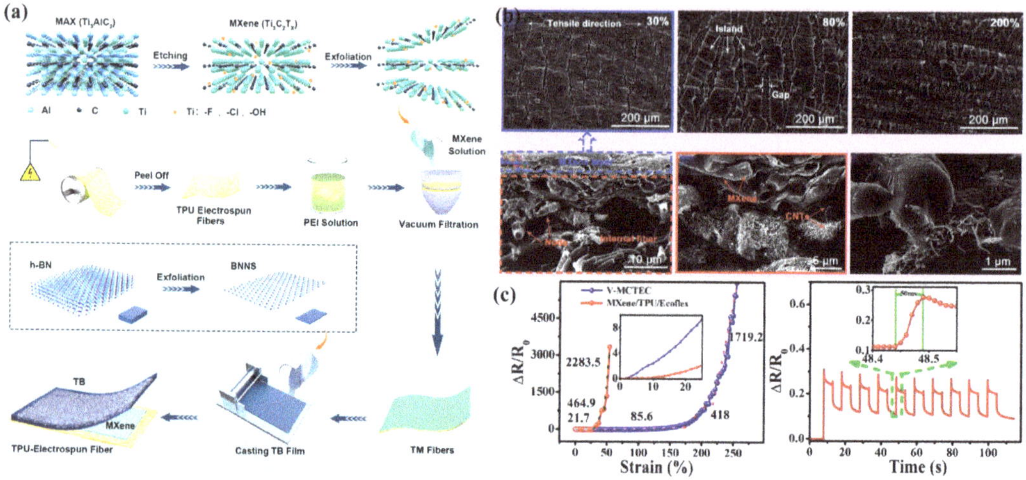

Figure 7. (**a**) Schematic illustration of the whole preparation process of the TPU/MXene/TPU-BNNS strain sensor. Reprinted with permission from Ref. [83]. Copyright 2024, American Chemical Society. (**b**) Crack morphology changes of MXene/CNT/TPU strain sensor under stretching strain of 30%, 80%, and 200%, respectively; cross-section SEM images of MXene/CNT/TPU composite film at different magnifications. Reprinted with permission from Ref. [84]. Copyright 2022, American Chemical Society. (**c**) Resistance changes with the strain of the vertical MXene-CNT/TPU/Ecoflex composites (V-MCTEC) and MXene/TPU/Ecoflex composites; the response time of V-MCTEC strain sensor under 0.1% strain. Reprinted with permission from Ref. [85]. Copyright 2024, Elsevier.

For MXene/TPU composite nanofiber membranes, because of the huge Young's modulus difference between the MXene and TPU matrices, the MXene conductive layer can be easily crushed into large fragments after the pre-stretching treatment to build an ultrasensitive microcracked structure [125]. However, the collapse of the microcracks under high strain leads to severe distortion and disintegration of the conductive coating. This causes the flexible strain sensors based on MXene/TPU composite nanofiber membranes to exhibit relatively high GF values and narrow strain sensing ranges. To address this issue, a common strategy is to utilize the synergistic effect of MXene and CNTs to construct excellent conductive networks with both high conductivity and strong tensile resistance in the TPU matrix [126]. The introduction of CNTs changes the direction of tensile stress transfer in MXene conductive coatings, resulting in smaller MXene fragments and denser microcrack structures. Meanwhile, CNTs can effectively bridge the MXene microcrack fragments, weakening the microcrack effect under tension and lengthening the separation process of neighboring microcrack fragments [127]. A research group has developed a MXene/CNT/TPU flexible strain sensor with a typical bilayer-conductive structure by self-assembling MXene and CNTs onto porous electrospun TPU nanofiber membranes via a vacuum filtration deposition process [84] (Figure 7b). During stretching, the brittle, densely stacked MXene upper lamellae crack, while the flexible MXene/CNTs-decorated TPU lower fibers connect the crack gaps, ensuring that the synergistic conductive network remains intact. This enables the MXene/CNT/TPU flexible strain sensors to exhibit high sensitivity and a wide operating range (GF values of 37.5, 289.5, 1070.1, and 2911 in the 0–50%, 50–150%, 150–250%, and 250–330% strain ranges, respectively), as well as a fast response time (80 ms) and superior durability (over 2600 cycles at 50% strain). On this basis, another research group developed a sandwich-structured MXene-CNT/TPU/Ecoflex flexible strain sensor by simultaneously coupling CNTs and MXene on wave-structured electrospun TPU nanofiber membranes through vacuum suction filtration and Ecoflex encapsulation [85] (Figure 7c). The low viscoelasticity of Ecoflex, combined with the wave-structured aligned TPU nanofiber network, was utilized to reduce the hysteresis (as low as 50 ms) of the prepared sensor when stretched in the vertical fiber direction. The MXene-CNT/TPU/Ecoflex flexible strain sensor also exhibits a low detection limit (0.1% strain), high sensitivity (GF of 1719.2 at 251% strain), and a wide sensing range (0.1–251% strain). Inspired by the pearl multilayer interaction structure, some researchers have developed a pearl-simulated multilayer structure of MGO/TPU flexible strain sensors by adding MXene/rGO (MGO) solution to multilayer TPU electrospun mats [128]. The Ti-O-C covalent bond formed by dehydration and nucleophilic substitution reactions between MXene and rGO strengthens the connection between MXene and rGO, enhancing the stretchability and sensing stability of the MGO/TPU flexible strain sensor (more than 5000 tensile cycles). Meanwhile, the pearl-like simulated multilayer structure allows the tuning of the operating range of the MGO/TPU flexible strain sensor (from 80% to 220%) by varying the mass ratio of MXene to rGO and the number of layers of TPU electrospun felts. Thanks to the synergy between MXene and rGO and the special design of the pearl-simulated multilayer structure, the MGO/TPU flexible strain sensor also demonstrates high sensitivity (GF > 84,326), fast response time (70 ms), and a low detection limit (0.05% strain).

MXene, a two-dimensional transition metal carbide, is discussed in this section for its role in electrospun TPU flexible strain sensors. MXene's conductivity, hydrophilicity, and mechanical properties make it suitable for creating composite nanofibers with enhanced conductivity and ductility. The section outlines strategies for depositing MXene layers and enhancing their dispersion in TPU, leading to sensors with high sensitivity, a wide strain range, and improved thermal regulation.

3.3. Metallic Materials

3.3.1. Silver Nanomaterials

Conventional metallic materials are not suitable for direct use in flexible strain sensors due to their hard texture and poor ductility; however, metallic nanomaterials are explored as conductive fillers due to their nanoscale size and are immobilized on electrospun TPU substrates through various processes (intercalation, coating, etc.) to create flexible strain sensors with excellent performance [129,130]. Among many metal nanomaterials, silver nanomaterials, such as silver nanowires (AgNWs) and silver nanoparticles (AgNPs), have been widely explored for the creation of high-performance flexible strain sensors by constructing various structures or synergizing with other conductive fillers (CNTs, MXene, etc.) in electrospun TPU substrates due to their high electrical conductivity (6.3×10^7 S/cm), excellent optical transparency, and appropriate yield strength [131–133].

Compared with one- or two-dimensional conductive materials with high aspect ratios, the nanoparticle morphology of AgNPs offers greater flexibility, which makes it more conducive to achieving reversible nanoscale recovery [134]. However, the high density of AgNPs and the poor adhesion between them and the electrospun TPU substrate make them susceptible to detachment by external forces, which affects the tensile and sensing stability of the prepared flexible strain sensors [135]. To solve this problem, some researchers developed a TPU/PDA/AgNP/SR flexible strain sensor by electroless deposition (ELD) of AgNPs on the surface of wavy electrospun TPU mats obtained by ethanol treatment with PDA and silicone rubber (SR) assistance [136]. The ethanol-treated electrospun TPU mats spontaneously curled to form a wavy structure, which, combined with the modification and assistance of PDA and silicone rubber (SR), significantly enhanced the deposition efficiency and fastness of AgNPs and improved the tensile and sensing stability of the sensors. Considering the high cost of DA, some other researchers suggested depositing AgNPs on the surface of electrospun TPU nanofiber felt with the aid of a mixed coating of tannic acid (TA) and hydrolyzable 3-aminopropyltriethoxysilane (APTES) (TA-APTES coating) to construct a composite conductive nanofiber material with a hierarchical AgNP shell and TPU nanofiber core microstructure [137]. The modification of the TA-APTES hybrid coating not only promotes the adsorption and immobilization of AgNPs on the surface of the electrospun TPU nanofiber mats but also enhances the interactions between the AgNPs and the electrospun TPU substrate, resulting in TPU@(TA-APTES)@AgNP flexible strain sensors that exhibit both high tensile properties (maximum strain of 565%) and high sensitivity (GF up to 6886). Moreover, some researchers have explored the use of the synergistic effect of zero-dimensional AgNPs and one-dimensional CNTs, two-dimensional MXene, to construct hybrid multidimensional conductive networks on electrospun TPU substrates, aiming to develop flexible strain sensors with excellent conductivity and stretchability [138,139]. A research group prepared a bilayer conductive TPU/Ag/CNT flexible strain sensor comprising a conductive layer of AgNPs and a conductive layer of CNTs by simultaneously anchoring AgNPs and CNTs on electrospun TPU nanofiber membranes via magnetron sputtering and ultrasonic treatment [86] (Figure 8a). During the stretching process, the rigid AgNP conductive layer continues to produce cracks until it deforms to form an "island structure"; and the high aspect ratio CNTs act as a "bridge" to connect the cracks in the "island structure" of the AgNP conductive layer, synergistically forming a more stable and stretch-resistant conductive network, thereby expanding the response range of the prepared strain sensor (up to 604% strain). The synergistic effect of AgNPs and CNTs also endows the TPU/Ag/CNT flexible strain sensors with excellent sensing performance, including high sensitivity (strain coefficient up to 6834), fast response and recovery time (122 ms/164 ms), and extremely low detection limit (0.1%). Notably, the oxidation of silver nanomaterials in air affects the stability and durability of flexible strain sensors during their application [140]. Therefore, a research group designed a dual conductive layer of AgNP/CNT/PDA/TPU flexible strain sensor leveraging the synergistic effect of AgNPs and CNTs by anchoring PDA on the surface of electrospun TPU fiber mats and then decorating CNTs and AgNPs by ultrasonication and in situ polymerization, respectively [141] (Figure 8b). The introduction of PDA not only improves the interaction between CNTs, AgNPs, and the interface with the electrospun TPU substrate, thereby broadening the detection range of the prepared sensors (up to 640%); it also effectively protects the AgNPs from oxidation during the working process, ensuring the response stability and long-term durability of the prepared sensors. Another research group modified the electrospun TPU nanofiber membrane by ultrasonic coating with MXene ($Ti_3C_2T_x$) followed by loading AgNPs by chemical deposition to prepare an AgNP/MXene/TPU flexible strain sensor utilizing the synergistic effect of AgNPs and MXene [142]. MXene ($Ti_3C_2T_x$) not only enhances the bonding stability and homogeneity between AgNPs and the electrospun TPU substrate through the rich chemical bonding on its surface but also disperses in the conductive network of AgNPs, acting as a "bridge" to enhance the electrical conductivity of the conductive network of AgNPs and its resistance to stretching. This enables AgNP/MXene/TPU flexible strain sensors to exhibit excellent EMI shielding performance (up to 87.31 dB at 8–12 GHz) as well as antimicrobial and hydrophobic properties, indicating great potential for applications in complex environments.

AgNWs have excellent thermal conductivity and chemical stability, but their unsatisfactory stretchability and poor adhesion to flexible substrates are important limitations to be addressed when applying them to electrospun TPU flexible strain sensors [143]. To address these limitations, and inspired by the convoluted structure of climbing plants, some researchers prepared a layered interlaced helical conductive yarn with both high mechanical elasticity and electrical conductivity by entangling the AgNW/MXene dimensional synergistic conductive network hierarchically with a network of electrospun TPU nanofibers [87] (Figure 8c). The high-aspect-ratio 1D AgNWs bridge the 2D MXene into a multidimensional synergistically interconnected conductive network, which intersperses with the electrospun TPU nanofiber network to form a unique hierarchical interlocking effect. This provides a robust channel for efficient electron/phonon transport in the layered helical conductive yarns, enabling the conductive yarns to exhibit high electrical conductivity (1.12×10^5 S/m) even under large mechanical strain (300%). The AgNW/MXene/TPU flexible strain sensors developed based on this layered helical conductive yarn have excellent electromechanical properties with a stable GF (1.7) even in the 600–1000% strain range.

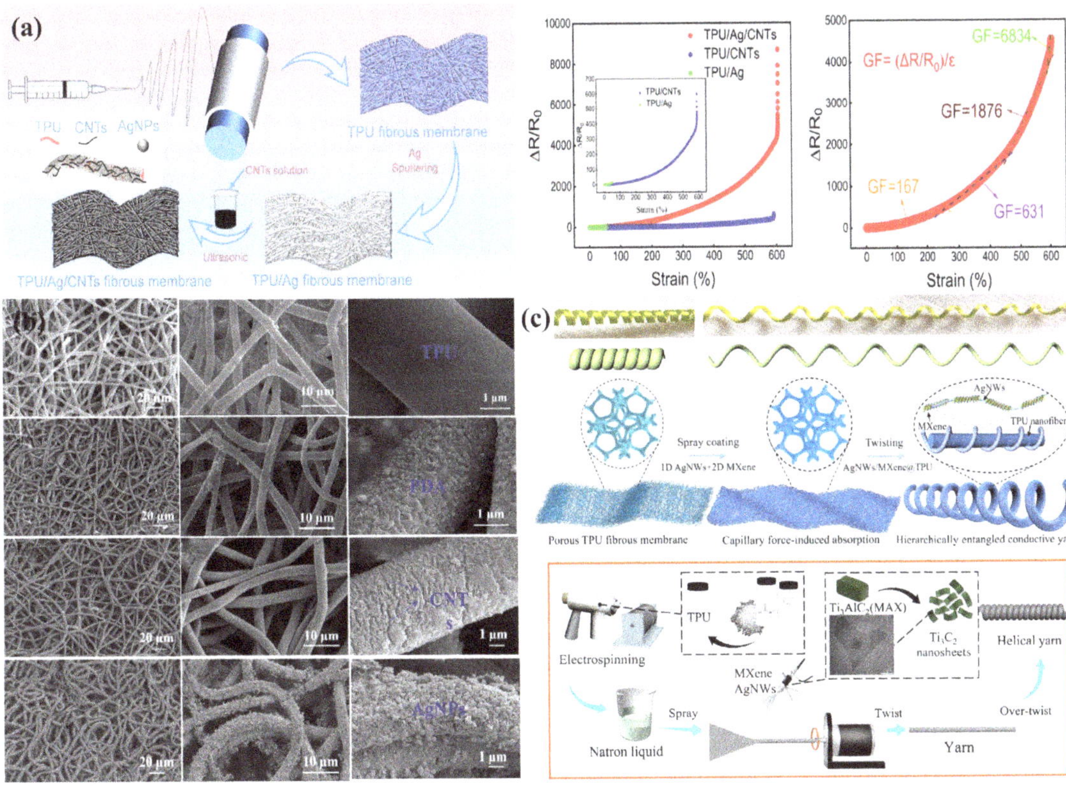

Figure 8. (**a**) Schematic illustration of the whole preparation process of the TPU/Ag/CNT fibrous membrane; relative resistance–strain sensing behaviors' comparison of the TPU/Ag/CNT strain sensor, TPU/CNT strain sensor, and TPU/Ag strain sensor; sensitivity of the TPU/Ag/CNT strain sensor in four regions from a 0 to 604% strain. Reprinted with permission from Ref. [86]. Copyright 2024, American Chemical Society. (**b**) SEM images of the surface for pure TPU mat (line 1), PDA/TPU mat (line 2), CNT/PDA/TPU mat (line 3), and AgNP/CNT/PDA/TPU composite (line 4), respectively. Reprinted with permission from Ref. [141]. Copyright 2023, Elsevier. (**c**) Photograph and schematic diagram of stretchable curling structure of climbing plant under original and stretching state; the structure evolution of hierarchically interlocked helical conductive yarn; schematic diagram of fabricating hierarchically interlocked helical conductive yarn via overtwisting the AgNWs/MXene multi-dimensional synergistic conductive network that entangled with elastic TPU nanofibers. Reprinted with permission from Ref. [87]. Copyright 2023, Elsevier.

This section focuses on silver nanomaterials, such as silver nanowires and nanoparticles, as conductive fillers in flexible strain sensors. The high electrical conductivity and optical transparency of silver nanomaterials make them ideal for sensor construction. The section discusses methods to improve the adhesion and dispersion of silver nanomaterials in TPU substrates, as well as the use of synergistic effects with other conductive materials to create multidimensional conductive networks for enhanced sensor performance.

3.3.2. Liquid Metal

Liquid metals (LM), such as eutectic gallium-indium and gallium-indium-tin alloys (GaIn and GaInSn), are cohesive, electrically conductive liquids with melting points close to room temperature [144]. Liquid metal not only overcomes the limitations of traditional metals, such as high Young's modulus and poor ductility, but also avoids the problems of high toxicity, low conductivity, and low vapor pressure associated with ionic liquid solvents [145]. It exhibits excellent properties such as low Young's modulus, fluidity, low viscosity, high conductivity, and low toxicity, and has significant potential for applications in electrospun TPU flexible strain sensors [146].

A commonly used approach is to create deformable conductive fibers (DCFs) by injecting LM into hollow electrospun TPU nanofibers [147]. Due to the thermoplasticity of TPU and the high electrical conductivity of LM, DCFs can be molded into complex 2D or 3D shapes by heating at low voltage to the two thermoplastic transition points of TPU. Additionally, leveraging LM's solid-liquid phase transition at its melting point, DCFs demonstrate variable shape memory properties at these transition points. The manufactured spiral DCF flexible strain sensors show excellent electrical performance (resistance change <2%) at a maximum strain of 3300%. However, LM potentially detaches from the hollow of the electrospun TPU nanofibers under external force, thus affecting the electrical conductivity of the composite fibers. Therefore, some researchers explored constructing a flexible 3D conductive network by coating LM droplets onto electrospun TPU nanofiber membranes via mechanical stretching to prepare an LM–TPU flexible strain sensor [88]. LM droplets are suspended between electrospun TPU nanofibers and coalesce into a conductive network. This enables the LM–TPU flexible strain sensors to exhibit an array of desirable properties, including a wide strain range (0–200%), good biocompatibility (can be used directly on the skin), and excellent stability and durability (9000 cycles). However, the simple coating renders the sensor less sensitive (GF of only 0.2 at 200% strain), and the lack of protection makes it susceptible to damage by the external environment, limiting its mass production and application. To solve these problems, some researchers proposed preparing ultra-high stretch, low-cost TPU/LM/NFs nanoyarns by printing LM onto the surface of electrospun TPU nanoyarns followed by encapsulation with nanofibers to form a protective layer [89]. The patterned printing of LM onto the surface of electrospun TPU nanoyarns makes it smooth, enhances interfacial interactions between the LM and TPU substrate, and improves the conductivity of the prepared TPU/LM/NFs flexible strain sensor (GF increased to 6 at 200% strain), while the protection provided by the outer nanofibers allows the sensor to exhibit high mechanical strength (40 MPa) and a wide sensing range (548%).

Liquid metals, with their low melting points and high conductivity, are explored in this section for their potential in electrospun TPU flexible strain sensors. The section describes approaches to creating deformable conductive fibers and flexible strain sensors using liquid metals, highlighting their biocompatibility, stability, and durability. It also discusses the challenges of maintaining electrical conductivity and environmental protection in liquid metal-based sensors.

3.4. Conductive Polymers

Conductive polymers belong to a class of polymers with exceptional electrical properties, showing tunable conductivity ranging from semiconductors to metallic conductors due to their intrinsic delocalized π-electrons [148]. Conductive polymers are also being explored to improve the sensitivity, sensing range, and durability of electrospun TPU flexible strain sensors because of their low cost, ease of synthesis, long-term stability, and non-toxicity [149]. Currently, typical conductive polymers like polypyrrole (PPy), poly(3,4-ethylenedioxythiophene):poly(styrenesulfonic acid) (PEDOT:PSS)), and polyaniline (PANI) are extensively used for the preparation of high-performance electrospun TPU flexible strain sensors.

PPy possesses excellent electrical conductivity, thermal stability, ease of preparation, and biocompatibility, making it ideal for direct use as a conductive filler in electrospun TPU flexible strain sensors [150]. Some researchers proposed preparing a fluorescent visible yarn strain sensor (SCFY strain sensor) based on a cured bead and crack structure by loading polydopamine (PDA) and polypyrrole (PPy) to create in situ designed cracks after heterostructurally designing a PDMS cured bead on the surface of the fluorescent fiber of TPU-TPE [90] (Figure 9a). The functional groups (e.g., ketone group and hydroxyl group) in PDA can form hydrogen bonds with N atoms in PPy as hydrogen bond donors or acceptors, which enhances the fastness of the conductive layer, thereby widening the working strain range (0–184%) of the prepared sensors. The PDMS solidified liquid beads and fibers exhibited stress mutation at the interface, causing more cracks in the PPy conductive layer under stress, thereby effectively enhancing the sensitivity of the sensor (GF value of 58.9 in the strain range of 143–184%). Moreover, the presence of TPE caused the cracks to emit blue fluorescence under UV lamp irradiation, enabling fluorescence visualization of the cracks.

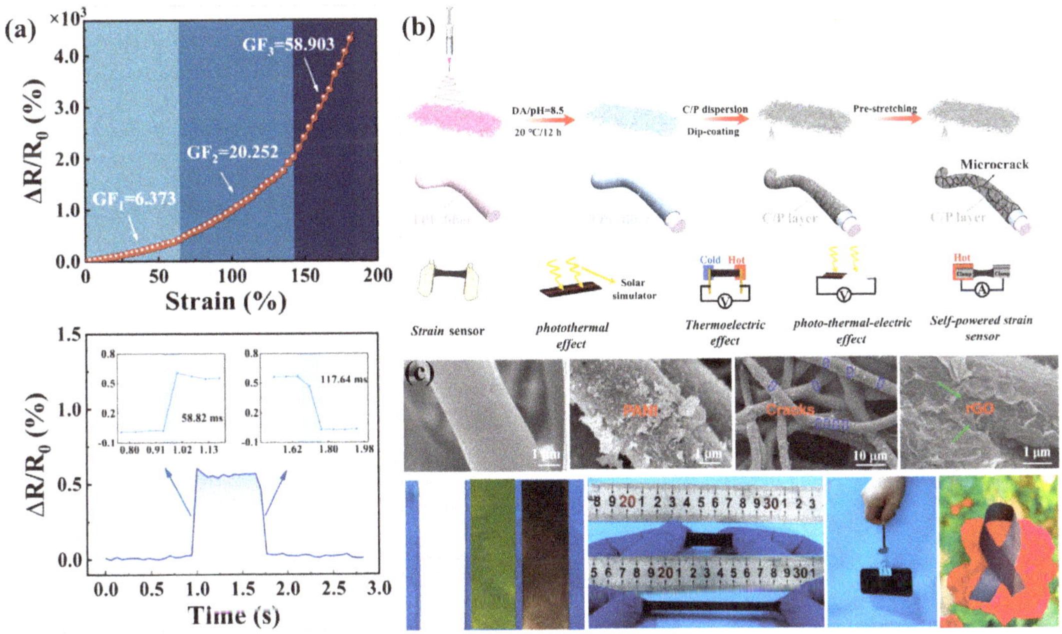

Figure 9. (**a**) GF of SCFY strain sensor toward the applied strain; the response time of SCFY strain sensor under 1% strain. Reprinted with permission from Ref. [90]. Copyright 2023, American Chemical Society. (**b**) Preparation process and characterization of CPT conductive fibrous membrane. Reprinted with permission from Ref. [91]. Copyright 2024, Elsevier. (**c**) SEM images of virgin TPU, PANI/TPU mats, and PANI/TPU mat after pre−stretching and GPTM; digital images of TPU, PANI/TPU, and the GPTM; the stretchability of the GPTM; photograph of GPTM hanging a phone; lightweight and flexibility of samples. Reprinted with permission from Ref. [92]. Copyright 2022, Elsevier.

Poly(3,4-ethylenedioxythiophene) (PEDOT), prepared by substituting thiophene rings at positions 3 and 4, demonstrates high corrosion resistance and stability in complex, harsh environments [151]. PEDOT also exhibits cost-effectiveness, superior electrical conductivity, and higher transparency than other conductive polymers [152]. However, this conductive polymer is insoluble in water or organic solvents, making it difficult to apply to electrospun TPU flexible strain sensors via impregnation or spin-coating techniques. Therefore, a negatively charged water-soluble component (PSS) was used to stabilize the positively charged PEDOT dispersion, resulting in the preparation of poly(3,4-ethylenedioxythiophene):poly(styrenesulfonic acid) (PEDOT:PSS)), which is easily applied to electrospun TPU flexible strain sensors [153]. A research team prepared a composite nanofiber membrane with a three-dimensional conductive network skeleton by dip-coating a mixed conductive filler of carboxylated multi-walled carbon nanotubes (C-MWCNT) and poly(3,4-ethylenedioxythiophene):poly(styrenesulfonic acid) (PEDOT:PSS)) with synergistic conductive effects onto an electrospun TPU nanofiber membrane substrate [91] (Figure 9b). Due to the Young's modulus difference between the hybrid conductive layer and the electrospun TPU substrate, a microcrack structure was constructed on the surface of the composite nanofiber membrane by pre-stretching, and the strain sensitivity of the glycol-treated e-CPT9/1 flexible strain sensor (mass ratio of C-MWCNT to PEDOT:PSS is 9/1) was efficiently tuned by varying the density of the microcracks (GF = 6008.3). Meanwhile, the detection range (0.5–680%) of the e-CPT9/1 flexible strain sensor was broadened due to the synergistic effect of C-MWCNT bridging adjacent PEDOT crack fragments. In addition, the prepared e-CPT9/1 flexible strain sensor also exhibits remarkable "photo-thermal-electrical" conversion capability due to the excellent photo-thermal and thermoelectric effects of C-MWCNT and PEDOT.

Polyaniline (PANI) is currently the most studied conductive polymer due to its strong polarity, low cost, high processability, environmental stability, and unique doping/de-doping mechanism [3]. The strong polarity of PANI induces high electrical conductivity, but it also leads to unsatisfactory elasticity of PANI [154]. Combining PANI with TPU improves the elasticity, but the strong polarity limits the ease of combination. In situ polymerization is a good method to combine electrospun TPU nanofiber membrane and PANI explored by many researchers [155]. A research group explored the simultaneous in situ polymerization of MWCNT and PANI on electrospun TPU nanofiber membranes to prepare a PANI-TPU/MWCNT flexible strain sensor with excellent performance [33]. Due to the loading of PANI and the good dispersion of MWCNT in the TPU matrix, PANI and MWCNT synergistically constructed a multidimensional hybrid conductive network with both high conductivity and strong stretchability. In addition, another research group assembled a flexible wearable rGO/PANI/TPU strain sensor (GPTSS) prepared by decorating polyaniline (PANI) nanoparticles bridged by reduced graphene oxide (rGO) nanosheets onto electrospun TPU nanofiber mats via in situ polymerization and ultrasonic treatment [92] (Figure 9c). The in situ polymerization of PANI nanoparticles coated on the electrospun TPU pads improved the interfacial adhesion to some extent and enhanced the deposition efficiency of rGO. The ultrasonic treatment introduces rGO nanosheets to bridge the PANI nanoparticles so that the synergistically constructed hybrid conductive network is easy to recover in time after damage. This enabled the preparation of a flexible GPTSS that simultaneously exhibits a wide strain range (0.1–300% strain), high sensitivity (GF up to 3000.2), fast response time (90 ms), and excellent long-term durability. In addition, the GPTSS is also able to accurately monitor NH3 (5 ppm) due to the H+-doped structure of the PANI nanoparticles.

Conductive polymers, such as polypyrrole, PEDOT:PSS, and polyaniline, are examined in this section for their use in enhancing the performance of electrospun TPU flexible strain sensors. These polymers offer tunable conductivity, low cost, and stability. The section details methods of incorporating conductive polymers into TPU substrates, such as in situ polymerization and the creation of synergistic conductive networks, resulting in sensors with improved sensitivity, response times, and durability.

4. The Application of Electrospun TPU Strain Sensors

Overall, flexible strain sensors based on electrospun TPU have potential applications in motion detection, health monitoring, and human–computer interaction due to their excellent sensitivity, wide strain operating range, and fast response time with excellent stability and durability [156–158].

4.1. Motion Detection

In the era of IoT, Industry 4.0, and big data analytics, there is an increasing demand for real-time monitoring of human activities, which has spurred the development and application of more flexible and sophisticated wearable strain sensors in the domain of motion monitoring [71,97,128]. These sensors are designed to cater to the diverse spectrum of human movement, ranging from large-scale joint movements such as those of the hands, arms, and legs, to the minute details of facial expressions and physiological activities like smiling, blinking, and swallowing [159]. A research team, leveraging the synergies of IoT and big data, has engineered a CNT/F-TPU flexible strain sensor that boasts exceptional stretchability and heightened sensitivity [68]. This sensor, as depicted in Figure 10a, when affixed to the wrist, elbow, and knee of a subject, captures continuous, periodic, and stable sensing signals even as these joints undergo bending. This capability underscores the sensor's proficiency in monitoring a broad spectrum of human body movements. Furthermore, the CNT/F-TPU strain sensor, when attached to the subject's finger, demonstrates its ability to monitor even subtle movements by producing a regular response curve as the bending angle incrementally increases from 0° to 30°. Additionally, when positioned on the face and neck, the sensor mirrors the frequency of the tester's facial expressions and swallowing actions, with a rapid increase in the $\Delta R/R_0$ value, thereby showcasing its potential to capture minute physiological changes. Meanwhile, another research collective has fabricated a highly stretchable and conductive CNT/TPU composite nanofiber yarn, ingeniously integrating it into an elastic self-adhesive bandage [25]. This smart sports bandage, as illustrated in Figure 10b, not only provides real-time guidance to beginners in sports like badminton by analyzing the peak characteristics of the relative resistance change (RRC) but also aids basketball players in refining their shooting postures and improving their scoring rates by sensing the elbow's bending during the shooting motion. Moreover, when wrapped around the knee, the bandage's sensor can monitor the strain and calculate the frequency and amplitude of leg movements, offering insights into the athlete's performance and travel status. When applied to the chest and abdomen, the sensor functions as a real-time health monitor, tracking heart rate, respiratory rate, and exercise intensity, thereby ensuring safety during physical activities and preventing potential risks associated with arrhythmia or respiratory distress. These innovations, powered by IoT connectivity, are a testament to the seamless integration of advanced materials, sensor technology, and data analytics, paving the way for a new horizon in personalized health monitoring and sports performance enhancement in the context of Industry 4.0 [113,125,144].

4.2. Health Monitoring

In the burgeoning field of digital health, the advancement of electrospun TPU flexible strain sensors plays a pivotal role in health monitoring [91,96,131]. These sensors are engineered to exhibit heightened sensitivity to the nuanced physiological signals that are often encountered in diverse and challenging environmental conditions [161]. Given the prolonged skin contact of such wearable devices, their design must also prioritize breathability and biocompatibility to ensure comfort and safety during extended use [162]. Several researchers have delved into the innovation of a 3D woven structure of MXene/AgNW/TPU core yarn sensors, known as KMYS, which have been seamlessly integrated into an array of apparel through the application of inlay yarn guide technology [129]. As shown in Figure 10c, the KMYS has been skillfully incorporated into various flexible wearable devices such as pendants, elbow pads, wristbands, finger cuffs, socks, and face masks. These devices are capable of sensing physiological rhythms, including the pulses of the carotid, brachial, radial, digital, and ankle arteries, as well as respiratory movements. The KMYS demonstrates its versatility by not only detecting the pronounced pulse signals of the carotid, brachial, and radial arteries but also by accurately capturing the subtler pulse waves of the finger and ankle arteries. A 31-day pulse monitoring experiment conducted on a 30-year-old female subject revealed that the KMYS could reliably capture daily pulse waveforms over an extended period, providing a precise indication of cardiovascular health. Furthermore, the KMYS, when worn on the abdomen, can monitor breathing and pulse signals during sleep, offering a proactive approach to preventing life-threatening conditions such as sleep apnea. The sensor's resilience is also evident in its ability to maintain stable pulse signal monitoring even when submerged in water, as demonstrated by tests where the KMYS was worn on the wrist and immersed. This capability underscores the sensor's potential for capturing high-fidelity physiological signals for health detection across a multitude of complex application scenarios, thereby reinforcing its role in the realm of digital health monitoring [103,114,147].

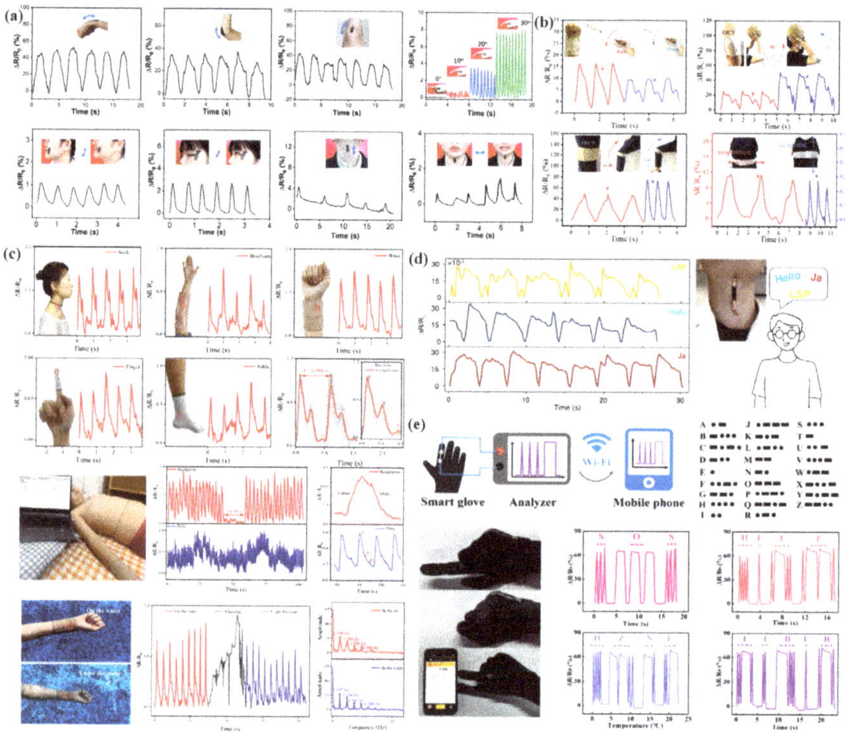

Figure 10. (**a**) Monitoring human body movement via a corresponding relative resistance change ($\Delta R/R_0$) signal of the CNT/F-TPU strain sensor: wrist movement; elbow movement; knee bending; finger movement with different bending degrees (0–30°); microexpressions of smiling; blinking; swallowing; pronouncing the word "work". Reprinted with permission from Ref. [68]. Copyright 2023, American Chemical Society. (**b**) Application of the smart sports bandage in badminton, basketball, walking and running, monitoring respiration and heartbeat. Reprinted with permission from Ref. [25]. Copyright 2022, Elsevier. (**c**) Demonstration of the KMYS for physiological signal detection under different conditions: pulse waveform at the carotid artery, brachial artery, radial artery digital artery, and ankle artery; the pulse signals (gray) and averaged pattern (red) using the KMYS; photograph showing the KMYS was worn against the abdomen of an overweight 10-year-old child for respiration monitoring; the respiration and pulse signals from the children during sleeping; the magnified views of respiration and pulse signals; photograph showing the KMYS was attached to the wrist on and under the water, respectively; pulse signals in different states; spectral information embedded within the pulse wave. Reprinted with permission from Ref. [129]. Copyright 2024, Elsevier. (**d**) Sensing application of the TPU/CB strain sensor in human motion: speaking different words with sample fixing at the neck. Reprinted with permission from Ref. [64]. Copyright 2021, Springer. (**e**) Information encryption and transmission based on Morse code realized by the IL/TPU ionogel fiber-shaped strain sensor; a scheme for connecting the sensor to a wireless transmitter; schematic diagram of Morse code; photograph of the sensor attached to a finger to send information; encryption and translation of "SOS", "HELP", "HZNU", and "FIBER". Reprinted with permission from Ref. [160]. Copyright 2022, American Chemical Society.

4.3. Human–Computer Interaction

In the burgeoning landscape of AI and robotics, the quest for more sensitive and intelligent human–computer interaction is driving innovation. Electrospun TPU flexible strain sensors, known for their high conductive sensitivity and exceptional tensile recovery performance, are being explored as cutting-edge wearable devices that revolutionize the domain of human–computer interaction [163]. As depicted in Figure 10d, a TPU/CB flexible strain sensor has been developed for voice interaction [64]. This sensor, when mounted on the neck, adeptly captures the minute vibrations of the user's vocal cords during speech. By analyzing the resultant response signal, it is possible to calculate the number of words spoken and the frequency of speech, thereby aiding the computer in speech recognition and enhancing the efficiency of human–computer interaction. Figure 10e introduces a TPU/IL ionogel fiber-like sensor designed for information encryption and transmission [160]. This sensor is integrated into a glove, transforming the human hand into a conduit for encrypted communication. By defining short and long finger bends as dots and dashes, respectively, Morse code is employed to encrypt letters or numbers [164]. These encrypted messages are then transmitted wirelessly to a cell phone, thereby broadening the horizons of advanced human–computer interaction. The integration of AI and robotics in these wearable sensors exemplifies the seamless fusion of technology and human capability, offering a new paradigm for secure, efficient, and intuitive communication between humans and machines. This innovation not only propels the field of human–computer interaction forward but also underscores the potential of AI and robotics in creating more responsive, intelligent, and interactive devices that can adapt to the user's needs.

5. Conclusions and Outlook

In recent years, flexible strain sensors have been increasingly developed for applications in motion monitoring, health detection, human–computer interaction, and other areas of high interest. The TPU nanofiber material fabricated by electrospinning combines the advantages of both nanomaterials and fiber materials, with excellent characteristics such as extremely fine size, large specific surface area, high porosity, controllable structure, and functional modification by doping, making it very suitable for the preparation of flexible strain sensors as a substrate material in combination with various conductive fillers. Therefore, it is necessary to review the flexible strain sensors based on this special substrate material. This review paper introduces and discusses recent advances in electrospun TPU flexible strain sensors, focusing on summarizing strategies for introducing a variety of conductive fillers, including carbonaceous materials, MXene, metallic materials, and conductive polymers.

(1) As we stand on the precipice of a transformative era in the field of flexible strain sensors, the future holds a wealth of opportunities and challenges. Enhancing material properties is central to ongoing research, with scientists striving to achieve a more uniform distribution of conductive fillers within the microstructure of materials. Exploring novel composites and nanomaterials, such as carbon nanotubes, graphene, and metallic nanoparticles, is expected to significantly improve the sensitivity, response time, and durability of strain sensors.

(2) The integration of electrospun TPU flexible strain sensors with AI and IoT presents a particularly promising avenue. AI algorithms can enable real-time data processing and analysis, leading to more precise health monitoring and predictive maintenance in industrial settings. IoT integration facilitates seamless data transmission, which is essential for remote monitoring and control systems in various applications.

(3) Innovations in manufacturing techniques, such as the fusion of 3D printing with electrospinning, are set to revolutionize the development of electrospun TPU flexible strain sensors. This hybrid approach allows for the creation of sensors with complex geometries and tailored mechanical properties to meet specific application requirements.

(4) Despite the optimistic outlook, challenges remain. The long-term stability and durability of sensors under real-world conditions are significant concerns. Sensors must be

able to withstand environmental factors such as temperature fluctuations, humidity, and mechanical stress without performance degradation. Addressing these issues requires rigorous material testing and the development of robust encapsulation techniques.

(5) Biocompatibility and safety are paramount, especially in applications where sensors come into direct contact with the human body. Extensive research is needed to ensure that sensor materials do not induce adverse reactions and are safe for long-term use. This includes understanding the interaction between sensor materials and biological systems at the molecular level.

(6) Establishing standardized testing protocols and regulatory frameworks is critical as the technology evolves. Universally accepted standards are necessary to evaluate the performance, reliability, and safety of electrospun TPU flexible strain sensors, facilitating their commercialization and building consumer trust.

(7) Cost remains a significant barrier to widespread adoption. Research into scalable manufacturing processes and the use of cost-efficient materials is essential for making these sensors more accessible to a broader market.

(8) Overcoming these challenges necessitates a collaborative approach, involving experts from diverse fields such as materials science, electrical engineering, biomedical engineering, and data science. Interdisciplinary research will foster innovation and accelerate the development of electrospun TPU flexible strain sensors capable of meeting future demands.

In conclusion, while the future of electrospun TPU flexible strain sensors is promising, it requires a concerted effort to navigate the complexities of material development, technological integration, and practical implementation. By addressing the challenges head-on and fostering a collaborative research environment, we can unlock the full potential of these sensors and usher in a new era of smart and responsive technologies that enhance our lives and industries.

Author Contributions: Z.Z. and W.T.: writing–conceptualization, visualization, editing, and visualization; T.X. and W.Z.: reviewing, data curation, and editing; J.Z. and C.B.: conceptualization. All authors have read and agreed to the published version of the manuscript.

Funding: This research was funded by the Key Project of Research and Development in Hainan Province (Grant No. ZDYF2023GXJS159), the Natural Science Foundation of Hainan Province (Grant No. 222MS008), and the Education Department of Hainan Province (Grant No. Hnky2020-1).

Data Availability Statement: The data used to support the findings of this study are available from the corresponding author upon request.

Conflicts of Interest: The authors declare no conflicts of interest.

References

1. Luo, Y.F.; Abidian, M.R.; Ahn, J.H.; Akinwande, D.; Andrews, A.M.; Antonietti, M.; Bao, Z.N.; Berggren, M.; Berkey, C.A.; Bettinger, C.J.; et al. Technology Roadmap for Flexible Sensors. *ACS Nano* **2023**, *17*, 5211–5295. [CrossRef]
2. Li, H.; Chen, J.W.; Chang, X.H.; Xu, Y.Q.; Zhao, G.Y.; Zhu, Y.T.; Li, Y.J. A highly stretchable strain sensor with both an ultralow detection limit and an ultrawide sensing range. *J. Mater. Chem. A* **2021**, *9*, 1795–1802. [CrossRef]
3. Liu, D.; Zhou, H.H.; Zhao, Y.Y.; Huyan, C.X.; Wang, Z.B.; Torun, H.; Guo, Z.H.; Dai, S.; Xu, B.B.; Chen, F. A Strand Entangled Supramolecular PANI/PAA Hydrogel Enabled Ultra-Stretchable Strain Sensor. *Small* **2022**, *18*, 2203258. [CrossRef]
4. Wei, Q.K.; Chen, G.R.; Pan, H.; Ye, Z.B.; Au, C.; Chen, C.X.; Zhao, X.; Zhou, Y.H.; Xiao, X.; Tai, H.L.; et al. MXene-Sponge Based High-Performance Piezoresistive Sensor for Wearable Biomonitoring and Real-Time Tactile Sensing. *Small Methods* **2022**, *6*, e2101051. [CrossRef]
5. Wang, X.J.; Xue, R.; Li, M.Z.; Guo, X.Y.; Liu, B.; Xu, W.G.; Wang, Z.; Liu, Y.Q.; Wang, G.H. Strain and stress sensing properties of the MWCNT/TPU nanofiber film. *Surf. Interfaces* **2022**, *32*, 102132. [CrossRef]
6. He, Z.; Byun, J.H.; Zhou, G.H.; Park, B.J.; Kim, T.H.; Lee, S.B.; Yi, J.W.; Um, M.K.; Chou, T.W. Effect of MWCNT content on the mechanical and strain-sensing performance of Thermoplastic Polyurethane composite fibers. *Carbon* **2019**, *146*, 701–708. [CrossRef]
7. Liu, L.P.; Niu, S.C.; Zhang, J.Q.; Mu, Z.Z.; Li, J.; Li, B.; Meng, X.C.; Zhang, C.C.; Wang, Y.Q.; Hou, T.; et al. Bioinspired, Omnidirectional, and Hypersensitive Flexible Strain Sensors. *Adv. Mater.* **2022**, *34*, 2200823. [CrossRef]

8. Liu, E.Z.; Cai, Z.M.; Ye, Y.W.; Zhou, M.Y.; Liao, H.; Yi, Y. An Overview of Flexible Sensors: Development, Application, and Challenges. *Sensors* **2023**, *23*, 817. [CrossRef]
9. Wang, X.; Gao, Q.S.; Schubert, D.W.; Liu, X.H. Review on Electrospun Conductive Polymer Composites Strain Sensors. *Adv. Mater. Technol.* **2023**, *8*, 2300293. [CrossRef]
10. Zheng, Q.B.; Lee, J.H.; Shen, X.; Chen, X.D.; Kim, J.K. Graphene-based wearable piezoresistive physical sensors. *Mater. Today* **2020**, *36*, 158–179. [CrossRef]
11. Chen, T.J.; Xie, Y.T.; Wang, Z.Y.; Lou, J.X.; Liu, D.Y.; Xu, R.X.; Cui, Z.Y.; Li, S.M.; Panahi-Sarmad, M.; Xiao, X.L. Recent Advances of Flexible Strain Sensors Based on Conductive Fillers and Thermoplastic Polyurethane Matrixes. *ACS Appl. Polym. Mater.* **2021**, *3*, 5317–5338. [CrossRef]
12. Yu, H.J.; An, H.S.; Shin, J.H.; Brunetti, A.; Lee, J.S. A new dip-coating approach for plasticization-resistant polyimide hollow fiber membranes: In situ thermal imidization and cross-linking of polyamic acid. Chem. *Eng. J.* **2023**, *473*, 145378. [CrossRef]
13. Lu, Y.; Li, W.S.; Zhou, J.F.; Ren, Y.; Wang, X.F.; Li, J.; Zhu, S. Strengthening and toughening behaviours and mechanisms of carbon fiber reinforced polyetheretherketone composites (CF/PEEK). *Compos. Commun.* **2023**, *37*, 101397. [CrossRef]
14. Kyeong, M.; Chae, J.E.; Lee, S.Y.; Lim, T.H.; Kim, M.; Lee, S.S.; Song, K.H.; Kim, H.J. Development of Poly(Arylene ether Sulfone)-Based blend membranes containing aliphatic moieties for the low-temperature decal transfer method. *J. Membr. Sci.* **2022**, *660*, 120853. [CrossRef]
15. Wang, J.J.; Yang, C.; Zhang, L.L.; Cui, C.W.; Zeng, X.F.; Chen, J.F. Monodispersed zinc oxide nanoparticles as multifunctional additives for polycarbonate thermoplastic with high transparency and excellent comprehensive performance. *Adv. Compos. Hybrid Mater.* **2022**, *5*, 2936–2947. [CrossRef]
16. Chen, C.H.; Hu, F.; Su, Z.H.; Yu, Y.J.; Wang, K.L.; Shi, Y.R.; Chen, J.; Xia, Y.; Gao, X.Y.; Wang, Z.K.; et al. Spring-Like Ammonium Salt Assisting Stress Release for Low-Temperature Deposited $FAPbI_3$ Films Toward Flexible Photovoltaic Application. *Adv. Funct. Mater.* **2023**, *33*, 2213661. [CrossRef]
17. Ren, J.Y.; Liu, Y.H.; Wang, Z.Q.; Chen, S.Q.; Ma, Y.F.; Wei, H.; Lu, S.Y. An Anti-Swellable Hydrogel Strain Sensor for Underwater Motion Detection. *Adv. Funct. Mater.* **2022**, *32*, 2107404. [CrossRef]
18. Ma, J.W.; Fan, H.Q.; Li, Z.X.; Jia, Y.X.; Yadav, A.K.; Dong, G.Z.; Wang, W.J.; Dong, W.Q.; Wang, S.R. Multi-walled carbon nanotubes/polyaniline on the ethylenediamine modified polyethylene terephthalate fibers for a flexible room temperature ammonia gas sensor with high responses. *Sens. Actuator B-Chem.* **2021**, *334*, 129677. [CrossRef]
19. Veske, P.; Bossuyt, F.; Vanfleteren, J. Testing for Wearability and Reliability of TPU Lamination Method in E-Textiles. *Sensors* **2022**, *22*, 156. [CrossRef]
20. Fu, Y.T.; Li, J.; Li, Y.Q.; Fu, S.Y.; Guo, F.L. Full-process multi-scale morphological and mechanical analyses of 3D printed short carbon fiber reinforced polyetheretherketone composites. *Compos. Sci. Technol.* **2023**, *236*, 109999. [CrossRef]
21. Mahdavi, H.; Hosseini, F.; Heidari, A.A.; Karami, M. Polyethersulfone-TPU blend membrane coated with an environmentally friendly sabja seed mucilage-Cu2+cross-linked layer with outstanding separation performance and superior antifouling. *J. Ind. Eng. Chem.* **2023**, *121*, 421–433. [CrossRef]
22. Mao, S.H.; Liu, W.; Xie, Z.M.; Zhang, D.; Zhou, J.H.; Xu, Y.S.; Fu, B.P.; Zheng, S.Y.; Zhang, L.; Yang, J.T. In Situ Growth of Functional Hydrogel Coatings by a Reactive Polyurethane for Biomedical Devices. *ACS Appl. Mater. Interfaces* **2023**, *15*, 56652–56664. [CrossRef]
23. Guo, Y.H.; Guo, Y.C.; He, W.D.; Zhao, Y.B.; Shen, R.Q.; Liu, J.X.; Wang, J. PET/TPU nanofiber composite filters with high interfacial adhesion strength based on one-step co-electrospinning. *Powder Technol.* **2021**, *387*, 136–145. [CrossRef]
24. Tyagi, M.; Rawat, R.S.; Tyagi, V.K.; Talwar, M.; Kumar, M.; Das, M.; Roy, P. Study on mechanical, thermal and shape memory properties of polycarbonate/thermoplastic polyurethane blends. *J. Appl. Polym. Sci.* **2024**, *141*, e55259. [CrossRef]
25. Tang, J.; Wu, Y.T.; Ma, S.D.; Yan, T.; Pan, Z.J. Flexible strain sensor based on CNT/TPU composite nanofiber yarn for smart sports bandage. *Compos. Part B-Eng.* **2022**, *232*, 109605. [CrossRef]
26. Zhang, X.N.; Li, J.W.; Lin, J.; Li, W.; Chu, W.; Zhang, M.M.; Lu, Y.Y.; He, X.H.; Zhao, Q.L. Highly Stretchable Electronic-Skin Sensors with Porous Microstructure for Efficient Multimodal Sensing with Wearable Comfort. *Adv. Mater. Interfaces* **2023**, *10*, 2201958. [CrossRef]
27. Li, B.; Liang, W.; Zhang, L.; Ren, F.C.; Xuan, F.Z. TPU/CNTs flexible strain sensor with auxetic structure via a novel hybrid manufacturing process of fused deposition modeling 3D printing and ultrasonic cavitation-enabled treatment. *Sens. Actuator A-Phys.* **2022**, *340*, 113526. [CrossRef]
28. Mei, S.X.; Zhang, X.M.; Ding, B.W.; Wang, J.Q.; Yang, P.F.; She, H.B.; Cui, Z.; Liu, M.Y.; Pang, X.C.; Fu, P. 3D-Printed thermoplastic polyurethane/graphene composite with porous segregated structure: Toward ultralow percolation threshold and great strain sensitivity. *J. Appl. Polym. Sci.* **2021**, *138*, 50168. [CrossRef]
29. Tian, Z.; Qin, W.; Wang, Y.; Li, X.; Gu, C.; Chen, J.; Yang, M.; Feng, L.; Chen, J.; Qiao, H. Ultra-stable strain/humidity dual-functional flexible wearable sensor based on brush-like AgNPs@ CNTs@ TPU heterogeneous structure. *Colloids Surf. A: Physicochem. Eng. Asp.* **2023**, *670*, 131398. [CrossRef]
30. Chang, X.T.; Sun, S.H.; Sun, S.B.; Liu, T.; Xiong, X.; Lei, Y.H.; Dong, L.H.; Yin, Y.S. ZnO nanorods/carbon black-based flexible strain sensor for detecting human motions. *J. Alloys Compd.* **2018**, *738*, 111–117. [CrossRef]
31. Pan, W.; Wang, J.; Li, Y.P.; Sun, X.B.; Wang, J.P.; Wang, X.X.; Zhang, J.; You, H.D.; Yu, G.F.; Long, Y.Z. Facile Preparation of Highly Stretchable TPU/Ag Nanowire Strain Sensor with Spring-Like Configuration. *Polymers* **2020**, *12*, 339. [CrossRef]

32. Shi, W.H.; Han, G.Y.; Chang, Y.Z.; Song, H.; Hou, W.J.; Chen, Q. Using Stretchable PPy@PVA Composites as a High-Sensitivity Strain Sensor To Monitor Minute Motion. *ACS Appl. Mater. Interfaces* **2020**, *12*, 45373–45382. [CrossRef]
33. Rehman, M.H.U.; Nazar, R.; Yasin, S.; Ramzan, N.; Habib, M.S. Development of PANI-TPU/MWCNTs based nanocomposites for piezoresistive strain sensing applications. *Mater. Lett.* **2022**, *328*, 133110. [CrossRef]
34. Backes, E.H.; Harb, S.V.; Pinto, L.A.; de Moura, N.K.; Morgado, G.F.D.; Marini, J.; Passador, F.R.; Pessan, L.A. Thermoplastic polyurethanes: Synthesis, fabrication techniques, blends, composites, and applications. *J. Mater. Sci.* **2024**, *59*, 1123–1152. [CrossRef]
35. Yang, Q.S.; Liu, N.; Yin, J.J.; Tian, H.; Yang, Y.; Ren, T.L. Understanding the Origin of Tensile Response in a Graphene Textile Strain Sensor with Negative Differential Resistance. *ACS Nano* **2022**, *16*, 14230–14238. [CrossRef] [PubMed]
36. Kumar, N.J.; Johnson, A.; Roggen, D.; Münzenrieder, N. A Flexible Capacitance Strain Sensor with Stitched Contact Terminals. *Adv. Mater. Technol.* **2023**, *8*, 2200410. [CrossRef]
37. Nazari, P.; Bauerle, R.; Zimmermann, J.; Melzer, C.; Schwab, C.; Smith, A.; Kowalsky, W.; Aghassi-Hagmann, J.; Hernandez-Sosa, G.; Lemmer, U. Piezoresistive Free-standing Microfiber Strain Sensor for High-resolution Battery Thickness Monitoring. *Adv. Mater.* **2023**, *35*, 2212189. [CrossRef]
38. Luo, Y.C.; Yu, M.L.; Zhang, Y.T.; Wang, Y.Y.; Long, L.; Tan, H.H.; Li, N.; Xu, L.J.; Xu, J.X. Highly sensitive strain sensor and self-powered triboelectric nanogenerator using a fully physical crosslinked double-network conductive hydrogel. *Nano Energy* **2022**, *104*, 107955. [CrossRef]
39. Wu, R.; Ma, L.; Liu, S.; Patil, A.B.; Hou, C.; Zhang, Y.; Zhang, W.; Yu, R.; Yu, W.; Guo, W.; et al. Fibrous inductance strain sensors for passive inductance textile sensing. Mater. *Today Phys.* **2020**, *15*, 100243. [CrossRef]
40. Shu, Q.; Xu, Z.B.; Liu, S.; Wu, J.P.; Deng, H.X.; Gong, X.L.; Xuan, S.H. Magnetic flexible sensor with tension and bending discriminating detection. *Chem. Eng. J.* **2022**, *433*, 134424. [CrossRef]
41. Ma, S.D.; Tang, J.; Yan, T.; Pan, Z.J. Performance of Flexible Strain Sensors With Different Transition Mechanisms: A Review. *IEEE Sens. J.* **2022**, *22*, 7475–7498. [CrossRef]
42. Migliorini, L.; Santaniello, T.; Falqui, A.; Milani, P. Super-Stretchable Resistive Strain Sensor Based on Ecoflex-Gold Nanocomposites. *ACS Appl. Nano Mater.* **2023**, *6*, 8999–9007. [CrossRef]
43. Cao, H.L.; Chai, S.S.; Tan, Z.F.; Wu, H.; Mao, X.; Wei, L.; Zhou, F.L.; Sun, R.J.; Liu, C.K. Recent Advances in Physical Sensors Based on Electrospinning Technology. *ACS Mater. Lett.* **2023**, *5*, 1627–1648. [CrossRef]
44. Xia, X.D.; Zhao, S.J.; Zhang, J.J.; Fang, C.; Weng, G.J. A unified investigation into the tensile and compressive sensing performance in highly sensitive MWCNT/epoxy nanocomposite strain sensor through loading-dependent tunneling distance. *Compos. Sci. Technol.* **2022**, *230*, 109723. [CrossRef]
45. Yu, Q.H.; Ge, R.; Wen, J.; Du, T.; Zhai, J.Y.; Liu, S.H.; Wang, L.F.; Qin, Y. Highly sensitive strain sensors based on piezotronic tunneling junction. *Nat. Commun.* **2022**, *13*, 778. [CrossRef] [PubMed]
46. Li, N.; Huang, G.W.; Liu, Y.; Qu, C.B.; Li, M.; Xiao, H.M. Performance Deficiency Improvement of CNT-Based Strain Sensors by Magnetic-Induced Patterning. *ACS Appl. Mater. Interfaces* **2023**, *15*, 5774–5786. [CrossRef] [PubMed]
47. Veeramuthu, L.; Venkatesan, M.; Benas, J.S.; Cho, C.J.; Lee, C.C.; Lieu, F.K.; Lin, J.H.; Lee, R.H.; Kuo, C.C. Recent Progress in Conducting Polymer Composite/Nanofiber-Based Strain and Pressure Sensors. *Polymers* **2021**, *13*, 4281. [CrossRef] [PubMed]
48. Hwang, H.; Kim, Y.; Park, J.H.; Jeong, U. 2D Percolation Design with Conductive Microparticles for Low-Strain Detection in a Stretchable Sensor. *Adv. Funct. Mater.* **2020**, *30*, 1908514. [CrossRef]
49. Yuan, H.Z.; Li, P.X.; Wang, X.Y.; Yu, C.; Wang, X.; Sun, J.T. Stretchable, ultrasensitive strain sensor with high-linearity by constructing crack-based dual conductive network. *Chem. Eng. J.* **2024**, *480*, 148102. [CrossRef]
50. Xue, J.J.; Wu, T.; Dai, Y.Q.; Xia, Y.N. Electrospinning and Electrospun Nanofibers: Methods, Materials, and Applications. *Chem. Rev.* **2019**, *119*, 5298–5415. [CrossRef]
51. Su, Z.Q.; Ding, J.W.; Wei, G. Electrospinning: A facile technique for fabricating polymeric nanofibers doped with carbon nanotubes and metallic nanoparticles for sensor applications. *RSC Adv.* **2014**, *4*, 52598–52610. [CrossRef]
52. Song, W.; Tang, Y.; Qian, C.; Kim, B.J.; Liao, Y.; Yu, D.-G. Electrospinning spinneret: A bridge between the visible world and the invisible nanostructures. *Innovation* **2023**, *4*, 100381. [CrossRef]
53. Gong, X.B.; Jin, C.F.; Liu, X.Y.; Yu, J.Y.; Zhang, S.C.; Ding, B. Scalable Fabrication of Electrospun True-Nanoscale Fiber Membranes for Effective Selective Separation. *Nano Lett.* **2023**, *23*, 1044–1051. [CrossRef]
54. Chen, X.; Zhang, J.T.; Lin, H.B.; Tian, M.; Li, M.C.; Tian, Y. Development and application of electrospun fiber-based multifunctional sensors. *Chem. Eng. J.* **2024**, *486*, 150204. [CrossRef]
55. Yang, G.; Tang, X.C.; Zhao, G.D.; Li, Y.F.; Ma, C.Q.; Zhuang, X.P.; Yan, J. Highly sensitive, direction-aware, and transparent strain sensor based on oriented electrospun nanofibers for wearable electronic applications. *Chem. Eng. J.* **2022**, *435*, 135004. [CrossRef]
56. Li, D.; Cui, T.R.; Jian, J.M.; Yan, J.L.; Xu, J.D.; Li, X.; Li, Z.; Yan, A.Z.; Chen, Z.K.; Shao, W.C.; et al. Lantern-Inspired On-Skin Helical Interconnects for Epidermal Electronic Sensors. *Adv. Funct. Mater.* **2023**, *33*, 2213335. [CrossRef]
57. Al-Dhahebi, A.M.; Ling, J.K.; Krishnan, S.G.; Yousefzadeh, M.; Elumalai, N.K.; Saheed, M.S.M.; Ramakrishna, S.; Jose, R. Electrospinning research and products: The road and the way forward. *Appl. Phys. Rev.* **2022**, *9*, 44. [CrossRef]
58. Thenmozhi, S.; Dharmaraj, N.; Kadirvelu, K.; Kim, H.Y. Electrospun nanofibers: New generation materials for advanced applications. *Mater. Sci. Eng. B-Adv. Funct. Solid-State Mater.* **2017**, *217*, 36–48. [CrossRef]
59. Wang, Y.; Yokota, T.; Someya, T. Electrospun nanofiber-based soft electronics. *NPG Asia Mater.* **2021**, *13*, 22. [CrossRef]

60. Gao, Q.; Agarwal, S.; Greiner, A.; Zhang, T. Electrospun fiber-based flexible electronics: Fiber fabrication, device platform, functionality integration and applications. *Prog. Mater. Sci.* **2023**, *137*, 101139. [CrossRef]
61. Selver, E.; Karaca, N.; Onen, A.; Ucar, N.; Altay, P. Morphological/alignment properties of thermoplastic polyurethane nanofiber affected by processing parameters. *J. Elastomer Plast.* **2021**, *53*, 769–783. [CrossRef]
62. Gao, Z.Y.; Xiao, X.; Carlo, A.D.; Yin, J.Y.; Wang, Y.X.; Huang, L.J.; Tang, J.G.; Chen, J. Advances in Wearable Strain Sensors Based on Electrospun Fibers. *Adv. Funct. Mater.* **2023**, *33*, 2214265. [CrossRef]
63. Chen, J.; Rong, F.; Xie, Y.B. Fabrication, Microstructures and Sensor Applications of Highly Ordered Electrospun Nanofibers: A Review. *Materials* **2023**, *16*, 3310. [CrossRef] [PubMed]
64. Wang, X.; Liu, X.H.; Schubert, D.W. Highly Sensitive Ultrathin Flexible Thermoplastic Polyurethane/Carbon Black Fibrous Film Strain Sensor with Adjustable Scaffold Networks. *Nano-Micro Lett.* **2021**, *13*, 64. [CrossRef] [PubMed]
65. Zhao, Y.; Ren, M.N.; Shang, Y.; Li, J.N.; Wang, S.; Zhai, W.; Zheng, G.Q.; Dai, K.; Liu, C.T.; Shen, C.Y. Ultra-sensitive and durable strain sensor with sandwich structure and excellent anti-interference ability for wearable electronic skins. *Compos. Sci. Technol.* **2020**, *200*, 108448. [CrossRef]
66. Wang, Y.H.; Li, W.Y.; Li, C.C.; Zhou, B.Z.; Zhou, Y.F.; Jiang, L.; Wen, S.P.; Zhou, F.L. Fabrication of ultra-high working range strain sensor using carboxyl CNTs coated electrospun TPU assisted with dopamine. *Appl. Surf. Sci.* **2021**, *566*, 150705. [CrossRef]
67. Yu, X.; Wu, Z.J.; Weng, L.; Jiang, D.W.; Algadi, H.; Qin, Z.F.; Guo, Z.H.; Xu, B.B. Flexible Strain Sensor Enabled by Carbon Nanotubes-Decorated Electrospun TPU Membrane for Human Motion Monitoring. *Adv. Mater. Interfaces* **2023**, *10*, 2202292. [CrossRef]
68. Meng, Y.Y.; Cheng, J.; Zhou, C.L. Superhydrophobic and Stretchable Carbon Nanotube/Thermoplastic Urethane-Based Strain Sensor for Human Motion Detection. *ACS Appl. Nano Mater.* **2023**, *6*, 5871–5878. [CrossRef]
69. Xiang, D.; Liu, L.B.; Chen, X.Y.; Wu, Y.P.; Wang, M.H.; Zhang, J.; Zhao, C.X.; Li, H.; Li, Z.Y.; Wang, P.; et al. High-performance fiber strain sensor of carbon nanotube/thermoplastic polyurethane@styrene butadiene styrene with a double percolated structure. *Front. Mater. Sci.* **2022**, *16*, 12. [CrossRef]
70. Wang, X.; Qu, M.C.; Wu, K.Q.; Schubert, D.W.; Liu, X.H. High sensitive electrospun thermoplastic polyurethane/carbon nanotubes strain sensor fitting by a novel optimization empirical model. *Adv. Compos. Hybrid Mater.* **2023**, *6*, 220586. [CrossRef]
71. Li, C.C.; Zhou, B.Z.; Zhou, Y.F.; Ma, J.W.; Zhou, F.L.; Chen, S.J.; Jerrams, S.; Jiang, L. Carbon Nanotube Coated Fibrous Tubes for Highly Stretchable Strain Sensors Having High Linearity. *Nanomaterials* **2022**, *12*, 2458. [CrossRef]
72. Jia, Y.Y.; Yue, X.Y.; Wang, Y.L.; Yan, C.; Zheng, G.Q.; Dai, K.; Liu, C.T.; Shen, C.Y. Multifunctional stretchable strain sensor based on polydopamine/reduced graphene oxide/electrospun thermoplastic polyurethane fibrous mats for human motion detection and environment monitoring. *Compos. Part B-Eng.* **2020**, *183*, 107696. [CrossRef]
73. Li, Z.Y.; Zhai, W.; Yu, Y.F.; Li, G.J.; Zhan, P.F.; Xu, J.W.; Zheng, G.Q.; Dai, K.; Liu, C.T.; Shen, C.Y. An Ultrasensitive, Durable and Stretchable Strain Sensor with Crack-wrinkle Structure for Human Motion Monitoring. *Chin. J. Polym. Sci.* **2021**, *39*, 316–326. [CrossRef]
74. Kang, L.H.; Ma, J.; Wang, C.; Li, K.C.; Wu, H.Y.; Zhu, M.F. Highly Sensitive and Wide Detection Range Thermoplastic Polyurethane/Graphene Nanoplatelets Multifunctional Strain Sensor with a Porous and Crimped Network Structure. *ACS Appl. Mater. Interfaces* **2024**, *16*, 2814–2824. [CrossRef] [PubMed]
75. Gao, Z.Y.; Wang, Y.X.; Mao, X.H.; Yan, X.H.; Zhang, X.T.; Xing, W.; Li, L.X.; Wang, Z.L.; Huang, L.J.; Tang, J.G. Polydopamine/Carbon Black/Carbon Nanofiber/Thermoplastic Polyurethane Composite Nanofiber Strain Sensor with Ultrahigh Loading Rate for Human Activity Monitoring. *ACS Appl. Nano Mater.* **2024**, *7*, 11690–11703. [CrossRef]
76. Zhang, Z.L.; Innocent, M.T.; Tang, N.; Li, R.Y.; Hu, Z.X.; Zhai, M.; Yang, L.J.; Ma, W.J.; Xiang, H.X.; Zhu, M.F. Electromechanical Performance of Strain Sensors Based on Viscoelastic Conductive Composite Polymer Fibers. *ACS Appl. Mater. Interfaces* **2022**, *14*, 44832–44840. [CrossRef]
77. Li, S.S.; Fu, Q.; Pan, C.X. A multi-functional wearable sensor based on carbon nanomaterials reinforced TPU fiber with high sensitivity. *J. Alloys Compd.* **2022**, *927*, 167041. [CrossRef]
78. Xia, J.F.; He, L.; Lu, Z.L.; Liu, L.P.; Song, J.A.; Chen, S.Y.; Wang, Q.S.; Hammad, F.A.; Tian, Y.L. Stretchable and Sensitive Strain Sensors Based on CB/MWCNTs-TPU for Human Motion Capture and Health Monitoring. *ACS Appl. Nano Mater.* **2023**, *6*, 9736–9745. [CrossRef]
79. Ding, Y.R.; Xue, C.H.; Guo, X.J.; Wang, X.; Jia, S.T.; An, Q.F. Flexible Superamphiphobic Film with a 3D Conductive Network for Wearable Strain Sensors in Humid Conditions. *ACS Appl. Electron. Mater.* **2022**, *4*, 345–355. [CrossRef]
80. Zhang, Y.J.; Li, C.C.; Zhou, B.Z.; He, H.T.; Zhou, Y.F.; Jiang, L.; Zhou, F.L.; Chen, S.J. A flexible strain sensor based on conductive TPU/CNTs-Gr composites. *J. Appl. Polym. Sci.* **2022**, *139*, e52475. [CrossRef]
81. Tang, J.; Wu, Y.T.; Ma, S.D.; Yan, T.; Pan, Z.J. Sensing mechanism of a flexible strain sensor developed directly using electrospun composite nanofiber yarn with ternary carbon nanomaterials. *Iscience* **2022**, *25*, 20. [CrossRef] [PubMed]
82. Zhang, C.J.; Li, J.C.; Chen, Y.W.; Zou, Z.F.; Zhang, Y.B.; Liu, Y.; Wu, B. Stretchable, flexible, and breathable MXene/dopamine/thermoplastic polyurethane nanofiber membrane with outstanding strain sensing and electromagnetic interference shielding performances. *Mater. Today Commun.* **2024**, *38*, 107968. [CrossRef]
83. Pan, K.L.; Wang, J.; Li, Y.; Lu, X.Y.; Hu, D.C.; Jia, Z.X.; Lin, J. Sandwich-Like Flexible Breathable Strain Sensor with Tunable Thermal Regulation Capability for Human Motion Monitoring. *ACS Appl. Mater. Interfaces* **2024**, *16*, 10633–10645. [CrossRef] [PubMed]

84. Dong, H.; Sun, J.C.; Liu, X.M.; Jiang, X.D.; Lu, S.W. Highly Sensitive and Stretchable MXene/CNTs/TPU Composite Strain Sensor with Bilayer Conductive Structure for Human Motion Detection. *ACS Appl. Mater. Interfaces* **2022**, *14*, 15504–15516. [CrossRef] [PubMed]
85. Ren, M.N.; Li, J.N.; Zhao, Y.; Zhai, W.; Zhou, K.K.; Yu, Y.F.; Wang, S.; Dai, K.; Liu, C.T.; Shen, C.Y. Highly strain-sensitive and stretchable multilayer conductive composite based on aligned thermoplastic polyurethane fibrous mat for human motion monitoring. *Compos. Commun.* **2024**, *46*, 101840. [CrossRef]
86. Shi, J.R.; Wang, D.; Sarkodie, B.; Wu, D.S.; Mao, Z.; Liu, Z.Y.; Feng, Q.; Xu, W.Z. Strain Sensors for Human Movement Detection Based on Fibrous Membranes Comprising Thermoplastic Polyurethane, Ag Nanoparticles, and Carbon Nanotubes. *ACS Appl. Nano Mater.* **2024**, *7*, 2051–2061. [CrossRef]
87. Zhang, W.X.; Miao, J.L.; Tian, M.W.; Zhang, X.J.; Fan, T.T.; Qu, L.J. Hierarchically interlocked helical conductive yarn enables ultra-stretchable electronics and smart fabrics. *Chem. Eng. J.* **2023**, *462*, 142279. [CrossRef]
88. Wang, X.F.; Liu, J.; Zheng, Y.Q.; Shi, B.; Chen, A.B.; Wang, L.L.; Shen, G.Z. Biocompatible liquid metal coated stretchable electrospinning film for strain sensors monitoring system. *Sci. China-Mater.* **2022**, *65*, 2235–2243. [CrossRef]
89. Uzabakiriho, P.C.; Wang, M.; Wang, K.; Ma, C.; Zhao, G. High-Strength and Extensible Electrospun Yarn for Wearable Electronics. *ACS Appl. Mater. Interfaces* **2022**, *14*, 46068–46076. [CrossRef]
90. Ai, J.W.; Wang, Q.Q.; Li, Z.Q.; Lu, D.X.; Liao, S.Q.; Qiu, Y.Y.; Xia, X.; Wei, Q.F. Highly Stretchable and Fluorescent Visualizable Thermoplastic Polyurethane/Tetraphenylethylene Plied Yarn Strain Sensor with Heterogeneous and Cracked Structure for Human Health Monitoring. *ACS Appl. Mater. Interfaces* **2023**, *16*, 1428–1438. [CrossRef]
91. Tian, Y.; Huang, M.J.; Wang, Y.L.; Zheng, Y.J.; Yin, R.; Liu, H.; Liu, C.T.; Shen, C.Y. Ultra-stretchable, sensitive and breathable electronic skin based on TPU electrospinning fibrous membrane with microcrack structure for human motion monitoring and self-powered application. *Chem. Eng. J.* **2024**, *480*, 147899. [CrossRef]
92. Zhan, P.F.; Zhai, W.; Wei, W.Y.; Ding, P.; Zheng, G.Q.; Dai, K.; Liu, C.T.; Shen, C.Y. Stretchable strain sensor with high sensitivity, large workable range and excellent breathability for wearable electronic skins. *Compos. Sci. Technol.* **2022**, *229*, 109720. [CrossRef]
93. Bozyel, I.; Keser, Y.I.; Gokcen, D. Triple mode and multi-purpose flexible sensor fabrication based on carbon black and thermoplastic polyurethane composite with propolis. Sens. *Actuator A-Phys.* **2021**, *332*, 113056. [CrossRef]
94. Zhang, R.; Hu, H.L.; Sun, Z.J.; Li, S.Q.; Liu, Z.H.; Hu, S.F.; Fu, X.D.; Liu, Q.T.; Wong, C.P. Improved Sensitivity of Flexible Conductive Composites Throughout the Working Strain Range Based on Bioinspired Strain Redistribution. *ACS Appl. Polym. Mater.* **2022**, *4*, 1608–1616. [CrossRef]
95. Zhang, W.Y.; Liu, Q.; Chen, P. Flexible Strain Sensor Based on Carbon Black/Silver Nanoparticles Composite for Human Motion Detection. *Materials* **2018**, *11*, 1836. [CrossRef] [PubMed]
96. Li, B.; Zhang, S.H.; Zhang, L.; Gao, Y.; Xuan, F.Z. Strain sensing behavior of FDM 3D printed carbon black filled TPU with periodic configurations and flexible substrates. *J. Manuf. Process.* **2022**, *74*, 283–295. [CrossRef]
97. Qin, Y.J.; Qu, M.C.; Pan, Y.M.; Zhang, C.H.; Schubert, D.W. Fabrication, characterization and modelling of triple hierarchic PET/CB/TPU composite fibres for strain sensing. *Compos. Part A-Appl. Sci. Manuf.* **2020**, *129*, 105724. [CrossRef]
98. Chen, T.J.; Wu, G.Z.; Panahi-Sarmad, M.; Wu, Y.T.; Xu, R.X.; Cao, S.J.; Xiao, X.L. A novel flexible piezoresistive sensor using superelastic fabric coated with highly durable SEBS/TPU/CB/CNF nanocomposite for detection of human motions. *Compos. Sci. Technol.* **2022**, *227*, 109563. [CrossRef]
99. Qaiser, N.; Al-Modaf, F.; Khan, S.M.; Shaikh, S.F.; El-Atab, N.; Hussain, M.M. A Robust Wearable Point-of-Care CNT-Based Strain Sensor for Wirelessly Monitoring Throat-Related Illnesses. *Adv. Funct. Mater.* **2021**, *31*, 2103375. [CrossRef]
100. Zhu, G.X.; Li, H.; Peng, M.L.; Zhao, G.Y.; Chen, J.W.; Zhu, Y.T. Highly-stretchable porous thermoplastic polyurethane/carbon nanotubes composites as a multimodal sensor. *Carbon* **2022**, *195*, 364–371. [CrossRef]
101. Zhao, X.X.; Guo, H.; Ding, P.; Zhai, W.; Liu, C.T.; Shen, C.Y.; Dai, K. Hollow-porous fiber-shaped strain sensor with multiple wrinkle-crack microstructure for strain visualization and wind monitoring. *Nano Energy* **2023**, *108*, 108197. [CrossRef]
102. Li, J.Y.; Xiang, D.; Su, P.; Zhao, C.X.; Li, H.; Li, Z.Y.; Wang, B.; Wang, P.; Li, Y.T.; Wu, Y.P. High-performance flexible strain sensors prepared by biaxially stretching conductive polymer composites with a double-layer structure. *Mater. Today Commun.* **2023**, *36*, 106548. [CrossRef]
103. Liu, W.W.; Xue, C.; Long, X.Y.; Ren, Y.; Chen, Z.; Zhang, W. Highly flexible and multifunctional CNTs/TPU fiber strain sensor formed in one-step via wet spinning. *J. Alloys Compd.* **2023**, *948*, 169641. [CrossRef]
104. Wang, Y.H.; Li, W.Y.; Zhou, Y.F.; Jiang, L.; Ma, J.W.; Chen, S.J.; Jerrams, S.; Zhou, F.L. Fabrication of high-performance wearable strain sensors by using CNTs-coated electrospun polyurethane nanofibers. *J. Mater. Sci.* **2020**, *55*, 12592–12606. [CrossRef]
105. Xu, X.X.; Yuan, Y.; Zhang, T.; Li, K.; Wang, S.Q.; Liang, C.; Zhu, H.P. A silanized MCNT/TPU-based flexible strain sensor with high stretchability for deformation monitoring of elastomeric isolators for bridges. *Constr. Build. Mater.* **2022**, *338*, 127664. [CrossRef]
106. Xiang, D.; Liu, L.B.; Xu, F.X.; Li, Y.Q.; Harkin-Jones, E.; Wu, Y.P.; Zhao, C.X.; Li, H.; Li, Z.Y.; Wang, P.; et al. Highly Sensitive Flexible Strain Sensor Based on a Double-percolation Structured Elastic Fiber of Carbon Nanotube (CNT)/Styrene Butadiene Styrene (SBS) @ Thermoplastic Polyurethane (TPU) for Human Motion and Tactile Recognition. *Appl. Compos. Mater.* **2023**, *30*, 307–322. [CrossRef]

107. Liu, L.B.; Xiang, D.; Zhang, X.X.; Harkin-Jones, E.; Wang, J.J.; Zhao, C.X.; Li, H.; Li, Z.Y.; Wang, L.; Wang, P.; et al. Highly sensitive flexible strain sensor based on carbon nanotube/styrene butadiene styrene@ thermoplastic polyurethane fiber with a double percolated structure. *Polym. Eng. Sci.* **2023**, *63*, 206–218. [CrossRef]
108. Hou, K.H.; Shi, X.; Yuan, J.E.; Zhang, D.; Bai, Y.S.; Xu, J.; Guo, F.M.; Zhang, Y.J.; Cao, A.Y.; Shang, Y.Y. Stretchable PU@CNT/MXene Fiber Fabricated by External Tension for High Performance Strain Sensor and Supercapacitor. *Adv. Mater. Technol.* **2023**, *8*, 11. [CrossRef]
109. Danová, R.; Olejnik, R.; Slobodian, P.; Matyas, J. The Piezoresistive Highly Elastic Sensor Based on Carbon Nanotubes for the Detection of Breath. *Polymers* **2020**, *12*, 713. [CrossRef]
110. Zhou, Y.; Stewart, R. Highly flexible, durable, UV resistant, and electrically conductive graphene based TPU/textile composite sensor. *Polym. Adv. Technol.* **2022**, *33*, 4250–4264. [CrossRef]
111. Gunasekaran, H.B.; Ponnan, S.; Zheng, Y.L.; Laroui, A.; Wang, H.P.; Wu, L.X.; Wang, J.L. Facile Fabrication of Highly Sensitive Thermoplastic Polyurethane Sensors with Surface- and Interface-Impregnated 3D Conductive Networks. *ACS Appl. Mater. Interfaces* **2022**, *14*, 22615–22625. [CrossRef]
112. Chen, X.X.; Yin, Z.Z.; Deng, Y.T.; Li, Z.H.; Xue, M.S.; Chen, Y.H.; Xie, Y.; Liu, W.Q.; He, P.; Luo, Y.D.; et al. Harsh environment-tolerant and robust superhydrophobic graphene-based composite membrane for wearable strain sensor. *Sens. Actuator A-Phys.* **2023**, *362*, 114630. [CrossRef]
113. Yang, J.L.; Ling, K.; Liu, L.H.; Zeng, X.H.; Xu, X.W.; Li, Z.L.; He, P. Printable and Wearable Graphene-Based Strain Sensor With High Sensitivity for Human Motion Monitoring. *IEEE Sens. J.* **2022**, *22*, 13937–13944. [CrossRef]
114. Li, B.; Luo, J.C.; Huang, X.W.; Lin, L.W.; Wang, L.; Hu, M.J.; Tang, L.C.; Xue, H.G.; Gao, J.F.; Mai, Y.W. A highly stretchable, super-hydrophobic strain sensor based on polydopamine and graphene reinforced nanofiber composite for human motion monitoring. *Compos. Part B-Eng.* **2020**, *181*, 107580. [CrossRef]
115. Yang, Y.J.; Yi, T.; Liu, Y.; Zhao, H.; Liang, C. Design of a Highly Sensitive Reduced Graphene Oxide/Graphene Oxide@Cellulose Acetate/Thermoplastic Polyurethane Flexible Sensor. *Sensors* **2022**, *22*, 3281. [CrossRef]
116. Zhang, X.Z.; Xiang, D.; Wu, Y.P.; Harkin-Jones, E.; Shen, J.B.; Ye, Y.; Tan, W.; Wang, J.J.; Wang, P.; Zhao, C.X.; et al. High-performance flexible strain sensors based on biaxially stretched conductive polymer composites with carbon nanotubes immobilized on reduced graphene oxide. *Compos. Part A-Appl. Sci. Manuf.* **2021**, *151*, 106665. [CrossRef]
117. Chen, X.Y.; Zhang, X.Z.; Xiang, D.; Wu, Y.P.; Zhao, C.X.; Li, H.; Li, Z.Y.; Wang, P.; Li, Y.T. 3D printed high-performance spider web-like flexible strain sensors with directional strain recognition based on conductive polymer composites. *Mater. Lett.* **2022**, *306*, 130935. [CrossRef]
118. Xu, W.H.; Ravichandran, D.; Jambhulkar, S.; Franklin, R.; Zhu, Y.X.; Song, K.N. Bioinspired, Mechanically Robust Chemiresistor for Inline Volatile Organic Compounds Sensing. *Adv. Mater. Technol.* **2020**, *5*, 2000440. [CrossRef]
119. Xiang, D.; Zhang, X.Z.; Han, Z.H.; Zhang, Z.X.; Zhou, Z.X.; Harkin-Jones, E.; Zhang, J.; Luo, X.; Wang, P.; Zhao, C.X.; et al. 3D printed high-performance flexible strain sensors based on carbon nanotube and graphene nanoplatelet filled polymer composites. *J. Mater. Sci.* **2020**, *55*, 15769–15786. [CrossRef]
120. Yang, K.; Yin, F.X.; Xia, D.; Peng, H.F.; Yang, J.Z.; Yuan, W.J. A highly flexible and multifunctional strain sensor based on a network-structured MXene/polyurethane mat with ultra-high sensitivity and a broad sensing range. *Nanoscale* **2019**, *11*, 9949–9957. [CrossRef]
121. Chen, Q.; Gao, Q.S.; Wang, X.; Schubert, D.W.; Liu, X.H. Flexible, conductive, and anisotropic thermoplastic polyurethane/polydopamine/MXene foam for piezoresistive sensors and motion monitoring. *Compos. Part A-Appl. Sci. Manuf.* **2022**, *155*, 106838. [CrossRef]
122. Zhang, Z.N.; Weng, L.; Guo, K.; Guan, L.Z.; Wang, X.M.; Wu, Z.J. Durable and highly sensitive flexible sensors for wearable electronic devices with PDMS-MXene/TPU composite films. *Ceram. Int.* **2022**, *48*, 4977–4985. [CrossRef]
123. Cui, M.J.; Wu, S.K.; Li, J.N.; Zhao, Y.; Zhai, W.; Dai, K.; Liu, C.T.; Shen, C.Y. An ultrasensitive flexible strain sensor based on CNC/CNTs/MXene/TPU fibrous mat for human motion, sound and visually personalized rehabilitation training monitoring. *Compos. Sci. Technol.* **2023**, *244*, 110309. [CrossRef]
124. Fang, F.Y.; Wang, H.; Wang, H.Q.; Gu, X.F.; Zeng, J.; Wang, Z.X.; Chen, X.D.; Chen, X.; Chen, M.Y. Stretchable MXene/Thermoplastic Polyurethanes Based Strain Sensor Fabricated Using a Combined Electrospinning and Electrostatic Spray Deposition Technique. *Micromachines* **2021**, *12*, 252. [CrossRef]
125. Zhang, D.B.; Yin, R.; Zheng, Y.J.; Li, Q.M.; Liu, H.; Liu, C.T.; Shen, C.Y. Multifunctional MXene/CNTs based flexible electronic textile with excellent strain sensing, electromagnetic interference shielding and Joule heating performances. *Chem. Eng. J.* **2022**, *438*, 135587. [CrossRef]
126. Wang, H.C.; Zhou, R.C.; Li, D.H.; Zhang, L.R.; Ren, G.Z.; Wang, L.; Liu, J.H.; Wang, D.Y.; Tang, Z.H.; Lu, G.; et al. High-Performance Foam-Shaped Strain Sensor Based on Carbon Nanotubes and $Ti_3C_2T_x$ MXene for the Monitoring of Human Activities. *ACS Nano* **2021**, *15*, 9690–9700. [CrossRef]
127. Cui, X.Y.; Miao, C.J.; Lu, S.W.; Liu, X.M.; Yang, Y.X.; Sun, J.C. Strain Sensors Made of MXene, CNTs, and TPU/PSF Asymmetric Structure Films with Large Tensile Recovery and Applied in Human Health Monitoring. *ACS Appl. Mater. Interfaces* **2023**, *15*, 59655–59670. [CrossRef]
128. Xu, W.J.; Hu, S.Y.; Zhao, Y.; Zhai, W.; Chen, Y.H.; Zheng, G.Q.; Dai, K.; Liu, C.T.; Shen, C.Y. Nacre-inspired tunable strain sensor with synergistic interfacial interaction for sign language interpretation. *Nano Energy* **2021**, *90*, 106606. [CrossRef]

129. Fan, W.J.; Li, C.; Wang, C.R.; Huang, C.X.; Ma, T.T.; Zhou, W.; Wan, Z. Kernmantle structured stretchable strain sensors for biomechanical sensing in amphibious environments. *Carbon* **2024**, *222*, 9. [CrossRef]
130. Liu, Y.M.; Lyu, P.; Hu, J.; Tang, W.Y.; Huang, Y.; Wang, B.S.; Xu, W.L.; Liu, X.; Zhang, R.F. Highly Sensitive and Color Adaptable Polyurethane-Based Strain Sensors with Embedded Silver Nanowires. *Adv. Mater. Technol.* **2023**, *8*, 118918. [CrossRef]
131. Zhang, C.; Ouyang, W.Y.; Zhang, L.; Li, D.C. A dual-mode fiber-shaped flexible capacitive strain sensor fabricated by direct ink writing technology for wearable and implantable health monitoring applications. *Microsyst. Nanoeng.* **2023**, *9*, 11. [CrossRef]
132. Sun, H.Y.; Zettl, J.; Willenbacher, N. Highly conductive and stretchable filament for flexible electronics. *Addit. Manuf.* **2023**, *78*, 158. [CrossRef]
133. Yin, R.; Yang, S.Y.; Li, Q.M.; Zhang, S.D.; Liu, H.; Han, J.; Liu, C.T.; Shen, C.Y. Flexible conductive Ag nanowire/cellulose nanofibril hybrid nanopaper for strain and temperature sensing applications. *Sci. Bull.* **2020**, *65*, 899–908. [CrossRef] [PubMed]
134. Huang, J.Y.; Li, D.W.; Zhao, M.; Mensah, A.R.; Lv, P.F.; Tian, X.J.; Huang, F.L.; Ke, H.Z.; Wei, Q.F. Highly Sensitive and Stretchable CNT-Bridged AgNP Strain Sensor Based on TPU Electrospun Membrane for Human Motion Detection. *Adv. Electron. Mater.* **2019**, *5*, 1900241. [CrossRef]
135. Li, G.; Xue, Y.S.; Peng, H.; Qin, W.F.; Zhou, B.; Zhao, X.; Liu, G.C.; Li, S.Y.; Guo, R.H. Wide strain range and high sensitivity sandwich structure CNTs/AgNWs/ CNTs/TPU strain sensors for human motion detection. *Sens. Actuator A-Phys.* **2024**, *366*, 114998. [CrossRef]
136. Li, Y.; Chen, Y.; Yang, Y.; Gu, J.D.; Ke, K.; Yin, B.; Yang, M.B. Aligned wave-like elastomer fibers with robust conductive layers via electroless deposition for stretchable electrode applications. *J. Mat. Chem. B* **2021**, *9*, 8801–8808. [CrossRef] [PubMed]
137. Yang, Y.Y.; Guo, Z.P.; Huang, J.J.; Zhang, S.Y.; Zhang, R.Q.; Gu, S.J.; Xu, J.; Cai, G.M.; Xu, W.L. Highly stretchable conductive elastomeric polyurethane nanofiber composite for human motion detection. *Mater. Lett.* **2021**, *293*, 129698. [CrossRef]
138. Zhang, S.J.; He, Z.L.; Zhou, G.H.; Jung, B.M.; Kim, T.H.; Park, B.J.; Byun, J.H.; Chou, T.W. High conductive free-written thermoplastic polyurethane composite fibers utilized as weight-strain sensors. *Compos. Sci. Technol.* **2020**, *189*, 108011. [CrossRef]
139. Wang, W.; Ma, Y.Y.; Wang, T.Y.; Ding, K.; Zhao, W.; Jiao, L.; Shu, D.K.; Li, C.Y.; Hua, F.G.; Jiang, H.; et al. Double-Layered Conductive Network Design of Flexible Strain Sensors for High Sensitivity and Wide Working Range. *ACS Appl. Mater. Interfaces* **2022**, *14*, 36611–36621. [CrossRef]
140. Sun, H.; Fang, X.D.; Fang, Z.Y.; Zhao, L.B.; Tian, B.; Verma, P.; Maeda, R.; Jiang, Z.D. An ultrasensitive and stretchable strain sensor based on a microcrack structure for motion monitoring. *Microsyst. Nanoeng.* **2022**, *8*, 111. [CrossRef]
141. Zhan, P.F.; Jia, Y.Y.; Zhai, W.; Zheng, G.Q.; Dai, K.; Liu, C.T.; Shen, C.Y. A fibrous flexible strain sensor with Ag nanoparticles and carbon nanotubes for synergetic high sensitivity and large response range. *Compos. Part A-Appl. Sci. Manuf.* **2023**, *167*, 107431. [CrossRef]
142. He, J.Q.; Li, A.; Wang, W.J.; Cui, C.; Jiang, S.; Chen, M.M.; Qin, W.F.; Tang, H.; Guo, R.H. Multifunctional Wearable Device Based on an Antibacterial and Hydrophobic Silver Nanoparticles/$Ti_3C_2T_x$ MXene/Thermoplastic Polyurethane Fibrous Membrane for Electromagnetic Shielding and Strain Sensing. *Ind. Eng. Chem. Res.* **2023**, *62*, 9221–9232. [CrossRef]
143. Li, J.N.; Zhao, Y.; Zhai, W.; Zhao, X.X.; Dai, K.; Liu, C.T.; Shen, C.Y. Highly aligned electrospun film with wave-like structure for multidirectional strain and visual sensing. *Chem. Eng. J.* **2024**, *485*, 149952. [CrossRef]
144. Wang, H.Z.; Li, R.F.; Cao, Y.J.; Chen, S.; Yuan, B.; Zhu, X.Y.; Cheng, J.S.; Duan, M.H.; Liu, J. Liquid Metal Fibers. *Adv. Fiber Mater.* **2022**, *4*, 987–1004. [CrossRef]
145. Park, Y.G.; Lee, G.Y.; Jang, J.; Yun, S.M.; Kim, E.; Park, J.U. Liquid Metal-Based Soft Electronics for Wearable Healthcare. *Adv. Healthc. Mater.* **2021**, *10*, 2002280. [CrossRef]
146. Peng, Y.; Peng, H.; Li, Y.; Zhou, J.C.; Zhang, J.Y. Thermoplastic and Electrically Conductive Fibers for Highly Stretchable and Sensitive Strain Sensors. *ACS Appl. Polym. Mater.* **2022**, *4*, 8795–8802. [CrossRef]
147. Xiong, Y.; Xiao, J.Y.; Chen, J.; Xu, D.; Zhao, S.S.; Chen, S.B.; Sheng, B. A multifunctional hollow TPU fiber filled with liquid metal exhibiting fast electrothermal deformation and recovery. *Soft Matter* **2021**, *17*, 10016–10024. [CrossRef] [PubMed]
148. Wang, Y.G.; Chao, M.Y.; Wan, P.B.; Zhang, L.Q. A wearable breathable pressure sensor from metal-organic framework derived nanocomposites for highly sensitive broad-range healthcare monitoring. *Nano Energy* **2020**, *70*, 104560. [CrossRef]
149. Jia, Z.X.; Li, Z.J.; Ma, S.F.; Zhang, W.Q.; Chen, Y.J.; Luo, Y.F.; Jia, D.M.; Zhong, B.C.; Razal, J.M.; Wang, X.G.; et al. Constructing conductive titanium carbide nanosheet (MXene) network on polyurethane/polyacrylonitrile fibre framework for flexible strain sensor. *J. Colloid Interface Sci.* **2021**, *584*, 1–10. [CrossRef]
150. Shen, H.Y.; Ke, H.Z.; Feng, J.D.; Jiang, C.Y.; Wei, Q.F.; Wang, Q.Q. Highly Sensitive and Stretchable c-MWCNTs/PPy Embedded Multidirectional Strain Sensor Based on Double Elastic Fabric for Human Motion Detection. *Nanomaterials* **2021**, *11*, 2333. [CrossRef]
151. Kayser, L.V.; Lipomi, D.J. Stretchable Conductive Polymers and Composites Based on PEDOT and PEDOT:PSS. *Adv. Mater.* **2019**, *31*, 1806133. [CrossRef]
152. Dauzon, E.; Lin, Y.B.; Faber, H.; Yengel, E.; Sallenave, X.; Plesse, C.; Goubard, F.; Amassian, A.; Anthopoulos, T.D. Stretchable and Transparent Conductive PEDOT:PSS-Based Electrodes for Organic Photovoltaics and Strain Sensors Applications. *Adv. Funct. Mater.* **2020**, *30*, 2001251. [CrossRef]
153. Shaker, A.; Hassanin, A.H.; Shaalan, N.M.; Hassan, M.A.; Abd El-Moneim, A. Micropatterned flexible strain gauge sensor based on wet electrospun polyurethane/PEDOT: PSS nanofibers. *Smart Mater. Struct.* **2019**, *28*, 075029. [CrossRef]

154. Jin, F.; Lv, D.W.; Shen, W.F.; Song, W.J.; Tan, R.Q. High performance flexible and wearable strain sensor based on rGO and PANI modified Lycra cotton e-textile. *Sens. Actuator A-Phys.* **2022**, *337*, 113412. [CrossRef]
155. Tong, L.; Wang, X.X.; He, X.X.; Nie, G.D.; Zhang, J.; Zhang, B.; Guo, W.Z.; Long, Y.Z. Electrically Conductive TPU Nanofibrous Composite with High Stretchability for Flexible Strain Sensor. *Nanoscale Res. Lett.* **2018**, *13*, 86. [CrossRef]
156. Lin, L.; Choi, Y.; Chen, T.; Kim, H.; Lee, K.S.; Kang, J.M.; Lyu, L.; Gao, J.F.; Piao, Y. Superhydrophobic and wearable TPU based nanofiber strain sensor with outstanding sensitivity for high-quality body motion monitoring. *Chem. Eng. J.* **2021**, *419*, 129513. [CrossRef]
157. Peng, M.L.; Li, X.; Liu, Y.F.; Chen, J.W.; Chang, X.H.; Zhu, Y.T. Flexible multisensory sensor based on hierarchically porous ionic liquids/ thermoplastic polyurethane composites. *Appl. Surf. Sci.* **2023**, *610*, 155516. [CrossRef]
158. Xiao, J.J.; He, J.Q.; Wang, W.J.; Chen, M.M.; Guo, R.H. A multifunctional flexible strain sensor based on an excellent sensing performance PDMS-MXene@CNT/TPU nanofiber membrane with hydrophobic and photothermal conversion performance. *New J. Chem.* **2023**, *47*, 19265–19275. [CrossRef]
159. Jiang, N.; Chang, X.H.; Hu, D.W.; Chen, L.R.; Wang, Y.P.; Chen, J.W.; Zhu, Y.T. Flexible, transparent, and antibacterial ionogels toward highly sensitive strain and temperature sensors. *Chem. Eng. J.* **2021**, *424*, 130418. [CrossRef]
160. Wang, F.; Chen, J.W.; Cui, X.H.; Liu, X.N.; Chang, X.H.; Zhu, Y.T. Wearable Ionogel-Based Fibers for Strain Sensors with Ultrawide Linear Response and Temperature Sensors Insensitive to Strain. *ACS Appl. Mater. Interfaces* **2022**, *14*, 30268–30278. [CrossRef]
161. Zhao, W.Y.; Lin, Z.F.; Wang, X.P.; Wang, Z.Y.; Sun, Z.L. Mechanically Interlocked Hydrogel-Elastomer Strain Sensor with Robust Interface and Enhanced Water-Retention Capacity. *Gels* **2022**, *8*, 625. [CrossRef] [PubMed]
162. Palicpic, C.M.; Khadka, R.; Yim, J.H. An effectively enhanced vapor phase hybridized conductive polymer based on graphene oxide and glycerol influence for strain sensor applications. *New J. Chem.* **2022**, *46*, 22162–22170. [CrossRef]
163. Qu, M.N.; Lv, Y.Q.; Ge, J.W.; Zhang, B.; Wu, Y.X.; Shen, L.; Liu, Q.H.; Yan, M.; He, J.M. Hydrophobic and multifunctional strain, pressure and temperature sensor based on TPU/SiO2-ILs ionogel for human motion monitoring, liquid drop monitoring, underwater applications. *Colloid Surf. A-Physicochem. Eng. Asp.* **2023**, *664*, 131103. [CrossRef]
164. Sun, J.Y.; Xiu, K.H.; Wang, Z.Y.; Hu, N.; Zhao, L.B.; Zhu, H.; Kong, F.Z.; Xiao, J.L.; Cheng, L.J.; Bi, X.Y. Multifunctional wearable humidity and pressure sensors based on biocompatible graphene/bacterial cellulose bioaerogel for wireless monitoring and early warning of sleep apnea syndrome. *Nano Energy* **2023**, *108*, 108215. [CrossRef]

sensors

MDPI

Review

A Review of Wearable Optical Fiber Sensors for Rehabilitation Monitoring

Xiangmeng Li *, Yongzhen Li, Huifen Wei, Chaohui Wang and Bo Liu

Shanxi Provincial Key Laboratory for Advanced Manufacturing Technology, North University of China, Taiyuan 030051, China; s202302017@st.nuc.edu.cn (Y.L.); huifenwei@nuc.edu.cn (H.W.); s202202014@st.nuc.edu.cn (C.W.); s202302019@st.nuc.edu.cn (B.L.)

* Correspondence: xmli123@nuc.edu.cn

Abstract: As the global aging population increases, the demand for rehabilitation of elderly hand conditions has attracted increased attention in the field of wearable sensors. Owing to their distinctive anti-electromagnetic interference properties, high sensitivity, and excellent biocompatibility, optical fiber sensors exhibit substantial potential for applications in monitoring finger movements, physiological parameters, and tactile responses during rehabilitation. This review provides a brief introduction to the principles and technologies of various fiber sensors, including the Fiber Bragg Grating sensor, self-luminescent stretchable optical fiber sensor, and optic fiber Fabry–Perot sensor. In addition, specific applications are discussed within the rehabilitation field. Furthermore, challenges inherent to current optical fiber sensing technology, such as enhancing the sensitivity and flexibility of the sensors, reducing their cost, and refining system integration, are also addressed. Due to technological developments and greater efforts by researchers, it is likely that wearable optical fiber sensors will become commercially available and extensively utilized for rehabilitation.

Keywords: wearable optical fiber sensor; rehabilitation; monitoring

1. Introduction

Based on the medium-variant projections from the United Nations' "World Population Prospects 2022", the share of the global population aged 65 and above is projected to rise from 10% to 16% by 2050. This suggests a global trend of population aging. By 2033, China, one of the most populous nations in the world, is expected to have over 400 million residents aged 60 and above, peaking at around 520 million by 2054. The trend of global population aging is evident both in the rising number of elderly individuals and their increasing share of the total population. For instance, countries like Italy, Finland, and Portugal have elderly populations exceeding 20% of their total populations [1]. As the elderly population ages, the risk of common diseases and accidents among this group is likely to rise significantly. For individuals with chronic hand conditions or disabilities, daily rehabilitation practices, including massage, acupuncture, and exercise therapy, are essential. Traditional rehabilitation treatments, confined to clinical settings, pose significant challenges for individuals with limited mobility. Currently, ongoing advancements in fields such as materials science, sensing technology, and communication engineering have led to the widespread adoption of wearable rehabilitation devices among patients with mobility impairments [2–5]. These devices offer patients flexible and continuous dynamic monitoring, supplying rehabilitation physicians with real-time data, thereby facilitating more convenient, remote treatment options during the rehabilitation process.

When the brain issues commands to move the fingers, it produces a range of biophysical signals, such as force, temperature, movement, and bioelectricity, as well as biochemical signals including hormones, lactic acid, cytokines, and changes in ion concentrations. Wearable sensors are capable of measuring and quantifying these signals, reflecting a range of hand physiological data, and aiding in the operation of wearable rehabilitation devices for hand treatment [6–8]. In the current market of wearable rehabilitation equipment, electronic sensors are predominantly utilized for detecting human physiological signals. Devices such as encoders, potentiometers, electrical goniometers, and linear variable differential transformers (LVDTs) are examples of electronic sensors that measure joint angular positions and segmental linear displacement [9]. However, the complexity of encoders necessitates meticulous manufacturing and precise installation. Additionally, they are susceptible to environmental factors like temperature, vibration, and dust, which can significantly increase measurement errors, thereby compromising the stability and reliability of wearable devices [10]. Potentiometers have a delicate mechanical structure that is vulnerable to both physical impacts and environmental humidity. Changes in humidity can degrade measurement precision and hasten the wear of the potentiometer materials, reducing its lifespan. Due to these sensitivities, potentiometers are ill-suited for wearable devices frequently experiencing motion or substantial stress [11]. Certain types of goniometers require frequent calibration to ensure data accuracy, which can increase the barriers to use and increase maintenance costs. Additionally, electric goniometers are susceptible to vibrations, which can cause output signal fluctuations and introduce uncertainty in the detection of human physiological signals for wearable devices requiring precise angular measurements [12]. LVDTs can suffer from issues like drift, reduced sensitivity, and limited resolution with prolonged use, necessitating regular calibration and maintenance. This requirement can escalate operational costs and system complexity, potentially rendering them unsuitable for applications demanding high precision and sensitivity [13]. Additionally, in the realm of wearable rehabilitation devices, resistive, capacitive, and piezoelectric sensors are predominantly utilized to measure physiological parameters, recognize gestures, monitor motion, and offer functionalities for data acquisition and user interaction. However, resistive, capacitive, and piezoelectric sensors are susceptible to damage. In particular, their sensitivity is highly affected by external environmental factors such as temperature and humidity fluctuations. An increase in temperature may cause thermal expansion of the sensor materials, affecting their resistance or capacitance values, and thereby reducing measurement accuracy. An increase in humidity may cause moisture condensation on the sensor surface or internally, which can not only lead to electrical short circuits but can also alter the electrical properties of the sensors, affecting the stability and reliability of the signal, and subsequently lead to issues such as increased system complexity, power consumption, and maintenance costs [9]. It is worth noting that when the actuator is operating the electromagnetic field it generates will simultaneously affect the real-time operation of the capacitive sensor, causing certain interference in the auxiliary control of wearable rehabilitation treatment devices for rehabilitation exercises and real-time monitoring of patient data [14]. Recent advancements in optical fiber sensing technology have established optical fiber sensors as viable alternatives to electronic sensors in wearable rehabilitation devices, emphasizing their key benefit: inherent immunity to electromagnetic interference. Furthermore, optical fiber sensors have found broad applications across the medical field [15–17], and optical fiber sensors are also utilized in sectors such as oil and gas [18], structural monitoring [19], environmental monitoring [20], aerospace [21,22], and industrial automation [23], benefited by their lightweight, compact size, high sensitivity, capability for remote measurement, ability to conduct multi-parameter measurements, and the capacity to integrate into sensor networks alongside other types of sensors.

To meet the specific application requirements of wearable rehabilitation equipment, optical fiber sensors can be integrated with a variety of materials to create devices that are compatible with the human body and exhibit high sensitivity [24–26]. This includes specialized sensors such as optical fiber temperature sensors, optical fiber angle sensors, and optical fiber strain sensors. These sensors can be embedded into wearable items such as gloves, wristbands, chest straps, and medical shirts to monitor patients' vital signs— including body temperature [27], pulse [28], respiratory rate [29], and exercise state [30,31]—in real time during rehabilitation. The data collected can then inform adjustments to the wearable rehabilitation equipment for optimal joint angles [32] and exercise modalities [33]. Representative applications of optical fiber sensors in the field of rehabilitation therapy are depicted in Figure 1. However, the widespread production and application of wearable optical fiber sensors in rehabilitation are currently limited due to their complex manufacturing processes and higher costs.

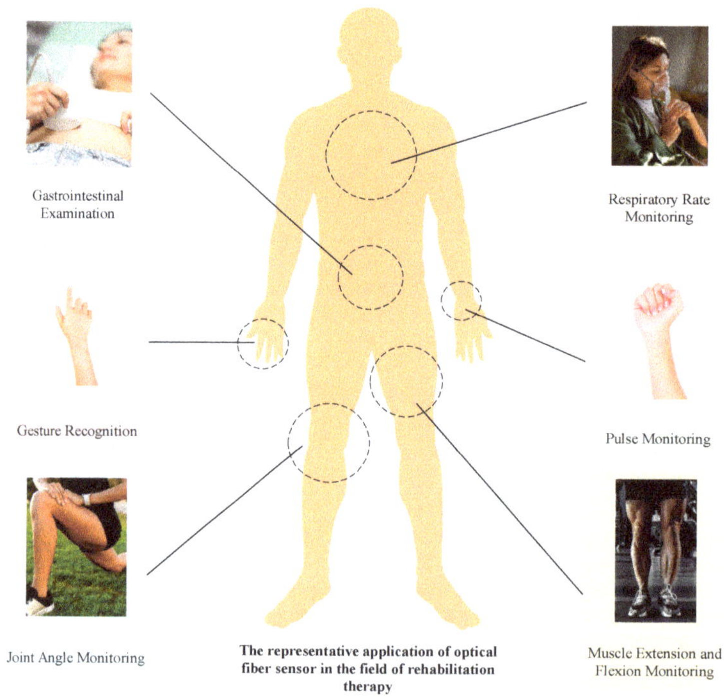

Figure 1. Representative applications for rehabilitation using wearable optical fiber sensors.

The detection of human finger parameters is crucial for both the recovery of muscle function and the assessment of hand motor skills during rehabilitation treatment. Wearable optical fiber sensors, offering a non-invasive, flexible approach with potential clinical applications, present a novel solution for rehabilitating hand function. However, applications in hand rehabilitation currently lack systematic analysis and the integration of key technologies. Thus, this paper reviews the technological advancements and potential applications within this domain. This paper discusses and summarizes the latest advancements in technology and finger parameter detection, drawing on the principles, fabrication methods, and materials of wearable optical fiber sensors. Section 2 primarily outlines the principles and fabrication methods of wearable optical fiber sensors. Section 3 presents case studies on the application of wearable optical fiber sensors for measuring human finger parameters. Section 4 discusses the challenges and prospective research directions for the implementation of wearable optical fiber sensors within the realm of finger parameter measurement. Section 5 offers a comprehensive summary of the entire review.

2. Working Principle of Wearable Optical Fiber Sensor

During the measurement of object parameters, wearable optical fiber sensors are subject to external influences, such as temperature and pressure. These influences can cause physical deformations like bending and stretching, which, in turn, alter the light transmission properties within the fibers. The sensor then detects these changes in the optical signal, which, upon processing using a photodetector and demodulator, are translated into the necessary measurement parameters. Presently, wearable optical fiber sensors primarily utilize fundamental physical properties like light intensity, wavelength, frequency, phase, and polarization for analyzing the characteristics of the objects under measurement. The diversity of optical fiber sensors is driving advancements in wearable rehabilitation equipment, offering patients a more user-friendly experience.

2.1. Fiber Bragg Grating Sensor

The Fiber Bragg Grating (FBG) sensor utilizes an internal grating etched into the optical fiber to convert optical signals into measurements indicative of stress, strain, or temperature. To mitigate the impact of environmental factors on the monitoring parameters, FBG sensors are typically encapsulated. Common encapsulation materials for FBG sensors include polymers [34,35], metals [36], and ceramics [37,38]. The choice of suitable encapsulation material is contingent upon the specific application environment and the performance demands of the sensor.

The working principle of the FBGs is illustrated in Figure 2a. When light propagates through an optical fiber, light at a specific central wavelength (the Bragg reflected light) is reflected at the Bragg grating region, while the rest of the wavelengths remain unaffected and continue to transmit. The reflected wavelength can be expressed as:

$$\lambda = 2 \times n_{eff} \times \Lambda. \quad (1)$$

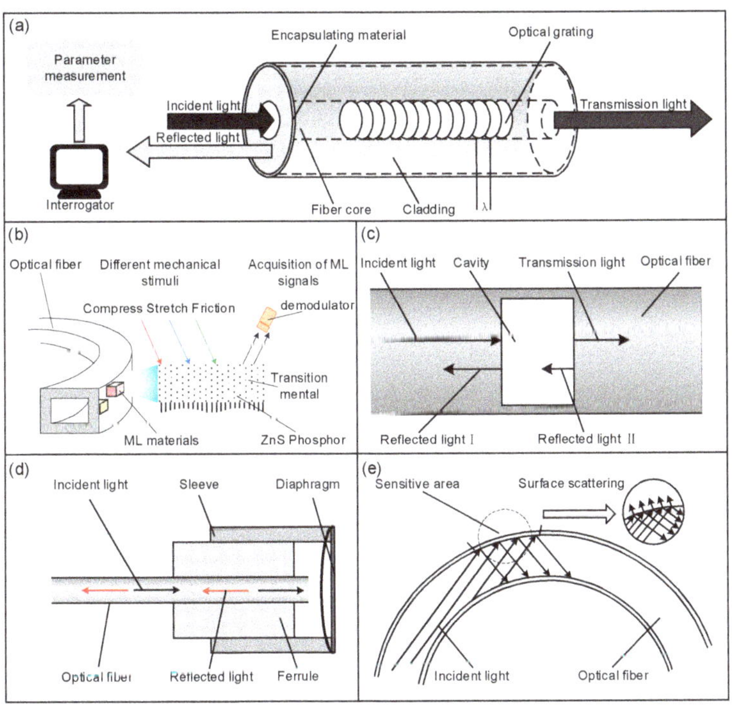

Figure 2. Schematic illustration of working principle for typical optical fiber sensors: (**a**) Fiber Bragg Grating sensor; (**b**) self-luminescent stretchable optical fiber sensor with mechanoluminescence (ML) materials; (**c**) optical fiber Fabry–Perot sensor; (**d**) diaphragm-type optical fiber sensor; and (**e**) polymer optical fiber sensor.

Among these parameters, λ represents the reflection wavelength of the FBGs, n_{eff} is the effective refractive index of the grating, and Λ denotes the period of the grating fringe. Strain ε and temperature T can influence the central wavelength of the FBGs. The relationship between strain ε, temperature T, and the wavelength shift of the FBGs is described by the subsequent equation:

$$\Delta\lambda = [(1 - P_e)\varepsilon + (\alpha + \xi)\Delta T]\lambda. \quad (2)$$

The changes encompass the central wavelength shift $\Delta\lambda$, strain ε, and temperature change ΔT, along with the photoelastic constant P_e, the thermal expansion coefficient α, and the thermo-optic coefficient ξ. FBGs offer distinct advantages for applications in structural, medical, and environmental monitoring, as well as in industrial process control, due to their adaptability to various environments and high measurement precision. They are especially well-suited for engineering applications that necessitate long-term stable monitoring. This establishes FBGs as a critical component of contemporary intelligent monitoring systems.

2.2. Self-Luminescent, Stretchable Optical Fiber Sensor

Currently, self-luminescent stretchable optical fiber sensors are typically fabricated using polymer casting and demolding techniques. Mechanoluminescence (ML) materials are strategically positioned within the fibers, with a demodulator attached at their extremities. Upon mechanical stimulation, the inherent self-luminescence of these materials is harnessed to produce internal sensor illumination, as depicted in Figure 2b.

Consider transition-metal-doped zinc sulfide phosphor, a common example of an ML material [39]. Upon mechanical stimulation of zinc sulfide phosphor, the doped transition metal ions are integrated into the crystal lattice of zinc sulfide, substituting for some zinc ions. Transition metal ions typically possess unpaired d-electron configurations. As the transition metal ions become excited, d-electrons transition from higher to lower energy levels, emitting photons in the process. The energy and wavelength of the emitted photons are determined by the energy level differences associated with the transition, resulting in varying emission wavelengths from the zinc sulfide material due to different transition metal ions. This emitted light propagates via total internal reflection along the optical fiber, with a demodulator at the terminus detecting the intensity and wavelength characteristics of light and generating a spectrum. Analysis of this spectrum can then indirectly infer the parameters of the physical quantity being measured.

2.3. Optic Fiber Fabry–Perot Sensor

The optical fiber Fabry–Perot sensor operates on the principle of Fabry–Perot interference. Current fabrication methods for Fabry–Perot sensors include the chemical etching method, which introduces bubbles into the fiber to create interference microcavities via a chemical reaction [40–42]; the laser processing method that employs femtosecond lasers to produce microcavities [43–45]; the direct welding method, which involves welding single-mode fibers with special fibers to form microcavities with gaps [46–49]; and the polymer-assisted filling method, which incorporates polymer materials into special fibers to create sealed air cavities during direct welding [50,51].

The working principle of the Fabry–Perot sensor, depicted in Figure 2c, involves using light propagation and reflection within the optical fiber to measure object parameters. A light source is connected to one end of the fiber, and reflections from the end faces of the cavity create an interference signal that varies with the length and refractive index of the cavity. Environmental changes can cause cavity deformation, resulting in a phase shift of the interference pattern. Consequently, Fabry–Perot sensors can measure a range of parameters, including humidity [52], temperature [53], curvature [54], acoustics [55], and strain [56].

A unique type of Fabry–Perot sensor, known as the diaphragm-type optical fiber sensor, features a fiber end that acts as the primary reflection point and a diaphragm that serves as the secondary reflection point. The diaphragm is constructed from materials like metal or polymers with desirable flexibility and deformability, which serve as sensing elements. This sensor responds to external stimuli to measure various physical quantities, including pressure, deformation, and temperature. The operating principle of the diaphragm-type optical fiber sensor is depicted in Figure 2d. The sensing principle involves the thin film making direct contact with the object under investigation by being attached to the end face of the optical fiber or encircling it. This vibration induces a change in the distance between the film and the end surface, thereby modulating the phase of the emitted light, which allows for the measurement of the physical quantity. This sensor responds to external stimuli to measure various physical quantities, including pressure, deformation, and temperature. During the measurement of physical quantities [57], the relationship between the deflection of the diaphragm h and the pressure P applied by a sound wave or pulse is given by

$$h = \frac{3(1 - v^2)Pa^4}{16Et^3}. \tag{3}$$

Here, v is the Poisson ratio of diaphragm, E is Young's modulus, P is the pressure induced by a pulse or sound signal, t is the thickness of the diaphragm, and a is the radius. Due to the pressure caused by the sound wave or pulse signal acting on the diaphragm, deformation occurs. This deformation modulates the phase of the light by altering the propagation path of the light at the end face of the optical fiber:

$$\Delta\varphi = \frac{4\pi nh}{\lambda} \tag{4}$$

where λ represents the central wavelength of the incident light and n denotes the refractive index of the fiber core.

Fabry–Perot sensors leverage interferometric principles to offer significant advantages in accurately measuring physical quantities, including temperature, pressure, and refractive index. The diaphragm-type optical fiber sensor, characterized by its flexibility, facilitates real-time monitoring and recording of acoustic cardiovascular conditions [57], imaging [58], and pulse waves [59] in patients during rehabilitation treatment. The fiber optic Fabry–Perot sensor holds broad application prospects in wearable rehabilitation devices, offering personalized and precise rehabilitation services to patients.

2.4. Polymer Optical Fiber Sensor

Polymer optical fiber sensors are constructed from polymer materials for the purpose of optical signal transmission and sensing. In comparison to traditional silica fibers, polymer fibers offer greater flexibility, lower refractive indices, and heightened sensitivity, thus providing distinct advantages for their use in optical fiber sensing applications [60]. The fabrication of these sensors generally encompasses three primary steps:

a. Polishing: Specialized equipment and techniques are employed to polish the polymer optical fiber, and 3D printing may be utilized [61] to create smooth and sensitive regions, ensuring a smooth fiber surface for effective interaction with external media. Within this process, parameters such as the polishing length, depth, and curvature radius can influence the sensitivity of the sensor.
b. Cladding Treatment: an appropriate cladding treatment is applied to the polished region to maintain optimal optical characteristics and mechanical integrity.
c. Fixed Encapsulation: the polished optical fiber is securely encapsulated to safeguard the polished region and establish a connection with the sensor system.

The operating principle of the polymer fiber sensor, as depicted in Figure 2e, is based on the principles of refraction and total internal reflection within the fiber. As light enters the polymer fiber and reaches the sensitive region, a portion is refracted into the external medium, while the remainder continues along the path of the fiber. Variations in the external refractive index of the medium, temperature, pressure, or other physical and chemical properties will result in changes to the intensities of both the refracted and transmitted light within the polished area. By measuring these variations in light intensity, the corresponding changes in the parameters of the object can be determined.

Presently, polymer optical fiber sensors are being utilized in the creation of curvature and joint motion sensors within the wearable rehabilitation treatment sector. These devices offer reliable medical data on physiological parameters, motion status, and muscle activity, which aids in the development and refinement of rehabilitation protocols [60].

2.5. Long-Period Fiber Grating Sensor

Long-Period Fiber Grating (LPFG) sensors are a type of optical sensor that utilizes optical fibers as the sensing medium. They achieve sensitive detection of external physical quantities—such as temperature, pressure, and bending—by introducing periodic refractive index variations within the optical fiber.

The working principle of LPFG sensors is based on the coupling effect between optical waveguide modes in the fiber. In an LPFG, periodic refractive index changes are introduced into the cladding of the optical fiber, forming a grating structure. As light propagates along the fiber and passes through the grating area, light of a specific wavelength resonates with the grating, leading to energy exchange between specific modes of the grating. This energy exchange results in an attenuation peak, known as the resonance peak, in the transmission spectrum for light of a particular wavelength. The resonant wavelength is related to the period of the grating, the effective refractive index of the fiber, and the refractive index of the surrounding environment. When environmental conditions change, such as temperature or pressure variations, the effective refractive index of the grating also changes, leading to a shift in the resonant wavelength. By monitoring the shift in the resonant wavelength, changes in external physical quantities can be detected.

The fabrication method for LPFG sensors typically involves the following steps [62,63]:

a. Fiber Selection: choose an appropriate single-mode or specialty optical fiber as the base material.
b. Grating Fabrication: use techniques such as ultraviolet light, CO_2 lasers, or femtosecond lasers to inscribe periodic refractive index variations on the fiber, creating the LPFG structure.
c. Chemical Etching: sometimes, chemical etching is required to adjust the parameters of the grating, such as by etching the outer cladding to enhance sensitivity to environmental conditions.
d. Functional Coating: apply a specific material, like graphene oxide, onto the grating surface to provide chemical functionality, facilitating subsequent immobilization of biomolecules or other types of sensing applications.

LPFG sensors, due to their high sensitivity, compatibility, and integrability, have a wide range of applications in various fields such as biomedical science and rehabilitation therapy [64–66].

2.6. Distributed Optical Fiber Sensor

Distributed Optical Fiber Sensors (DOFSs) are a type of sensor technology that utilizes optical fibers as the sensing medium to continuously, and in real time, monitor physical quantities (such as temperature, strain, vibration, etc.) along the length of the optical fiber.

The working principle of DOFS is mainly based on the light scattering effects in the optical fiber, including Rayleigh scattering, Brillouin scattering, and Raman scattering. These scattering effects interact with the light waves propagating in the optical fiber, causing changes in the frequency, phase, intensity, etc., of the light waves. These changes are related to the changes in physical quantities along the fiber. By measuring these changes, the distribution of physical quantities along the fiber can be inferred [67,68].

Rayleigh scattering: elastic scattering caused by tiny density fluctuations in the optical fiber, with the scattered light frequency being the same as the incident light, is commonly used for distributed temperature and vibration monitoring.

Brillouin scattering: Inelastic scattering caused by the interaction of sound waves in the optical fiber, resulting in frequency changes, is used for measuring temperature and strain.

Raman scattering: inelastic scattering caused by molecular vibrations in the optical fiber material, producing Stokes and anti-Stokes light, is used for temperature monitoring.

The fabrication of Distributed Optical Fiber Sensors involves the selection of optical fibers, the preparation of the fibers, and the integration of the sensor system. The choice of optical fiber depends on the monitoring parameters required and the environmental conditions. For example, single-mode fibers are typically used for Rayleigh and Raman scattering sensors, while multi-mode or few-mode fibers may be used for Brillouin scattering sensors. The preparation of optical fibers includes drawing the fibers, coating, and possible special treatments such as doping or the creation of FBGs [69].

Distributed Optical Fiber Sensors are widely used in various fields such as healthcare and environmental monitoring, but they still face some challenges, including cost, spatial resolution, measurement range, and the capability for multi-parameter measurement [18,70].

2.7. Micro-/Nanofibers Sensor

Micro-/Nanofiber (MNF) sensors are a highly sensitive sensing technology that utilizes micro/nanoscale optical fibers as the sensing medium to achieve continuous, real-time monitoring of physical quantities (such as temperature, strain, pressure, humidity, etc.) along the length of the fiber. The working principle of MNF primarily relies on various effects produced by the interaction of light with matter within the optical fiber, including intensity, scattering, absorption, interference, and changes in waveguide modes. These effects are associated with changes in the physical quantities along the fiber, and by precisely measuring these changes, the distribution of physical quantities along the fiber can be monitored [71–73].

In terms of manufacturing methods, MNF sensors involve a variety of techniques [74,75]:

a. Electrospinning technology: This process uses high voltage to eject a polymer solution into fine fibers, creating fibers at the micro/nano-scale.
b. Flame heating method: used for preparing silicon dioxide MNFs, this method heats standard optical fibers with a flame, gradually stretching and reducing their diameter.
c. Laser heating method: A CO_2 laser beam serves as the heat source for preparing MNFs with excellent surface smoothness and uniform diameter within a micro furnace.
d. Chemical synthesis and nano-lithography: these techniques are used for fabricating polymer MNFs and include methods such as chemical synthesis, electrospinning, and physical drawing.

MNFs have been widely applied in fields such as healthcare, environmental monitoring, and industry [76–78].

2.8. Performance Comparison of Wearable Optical Fiber Sensors

We can explore three main aspects: a comparison of advantages and disadvantages, materials and costs, and production and manufacturing processes.

a. Comparison of Advantages and Disadvantages: Self-luminescent, stretchable optical fiber sensors are superior in applications requiring no external power, yet they come with higher production costs and design complexities. In contrast, FBGs and Fabry–Perot sensors excel in precision monitoring for structural health, temperature, and strain measurements, though they have high production costs and require precise manufacturing processes. Diaphragm-type and polymer fiber optic sensors, however, provide better cost-effectiveness owing to simplified production processes and cost-efficient materials. LPFGs and DOFs offer highly sensitive detection of external physical quantities, but their practical use may require complex signal processing and data analysis.
b. Materials and Costs: Material selection is crucial; self-luminescent, stretchable optical fiber sensors and FBG sensors often necessitate specific high-performance materials, influencing production costs. Fabry–Perot sensors utilize precision materials, yet their varied designs can accommodate cost control, to some extent. Diaphragm-type and polymer fiber optic sensors typically employ cost-effective, easily processed materials like polydimethylsiloxane (PDMS) and polycarbonate (PC). Material selection for LPFGs and DOFs hinges on sensing requirements and the ability to adapt

to environmental conditions. Overall, material selection and cost management are pivotal in determining the broad application of sensors.

c. Production and Manufacturing Processes: The unique nature of self-luminescent, stretchable optical fiber sensors demands intricate mechanical design and meticulous manufacturing processes. Producing FBGs and Fabry–Perot sensors requires sophisticated laser writing and precise controls, somewhat restricting their large-scale manufacturing. Diaphragm-type and polymer fiber optic sensors can utilize industrial polishing, 3D printing, and injection molding, thus lowering production barriers and costs. LPFGs fabrication entails creating periodic refractive index changes, potentially using UV exposure, CO_2 lasers, or chemical etching. DOF production hinges on optical fiber selection, preparation, and sensor system integration and may include special treatments like doping or FBG fabrication.

3. The Application of Penetrable Optical Fiber Sensors in Finger Parameter Detection

Hand diseases are prevalent health issues that can significantly impact one's quality of life. Yet, accurate motion guidance and detection are especially critical during rehabilitation sessions. Proper exercise guidance can prevent sports injuries and expedite the rehabilitation process for patients with hand diseases. By monitoring patients' exercises, medical personnel can promptly identify irregularities, providing timely adjustments and guidance to prevent unnecessary injuries resulting from improper exercise techniques during rehabilitation. Hand rehabilitation often necessitates the use of various assistive devices and technologies for movement monitoring and guidance. Nevertheless, advancements in science and technology have introduced optical fiber sensors as a novel solution for hand rehabilitation challenges. Optical fiber sensors can be adeptly integrated into hand rehabilitation for the monitoring and guidance of hand movements. For instance, embedding optical fiber sensors within rehabilitation aids allows for real-time monitoring of hand movements and targeted guidance based on data analysis.

3.1. *Detection of Finger Movement*

3.1.1. Gesture Recognition

Gesture recognition technology is becoming increasingly vital in rehabilitation, aiding in the recovery of motor function for stroke and hand injury patients. Conventional rehabilitation typically depends on manual guidance and assessment from physiotherapists, a process that is time-consuming, labor-intensive, and challenges the feasibility of continuous monitoring and tailored training. The advent of wearable optical fiber sensors has revolutionized rehabilitation treatment. These sensors can accurately capture and record finger movements and joint angle changes, enabling real-time gesture recognition through computer-aided analysis. Furthermore, when integrated with virtual reality and game-based systems, wearable optical fiber sensors can offer a more diverse and engaging range of training modalities, enhancing rehabilitation efficiency.

In the realm of gesture recognition, wearable optical fiber sensors leverage their high sensitivity to accurately detect hand movements and joint changes, enabling precise identification of complex gestures. This capability for discerning nuanced movements not only underpins applications in virtual reality, gaming interaction, and assistive communication but also introduces innovative approaches to rehabilitation medicine and athletic training [79–82]. Bai et al. introduced a novel, stretchable lightguide for multimodal sensing (SLIMS) [31], consisting of a series of parallel elastomeric optical fibers featuring continuous or discrete ribbon patterns. SLIMS can discern and quantify the position, extent, and nature of mechanical deformations—such as stretching, bending, or pressing—utilizing the principles of frustrated total internal reflection and absorption. By integrating SLIMS sensors into a wireless glove, they enabled multi-point and multi-mode deformation decoupling, allowing for real-time tracking of a comprehensive range of finger joint movements and external pressures.

Liang et al. introduced a self-powered, stretchable mechanoluminescent optical fiber strain sensor [83], as shown in Figure 3a. By integrating a mechanoluminescent (ML) phosphor with an elastomeric optical fiber, the sensor emits visible light upon external stretching or release, eliminating the need for an external light source or power supply. The sensor demonstrates a linear strain response of up to 50% and a high precision strain measurement capability, achieving an accuracy of 1%. Additionally, they showcased the application of the sensors in wearable gloves and implantable devices for detecting human motion. For instance, integrating the sensors into gloves allows for monitoring finger joint movements, while those implanted in tissue, such as that of pig, can track muscle movements. Yang et al. introduced a self-powered photoelectric cooperative fiber sensor (SOEFSs), an optical fiber electronic device capable of visualizing and digitizing mechanical stimulation without the need for an external power supply [84]. The sensor responds to mechanical stimulation through a combination of mechanoluminescence (ML) and the triboelectric effect. SOEFSs feature a unique sheath–core structure, with the sheath composed of an optically active composite material and the core made of electrically active material. Utilizing a thermoplastic ML material system and spinning process, an industrial-grade, continuous manufacturing and recycling process for SOEFSs has been established, realizing sustainable production.

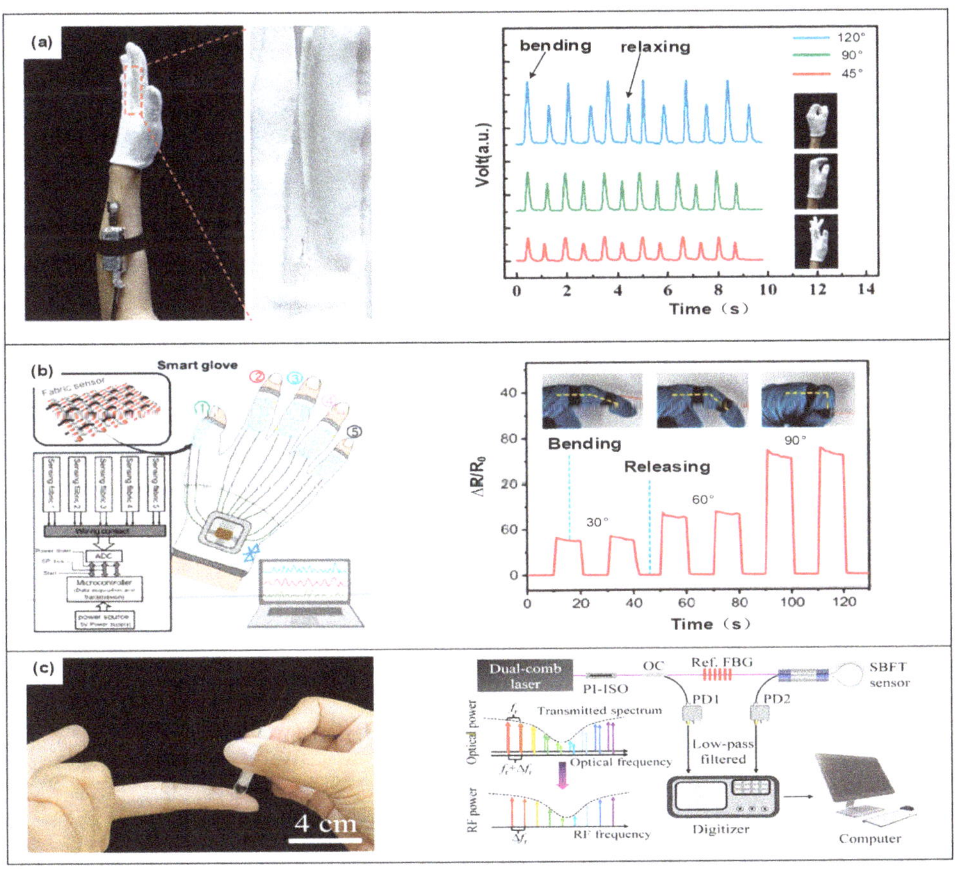

Figure 3. Application diagram of optical fiber sensor in detecting finger parameters. (**a**) Self-powered, stretchable mechanoluminescent optical fiber strain sensor, courtesy of Ref. [67]. (**b**) Structure diagram and detection diagram of stretchable optical strain sensor, courtesy of Ref. [74]. (**c**) Application and the schematic diagram of a soft, bio-inspired, fiber optic tactile sensor (SBFT), courtesy of Ref. [83].

The application of optical fiber sensors in gesture recognition demonstrates their unique advantages. The basic composition of a gesture recognition system typically encompasses three core components: an optical fiber sensor array, a signal acquisition and processing module, and a pattern recognition algorithm. Through the synergistic operation of these components, the system can accurately capture and recognize a wide array of gestures.

3.1.2. Measurement of Finger Joint Angle

In daily human activities, from delicate operations to routine typing, precise control and measurement of finger joint angles are essential. Detecting finger joint angles is crucial for assessing an individual's athletic capacity and health status and is also indispensable in fields such as rehabilitation medicine, physical training, ergonomics, and virtual reality. Sensors can accurately measure joint flexion by monitoring the propagation changes of light within optical fibers, thereby acquiring precise joint angle data [85–88]. A key advantage of this technology is its minimal susceptibility to environmental interference and non-irritating nature to the skin, allowing for prolonged wear and real-time monitoring. Li et al. introduced an optical neuron (MPON) utilizing a silicon-based microfiber probe for detecting and recognizing the motion of robotic fingers [89]. Silicon-based microfiber probes are fabricated by drawing biconical, silicon-based optical microfibers from standard single-mode fibers and embedding them within layers of PDMS. To safeguard the waveguide structure of the microfiber probe and protect against environmental contamination, a PDMS–Teflon–Microfiber–Teflon–PDMS composite structure was developed. With this composite structure, the MPON is capable of accurately measuring a wide range of bending angles with a rapid response.

Qu et al. introduced a high-sensitivity, stretchable optical strain sensor designed for monitoring human activity and providing medical care [90]. The structural and detection diagrams for this sensor are presented in Figure 3b. The sensor employed a simple, cost-effective design with unique mechanical and sensing properties, allowing for integration into clothing or direct attachment to the skin to monitor a range of human activities, from subtle physiological signals like wrist pulses to large joint movements and gestures. The sensor is capable of measuring large strains, up to 100%, with a low detection limit of 0.09%, a rapid response time of less than 12 ms, and high repeatability exceeding 6000 cycles.

The application of optical fiber sensors in detecting knuckle bending angles reveals their potential in biomedical engineering. The flexible optical fiber sensor, when placed close to the skin, can accurately detect minute changes in the knuckles. The signal processing unit converts the collected sensor data into an electrical signal and performs initial amplification and filtering. Data analysis software employs advanced algorithms to interpret the signals, enabling the precise calculation of knuckle bending angles.

3.2. *Detection of Finger Physiological Parameters*

Optical fiber sensors offer an effective means for the precise monitoring of physiological parameters, including finger movement, blood flow, and muscle activity. Furthermore, the wearability of optical fiber sensors allows for extended monitoring without disruption to patients' daily routines, which is crucial for enhancing rehabilitation efficiency and patient experience.

3.2.1. Blood Oxygen Saturation and Pulse Waveform

Oxygen saturation indicates the level of oxygenated hemoglobin in the blood, and the pulse waveform offers insights into the pumping function of the heart and the condition of the blood vessels. During rehabilitation treatment for conditions like heart disease, chronic obstructive pulmonary disease, and sleep apnea syndrome, monitoring blood oxygen saturation and pulse waveforms is crucial for guiding treatment and preventing potential complications [91]. For instance, continuous monitoring of blood oxygen saturation can assist physicians in evaluating the respiratory support requirements of the patient, and alterations in the pulse waveform may signal changes in cardiac load or enhancements in vascular elasticity. Optical fiber sensors are capable of continuously and accurately collecting physiological data without disrupting patients' daily activities. For instance, Shweta Pant et al. [92] introduced a novel, non-invasive finger plethysmography technique utilizing a Fiber Bragg Grating sensor to capture arterial pulse waveforms. This approach differs from the traditional pulse oximetry method, which assesses not only oxygen saturation in the blood but also records the actual pulse pressure waveform experienced by the arterial wall, offering a more comprehensive assessment of cardiovascular health.

Concurrently, the FBG Plethysmographic Pulse Recorder (FBGPPR), a device they designed, can dynamically record volume changes in the fingers due to pulsatile blood flow in the ulnar artery. Data from the FBGPPR were compared against the radial artery pulse waveform (RAPPW). as recorded using the FBG-based pulse recorder (FBGPR). Furthermore, the study illustrates the use of the FBGPPR and an electronic stethoscope to assess pulse transmission times and analyzes the captured pulse waveforms, which can reveal subtle changes and significant features within them. The characteristics are closely associated with factors such as vascular aging, arterial stiffness, the atherosclerosis index, and the effects of vasoactive drugs.

The application of optical fiber sensor technology enhances both the quality and efficiency of monitoring and offers patients a more comfortable and convenient rehabilitation experience. As technology advances, the use of optical fiber sensors in rehabilitation is expected to expand, offering increasingly personalized and efficient treatment options to patients.

3.2.2. Skin Temperature

In rehabilitation treatment, finger temperature serves as a physiological indicator that reflects the state of local blood circulation and tissue metabolic activity. Assessing normal finger temperature indices can facilitate wound healing, alleviate pain, prevent infections, and accelerate the rehabilitation process. For instance, during hand trauma or post-surgical rehabilitation, monitoring finger temperature can help to identify circulatory disorders and prevent complications in a timely manner. Utilizing optical fiber sensors for temperature monitoring enables continuous, real-time tracking of finger temperatures and offers scientific data to support rehabilitation treatments [93,94].

Liu et al. [95] introduced a ratiometric fluorescence optical fiber sensor that utilizes integrated silicate and tellurite glass for real-time human thermal monitoring. Within this sensor, Er^{3+}/Yb^{3+} co-doped tellurite glass generates up-conversion fluorescence emissions associated with thermal coupling energy levels, enabling temperature measurements through near-infrared excitation. The molten tellurite glass was drawn into a silicate tube using a negative pressure technique, achieving tight adhesion between the dissimilar glass materials, which have notably different softening temperatures. The sensor can be placed in the mouth or on the skin of the hand to continuously monitor body temperature. Additionally, the sensor can be deployed in the oral exhalation and nasal breathing environment for real-time thermal activity monitoring, offering a novel and highly reliable approach for human health surveillance. The research indicates that the sensor boasts high sensitivity, rapid response, a compact design, robust packaging, low cost, and high reliability, making it potentially valuable for real-time life and health monitoring applications.

Fiber optic sensors have demonstrated considerable value in the detection of finger physiological parameters, particularly in the real-time tracking of critical indicators like blood oxygen saturation, pulse waveform, and body temperature. These sensors provide physicians with crucial health data through high-precision measurements, aiding in the assessment of treatment efficacy and the adjustment of rehabilitation strategies. Optical fiber sensor technology is progressing towards greater sensitivity, enhanced durability, and reduced costs.

3.3. Detection of Finger Touch

In the rehabilitation field, advancements in optical fiber sensor technology have introduced new possibilities for tactile detection. By quantifying the effects of temperature, pressure, and slip on the fingers, these sensors can offer comprehensive tactile feedback [96–99]. This multi-parameter monitoring assists healthcare professionals in gaining a deeper understanding of patients' rehabilitation processes, thereby optimizing treatment plans and enhancing the personalization of care.

Shang and colleagues [100] introduced a soft, bio-inspired, fiber optic tactile sensor (SBFT) capable of sensing and differentiating between temperature and pressure akin to human skin, as shown in Figure 3c. The SBFT sensor is constructed by encapsulating a macroscopically curved Fiber Bragg Grating (FBG) within an elastomer droplet, generating two distinct types of optical resonances—FBG-related and whispering gallery modes (WGMs)—which propagate along the curvature. By leveraging the distinct thermo-optical and stress-optical properties of FBG and WGM resonances, the SBFT sensor can fully decouple pressure and temperature measurements, achieving an accuracy of 0.2 °C for temperature and 0.8 mN for pressure.

Li et al. introduced a skin-like optical fiber tactile sensor (SOFT), inspired by artificial intelligence, designed to emulate the tactile signal generation and processing functions of human skin [101]. It possesses multifunctional tactile interaction capabilities, including the detection of tactile amplitude, position, and tensile strain. The stretch unit of the SOFT features a unique "double X" configuration, enhancing the stretchability of the sensor, with tactile units aligned in a straight line for ease of array partitioning. The SOFT is composed of four FBG sensors embedded within a three-layer structure resembling skin, creating a flexible tactile sensing array capable of over 20% stretchability. Utilizing a two-stage cascade neural network, the SOFT can concurrently discern the position and magnitude of applied forces. Within the 0–3.5 N force range, the SOFT achieves a contact position recognition accuracy of 92.41%, with an error margin of less than 4.2%.

Optical fiber sensor technology mimics human tactile responses by precisely monitoring critical tactile indicators, like pressure, slip, and temperature, and sensitively detecting and quantifying external stimuli. This aids medical professionals in assessing treatment efficacy and refining treatment plans.

4. Challenges and Future Research Directions of Wearable Optical Fiber Sensors

We summarized the applications, types of optical fiber sensors, and performance of the various fiber optic sensors listed in Table 1. Currently, the majority of wearable optical fiber sensors utilized for finger parameter recognition are not yet extended for providing feedback in human health assessments or for on-demand robotic assistance. As noted in Ref. [102], the integration of wearable optical fiber sensing systems into exoskeleton assistive devices remains in the conceptual phase. Looking ahead, a key research direction for optical fiber sensors is to develop a compact, fully integrated, and cost-effective monitoring system suitable for clinical settings. However, current optical fiber sensing systems lack the necessary integration, making them less competitive than electronic sensors with regard to widespread clinical application. Numerous challenges persist, including sensor structural design, optimal measurement site selection, signal acquisition, multi-parameter measurement capabilities, and production costs, among others.

When designing wearable optical fiber sensors for finger parameter recognition, several key considerations should be addressed:

a. They must exhibit high sensitivity and flexibility.
b. The structure of the sensor must accurately detect changes at the measurement point, conform to the joints and muscles, accommodate the full range of human finger motion, and account for the effects of relative displacement and friction at the sensor-to-skin interface on measurement accuracy.

c. Employing flexible encapsulation or manufacturing techniques, along with a thoughtful arrangement of optical fiber sensors, can fulfill the durability requirements necessary for wearable applications.
d. It is essential to analyze the mechanical properties of flexible materials, ensure the stability and repeatability of their performance for long-term monitoring, and address the hysteresis inherent in flexible materials, which can affect measurement accuracy [103,104]. Hysteresis compensation techniques should be integrated to enhance the precision of the measured parameters.

Table 1. Measurement of human finger information based on optical fiber sensing.

Reference	Application	Type of Optical Fiber Sensors	Performance
[80]	Gesture recognition	SMF Sensor	Capable of recognizing up to 12 basic gestures with an accuracy rate as high as 94%.
[85]	Finger joint angle	POF Sensor	The sensitivity can reach 0.070 dB/°.
[87]	Pulse waveform	SCF Sensor	At pressures below 200 Pa, the sensitivity is 2.2 kPa^{-1}; within the pressure range of 200–600 Pa, the sensitivity is 0.91 kPa^{-1}.
[91]	Oxygen saturation and Heart rate	POF Sensor	Capable of rapidly responding to external pressure changes, with a response time of 5 milliseconds.
[93]	Temperature monitoring	MNF Sensor	Demonstrated a temperature sensitivity as high as −30 nm/°C. Capable of achieving a resolution of 0.0012 °C.
[96]	Tactile and temperature monitoring	FBG Sensor	Tactile sensitivity of 7.287 nm/MPa. Temperature sensitivity of 13 pm/°C.
[82]	Touch sensing and body temperature	FBG Sensor	A pressure sensitivity of 0.03 nm/kPa, a bending angle sensitivity of 0.19 nm/°, and a temperature sensitivity of 0.04 nm/°C.

At present, while fiber optic sensing technology, particularly FBG and Fabry–Perot sensors, excels in functional performance, it encounters challenges in practical application, such as high costs, large detection equipment size, and overall transportation difficulties, which limit their extensive use in clinical and rehabilitation settings. Nevertheless, as science and technology advance, we anticipate significant future transformations of these sensing systems. First, during deployment, beyond the typical issues like temperature, vibration, and dust, numerous environmental elements can impact the performance and precision of fiber optic sensors. These include humidity, electromagnetic interference, chemical contaminants, pressure variations, UV radiation, and air pollutants. These can irreversibly impair signal acquisition, transmission, measurement accuracy, sensitivity, and the integrity of the fiber material. Measures to combat these factors are mainly reflected in the protection with special coatings and encapsulation materials, the implementation of fiber optic isolation techniques to electrically insulate the optical sensor from subsequent electronic components, and the application of ultraviolet-resistant materials and coatings. Thus, employing innovative structures, materials, and design approaches is essential for sensor miniaturization, stable signal transfer, and improved detection capabilities. These innovations broaden the applicability of fiber optic sensors, facilitating precise measurements even in challenging conditions [105–107]. Second, combining multidimensional modeling with machine learning provides a novel approach to parameter decoupling. Machine learning has demonstrated significant potential in gait phase recognition and prediction. Integrating artificial intelligence and big data analysis allows for in-depth sensor data analysis, facilitating accurate diagnoses and personalized treatment plans [108–110]. Third, through the optimization and refinement of query algorithms, sensor performance can be enhanced while simultaneously reducing complexity and cost [111–114]. A notable advantage of wavelength-modulated optical fiber sensors is their capability to measure multiple parameters using a single optical fiber. This capability is particularly crucial for scenarios requiring the simultaneous monitoring of multiple biomechanical parameters. Additionally, phase- and intensity-modulated optical fiber sensors are capable of multi-parameter measurement, offering enhanced possibilities for temporal and spatial monitoring. The advancement of distributed optical fiber sensing technology, particularly two-dimensional sensing, offers novel solutions for high-resolution measurements of position and pressure. Compared to traditional quasi-distributed measurements, two-dimensional sensing technology exhibits superior spatial resolution and accuracy, providing significant advantages for monitoring human biomechanical parameters. Concurrently, advances in wireless technology are expected to improve the convenience and real-time capabilities of fiber optic sensor data collection [110,115]. These technologies enable comprehensive patient analysis and assessment, promoting the development of more precise treatment plans.

Currently, GaAs-based, phosphor, MEMS, and Distributed Optical Fiber sensors have been commercially available for years and hold substantial market value. In the future, the design and manufacturing of wearable fiber optic sensors must balance performance with cost-effectiveness. Advancements and innovations in technology will likely make future fiber optic sensors more efficient and cost-effective, expanding their use in clinical and rehabilitation fields [116–120].

5. Conclusions

To summarize, this short review provided a comprehensive overview of the applications and advancements of wearable optical fiber sensors within rehabilitation, aligning with the requirements of wearable sensor technology in the rehabilitation field and leveraging the benefits of optical fiber sensors. Optical fiber sensors hold significant promise in rehabilitation treatment due to their high sensitivity, interference resistance, and portability. This review has described the operating principles and manufacturing techniques of various optical fiber sensors, including self-luminous stretchable sensors and FBG sensors, and discussed their applications in monitoring finger movements, physiological parameters, and tactile sensations. Additionally, the review has highlighted the challenges of current technology, including higher sensitivity and lower costs, and outlined potential future research directions. It is anticipated that wearable optical fiber sensors will become commercialized and find widespread application in rehabilitation therapy.

Author Contributions: X.L. and Y.L. conceived the idea of the investigation, H.W. supervised the investigation, Y.L., C.W. and B.L. conducted the literature reviewing, X.L. and Y.L. wrote the manuscript, and all the authors edited and approved the manuscript. All authors have read and agreed to the published version of the manuscript.

Funding: This work was financially supported by the Science and Technology Cooperation and Exchange Special Project of Shanxi Province (Grant No. 202204041101006), Fundamental Research Program of Shanxi Province (Grant Nos. 20210302123013, 202203021222077), and research project supported by Shanxi Scholarship Council of China (Grant No. 2023-130).

Data Availability Statement: No new data were created or analyzed in this study. Data sharing is not applicable to this article.

Conflicts of Interest: The authors declare no conflicts of interest.

References

1. 2022 Revision of World Population Prospects. Available online: https://population.un.org/wpp/ (accessed on 20 May 2024).
2. Maceira-Elvira, P.; Popa, T.; Schmid, A.C.; Hummel, F.C. Wearable technology in stroke rehabilitation: Towards improved diagnosis and treatment of upper-limb motor impairment. *J. NeuroEng. Rehabil.* **2019**, *16*, 142–159. [CrossRef]
3. Lin, Y.; Shull, P.B. Novel, Soft, Water-Filled Acoustic Waveguides for Simultaneous Tactile Force and Location Sensing. *IEEE Trans. Ind. Electron.* **2023**, *71*, 8141–8155. [CrossRef]
4. Nelson, B.W.; Low, C.A.; Jacobson, N.; Arean, P.; Torous, J.; Allen, N.B. Guidelines for wrist-worn consumer wearable assessment of heart rate in biobehavioral research. *npj Digit. Med.* **2020**, *3*, 30. [CrossRef]

5. Feng, Y.; Ju, L.; Jia, H.; Liu, H.; Ding, X.; Zhang, W. Intentionally Light-Loss Carbon-Optic Fiber (COF) Twisted Sensor for Calf Strength Sensing via Monitoring Vastus Medialis. *IEEE Sens. J.* **2023**, *23*, 9271–9279. [CrossRef]
6. Luo, J.; Liu, H.; Xing, H.; Zhu, R.; Chen, Z.; Xu, F. Large Curvature Bending Measurable Fiber-Optic Neurons for Multi-Joint Bending Perception. *J. Light. Technol.* **2023**, *41*, 5780–5787. [CrossRef]
7. Zhang, L.; Tong, L. A bioinspired flexible optical sensor for force and orientation sensing. *Opto-Electron. Adv.* **2023**, *6*, 230051–230053. [CrossRef]
8. Wang, X.; Chen, G.; Zhang, K.; Li, R.; Jiang, Z.; Zhou, H.; Gan, J.; He, M. Self-Healing Multimodal Flexible Optoelectronic Fiber Sensors. *Chem. Mater.* **2023**, *35*, 1345–1354. [CrossRef]
9. Moreno, J.C.; Bueno, L. *Wearable Robot Technologies*, 1st ed.; Pons, J.L., Ed.; Wiley: Hoboken, NJ, USA, 2008; pp. 165–200.
10. Jaramillo, I.E.; Jeong, J.G.; Lopez, P.R.; Lee, C.-H.; Kang, D.-Y.; Ha, T.-J.; Oh, J.-H.; Jung, H.; Lee, J.H.; Lee, W.H.; et al. Real-Time Human Activity Recognition with IMU and Encoder Sensors in Wearable Exoskeleton Robot via Deep Learning Networks. *Sensors* **2022**, *22*, 9690. [CrossRef]
11. Othman, A.; Hamzah, N. Design and Development of an Adjustable Angle Sensor Based on Rotary Potentiometer for Measuring Finger Flexion. In Proceedings of the 2016 6th IEEE International Conference on Control System, Computing and Engineering (ICCSCE), Penang, Malaysia, 25–27 November 2016.
12. Lorussi, F.; Galatolo, S.; De Rossi, D.E. Textile-Based Electrogoniometers for Wearable Posture and Gesture Capture Systems. *IEEE Sens. J.* **2009**, *9*, 1014–1024. [CrossRef]
13. Yang, Y.S.; Bae, K.Y. The Modelling and Design of a Linear Variable Differential Transformer. *Int. J. Precis. Eng. Manuf.* **2022**, *23*, 153–162. [CrossRef]
14. Leal-Junior, A.G.; Frizera, A.; Vargas-Valencia, L.; dos Santos, W.M.; Bo, A.P.L.; Siqueira, A.A.G.; Pontes, M.J. Polymer Optical Fiber Sensors in Wearable Devices: Toward Novel Instrumentation Approaches for Gait Assistance Devices. *IEEE Sens. J.* **2018**, *18*, 7085–7092. [CrossRef]
15. Zu, L.; Wang, X.; Liu, P.; Xie, J.; Zhang, X.; Liu, W.; Li, Z.; Zhang, S.; Li, K.; Giannetti, A.; et al. Ultrasensitive and Multiple Biomarker Discrimination for Alzheimer's Disease via Plasmonic & Microfluidic Sensing Technologies. *Adv. Sci.* **2024**, e2308783. [CrossRef]
16. Bratash, O.; Buhot, A.; Leroy, L.; Engel, E. Optical fiber biosensors toward in vivo detection. *Biosens. Bioelectron.* **2024**, *251*, 116088. [CrossRef]
17. Lopes, R.N.; Pinto, P.H.S.; Vargas, J.D.L.; Dante, A.; Macrae, A.; Allil, R.C.B.; Werneck, M.M. Development of an Immunocapture-Based Polymeric Optical Fiber Sensor for Bacterial Detection in Water. *Polymers* **2024**, *16*, 861. [CrossRef] [PubMed]
18. Ashry, I.; Mao, Y.; Wang, B.; Hveding, F.; Bukhamsin, A.Y.; Ng, T.K.; Ooi, B.S. A Review of Distributed Fiber-Optic Sensing in the Oil and Gas Industry. *J. Light. Technol.* **2022**, *40*, 1407–1431. [CrossRef]
19. Du, C.; Dutta, S.; Kurup, P.; Yu, T.; Wang, X. A review of railway infrastructure monitoring using fiber optic sensors. *Sens. Actuators A Phys.* **2020**, *303*, 111728. [CrossRef]
20. Fan, L.; Bao, Y. Review of fiber optic sensors for corrosion monitoring in reinforced concrete. *Cem. Concr. Comps.* **2021**, *120*, 104029. [CrossRef]
21. Alvarez-Montoya, J.; Carvajal-Castrillón, A.; Sierra-Pérez, J. In-flight and wireless damage detection in a UAV composite wing using fiber optic sensors and strain field pattern recognition. *Mech. Syst. Signal Process.* **2020**, *136*, 106526. [CrossRef]
22. Cutolo, A.; Bernini, R.; Berruti, G.M.; Breglio, G.; Bruno, F.A.; Buontempo, S.; Catalano, E.; Consales, M.; Coscetta, A.; Cusano, A.; et al. Innovative Photonic Sensors for Safety and Security, Part II: Aerospace and Submarine Applications. *Sensors* **2023**, *23*, 2417. [CrossRef]
23. Liu, Y.; Li, X.; Zhang, Y.-N.; Zhao, Y. Fiber-optic sensors based on Vernier effect. *Measurement* **2021**, *167*, 108451. [CrossRef]
24. Zaltieri, M.; Massaroni, C.; Presti, D.L.; Bravi, M.; Sabbadini, R.; Miccinilli, S.; Sterzi, S.; Formica, D.; Schena, E. A Wearable Device Based on a Fiber Bragg Grating Sensor for Low Back Movements Monitoring. *Sensors* **2020**, *20*, 3825. [CrossRef] [PubMed]
25. Li, T.; Qiu, L.; Ren, H. Distributed Curvature Sensing and Shape Reconstruction for Soft Manipulators with Irregular Cross Sections Based on Parallel Dual-FBG Arrays. *IEEE/ASME Trans. Mechatron.* **2020**, *25*, 406–417. [CrossRef]
26. Zhang, W.; Jia, H.; Ju, L.; Shi, Y.; Ding, X.; Feng, Y. Bending-Sensitive Optical Waveguide Sensor with Carbon-Fiber Layer for Monitoring Grip Strength. *IEEE Trans. Neural Syst. Rehabil. Eng.* **2023**, *31*, 1922–1932. [CrossRef] [PubMed]
27. Pan, J.; Zhang, Z.; Jiang, C.; Zhang, L.; Tong, L. A multifunctional skin-like wearable optical sensor based on an optical micro-/nanofibre. *Nanoscale* **2020**, *12*, 17538–17544. [CrossRef] [PubMed]
28. Li, J.H.; Chen, J.H.; Xu, F. Sensitive and Wearable Optical Microfiber Sensor for Human Health Monitoring. *Adv. Mater. Technol.* **2018**, *3*, 1800296–1800303. [CrossRef]
29. Presti, D.L.; Massaroni, C.; Zaltieri, M.; Sabbadini, R.; Carnevale, A.; Di Tocco, J.; Longo, U.G.; Caponero, M.A.; D'Amato, R.; Schena, E.; et al. A Magnetic Resonance-Compatible Wearable Device Based on Functionalized Fiber Optic Sensor for Respiratory Monitoring. *IEEE Sens. J.* **2021**, *21*, 14418–14425. [CrossRef]
30. Leber, A.; Cholst, B. Stretchable Thermoplastic Elastomer Optical Fibers for Sensing of Extreme Deformations. *Adv. Funct. Mater.* **2018**, *29*, 1802629–1802636. [CrossRef]
31. Bai, H.D.; Li, S.; Barreiros, J.; Tu, Y.; Pollock, C.R.; Shepherd, R.F. Stretchable distributed fiber-optic sensors. *Science* **2020**, *370*, 848–852. [CrossRef] [PubMed]

32. Vargas-Valencia, L.S.; Schneider, F.B.A.; Leal-Junior, A.G.; Caicedo-Rodriguez, P.; Sierra-Arevalo, W.A.; Rodriguez-Cheu, L.E.; Bastos-Filho, T.F.; Frizera-Neto, A. Sleeve for Knee Angle Monitoring: An IMU-POF Sensor Fusion System. *IEEE J. Biomed. Health Inf.* **2021**, *25*, 465–474. [CrossRef]
33. Fajardo, J.; Neto, A.R.; Ferman, V.; Muñoz, A.; Andrade, D.; Rohmer, E. A Wearable Robotic Glove based on Optical FMG Driven Controller. In Proceedings of the 2019 IEEE 4th International Conference on Advanced Robotics and Mechatronics (ICARM), Toyonaka, Japan, 3–5 July 2019.
34. Lun, T.L.T.; Wang, K.; Ho, J.D.-L.; Lee, K.-H.; Sze, K.Y.; Kwok, K.-W. Real-Time Surface Shape Sensing for Soft and Flexible Structures Using Fiber Bragg Gratings. *IEEE Rob. Autom. Lett.* **2019**, *4*, 1454–1461. [CrossRef]
35. Wang, H.; Zhang, R.; Chen, W.; Liang, X.; Pfeifer, R. Shape Detection Algorithm for Soft Manipulator Based on Fiber Bragg Gratings. *IEEE/ASME Trans. Mechatron.* **2016**, *21*, 2977–2982. [CrossRef]
36. He, Q.; Zhang, Q.C. A flexible temperature sensing finger using optical fiber grating for soft robot application. *Optoelectron. Lett.* **2021**, *17*, 400–406. [CrossRef]
37. Wu, B.; Chen, H.; Xiao, S.; Wang, M.; Wang, Z.; Yan, F. High-Sensitivity Fiber-Optic Voltage Sensor Based on an Optoelectronic Oscillator Using a PZT-Stack and an Equivalent Phase-Shifted Fiber Bragg Grating. *IEEE Sens. J.* **2023**, *23*, 19332–19338. [CrossRef]
38. Dang, W.; Li, Z.; Dan, J.; Nan, P.; Xin, G.; Lim, K.-S.; Ahmad, H.; Yang, H. High sensitivity fiber Bragg grating (FBG) sensor based on hollow core silica tube (HCST) sensitization for gas pressure and temperature discrimination. *Opt. Fiber Technol.* **2023**, *75*, 103202. [CrossRef]
39. Ren, B.Y.; Chen, B.; Zhang, X.; Wu, H.; Fu, Y.; Peng, D. Mechanoluminescent optical fiber sensors for human-computer interaction. *Sci. Bull.* **2023**, *68*, 542–545. [CrossRef]
40. Pevec, S.; Donlagic, D. Miniature all-fiber Fabry-Perot sensor for simultaneous measurement of pressure and temperature. *Appl. Opt.* **2012**, *51*, 4536–4541. [CrossRef] [PubMed]
41. Liu, Y.; Wang, D.N.; Chen, W.P. Crescent shaped Fabry-Perot fiber cavity for ultra-sensitive strain measurement. *Sci. Rep.* **2016**, *6*, 38390. [CrossRef] [PubMed]
42. Cibula, E.; Donlagic, D. In-line short cavity Fabry-Perot strain sensor for quasi distributed measurement utilizing standard OTDR. *Opt. Express* **2007**, *15*, 8719–8730. [CrossRef] [PubMed]
43. Liao, C.R.; Hu, T.Y.; Wang, D.N. Optical fiber Fabry-Perot interferometer cavity fabricated by femtosecond laser micromachining and fusion splicing for refractive index sensing. *Opt. Express* **2012**, *20*, 22813–22818. [CrossRef]
44. Tian, M.; Lu, P.; Chen, L.; Liu, D.; Yang, M.; Zhang, J. Femtosecond laser fabricated in-line micro multicavity fiber FP interferometers sensor. *Opt. Commun.* **2014**, *316*, 80–85. [CrossRef]
45. Li, J.J.; Liu, Y.; Qu, S. Research Progress on Optical Fiber Functional Devices Fabricated by Femtosecond Laser Micro-Nano Processing. *Laser Optoelecton. Prog.* **2020**, *57*, 111402–111417.
46. Yang, F.; Tan, Y.; Jin, W.; Lin, Y.; Qi, Y.; Ho, H.L. Hollow-core fiber Fabry–Perot photothermal gas sensor. *Opt. Lett.* **2016**, *41*, 3025–3028. [CrossRef] [PubMed]
47. Jia, P.; Fang, G.; Liang, T.; Hong, Y.; Tan, Q.; Chen, X.; Liu, W.; Xue, C.; Liu, J.; Zhang, W.; et al. Temperature-compensated fiber-optic Fabry–Perot interferometric gas refractive-index sensor based on hollow silica tube for high-temperature application. *Sens. Actuators B Chem.* **2017**, *244*, 226–232. [CrossRef]
48. Yu, Y.Q.; Chen, X.; Huang, Q.; Du, C.; Ruan, S.; Wei, H. Enhancing the pressure sensitivity of a Fabry-Perot interferometer using a simplified hollow-core photonic crystal fiber with a microchannel. *Appl. Phys. B Laser Opt.* **2015**, *120*, 461–467. [CrossRef]
49. Lee, C.L.; Ho, H.Y.; Gu, J.-H.; Yeh, T.-Y.; Tseng, C.-H. Dual hollow core fiber-based Fabry-Perot interferometer for measuring the thermo-optic coefficients of liquids. *Opt. Lett.* **2015**, *40*, 459–462. [CrossRef] [PubMed]
50. Zhao, Y.; Chen, M.-Q.; Xia, F.; Lv, R.-Q. Small in-fiber Fabry-Perot low-frequency acoustic pressure sensor with PDMS diaphragm embedded in hollow-core fiber. *Sens. Actuators A Phys.* **2018**, *270*, 162–169. [CrossRef]
51. Li, M.; Liu, Y.; Gao, R.; Li, Y.; Zhao, X.; Qu, S. Ultracompact fiber sensor tip based on liquid polymer-filled Fabry-Perot cavity with high temperature sensitivity. *Sens. Actuators B Chem.* **2016**, *233*, 496–501. [CrossRef]
52. Zhu, C.; Huang, J. Highly Sensitive Humidity Sensor Based on Optical Fiber Fabry-Perot Interferometer. In Proceedings of the 2023 IEEE Sensors, Vienna, Austria, 29 October–1 November 2023.
53. Liu, Z.; Zhao, B.; Zhang, Y.; Zhang, Y.; Sha, C.; Yang, J.; Yuan, L. Optical fiber temperature sensor based on Fabry-Perot interferometer with photopolymer material. *Sens. Actuators A Phys.* **2022**, *347*, 113894. [CrossRef]
54. Monteiro, C.S.; Ferreira, M.S.; Silva, S.O.; Kobelke, J.; Schuster, K.; Bierlich, J.; Frazão, O. Fiber Fabry-Perot interferometer for curvature sensing. *Photonic Sens.* **2016**, *6*, 339–344. [CrossRef]
55. Wang, S.; Lu, P.; Zhang, J.; Xu, H.; Yao, Q.; Chen, Y.; Liu, L.; Liao, H.; Sun, Y.; Ni, W.; et al. An Infrasound Sensor Based on Extrinsic Fiber-Optic Fabry–Perot Interferometer Structure. *IEEE Photonics Technol. Lett.* **2016**, *28*, 1264–1267. [CrossRef]
56. Jiang, M.; Gerhard, E. A simple strain sensor using a thin film as a low-finesse fiber-optic Fabry–Perot interferometer. *Sens. Actuators A Phys.* **2001**, *88*, 41–46. [CrossRef]
57. Wang, D.; Fan, S.; Jin, W. Graphene diaphragm analysis for pressure or acoustic sensor applications. *Microsyst. Technol.* **2013**, *21*, 117–122. [CrossRef]
58. Wang, J.; Ai, F.; Sun, Q.; Liu, T.; Li, H.; Yan, Z.; Liu, D. Diaphragm-based optical fiber sensor array for multipoint acoustic detection. *Opt. Express* **2018**, *26*, 25293–25304. [CrossRef]

59. Wang, J.; Liu, K.; Sun, Q.; Ni, X.; Ai, F.; Wang, S.; Yan, Z.; Liu, D. Diaphragm-based optical fiber sensor for pulse wave monitoring and cardiovascular diseases diagnosis. *J. Biophotonics* **2019**, *12*, 201900084–201900093. [CrossRef]
60. Leal-Junior, A.G.; Frizera, A.; Pontes, M.J. Sensitive zone parameters and curvature radius evaluation for polymer optical fiber curvature sensors. *Opt. Laser Technol.* **2018**, *100*, 272–281. [CrossRef]
61. Li, C.; Liu, J.; Chi, Z.; Zhang, H.; Peng, X. Wearable Wrist Joint Movement Detection Using Two Sensitivity-Enhanced Plastic Optical Fibers. In Proceedings of the 2018 IEEE Sensors, New Delhi, India, 28–31 October 2018.
62. Wang, L.; Zhang, W.; Chen, L.; Bai, Z.; Liu, F.; Yan, T. Torsion sensor based on two cascaded long period fiber gratings fabricated by CO_2 laser pulse irradiation and HF etching technique respectively. *J. Mod. Opt.* **2016**, *64*, 541–545. [CrossRef]
63. Chan, K.-H.; Lin, H.-J.; Hwang, S.-J. Long-period fiber gratings based on blue phase liquid crystal/polymer composites. *Liq. Cryst.* **2021**, *49*, 50–58. [CrossRef]
64. Esposito, F.; Sansone, L.; Srivastava, A.; Cusano, A.M.; Campopiano, S.; Giordano, M.; Iadicicco, A. Label-free detection of vitamin D by optical biosensing based on long period fiber grating. *Sens. Actuators B Chem.* **2021**, *347*, 130637. [CrossRef]
65. Zhu, C.; Huang, S.; Tang, C.; Yang, G.; Zhao, Y.; Li, H. Miniature Temperature-Independent Curvature Sensor Based on a Phase-Shifted Long-Period Fiber Grating Using Deep Tapering. *IEEE Sens. J.* **2023**, *23*, 14174–14181. [CrossRef]
66. Ran, J.; Chen, Y.; Wang, G.; Zhong, Z.; Zhang, J.; Xu, O.; Huang, Q.; Lei, X. Mechanically Induced Long-Period Fiber Gratings and Applications. *Photonics* **2024**, *11*, 223. [CrossRef]
67. Li, J.; Zhang, M. Physics and applications of Raman distributed optical fiber sensing. *Light. Sci. Appl.* **2022**, *11*, 128. [CrossRef] [PubMed]
68. Bai, Q.; Wang, Q.; Wang, D.; Wang, Y.; Gao, Y.; Zhang, H.; Zhang, M.; Jin, B. Recent Advances in Brillouin Optical Time Domain Reflectometry. *Sensors* **2019**, *19*, 1862. [CrossRef] [PubMed]
69. Lu, P.; Lalam, N.; Badar, M.; Liu, B.; Chorpening, B.T.; Buric, M.P.; Ohodnicki, P.R. Distributed optical fiber sensing: Review and perspective. *Appl. Phys. Rev.* **2019**, *6*, 041302. [CrossRef]
70. Leal-Junior, A.G.; Ribeiro, D.; Avellar, L.M.; Silveira, M.; Díaz, C.A.R.; Frizera-Neto, A.; Blanc, W.; Rocon, E.; Marques, C. Wearable and Fully-Portable Smart Garment for Mechanical Perturbation Detection with Nanoparticles Optical Fibers. *IEEE Sens. J.* **2021**, *21*, 2995–3003. [CrossRef]
71. Yin, J.; Reddy, V.S.; Chinnappan, A.; Ramakrishna, S.; Xu, L. Electrospun Micro/Nanofiber with Various Structures and Functions for Wearable Physical Sensors. *Polym. Rev.* **2022**, *63*, 715–762. [CrossRef]
72. Ma, S.; Wang, X.; Li, P.; Yao, N.; Xiao, J.; Liu, H.; Zhang, Z.; Yu, L.; Tao, G.; Li, X.; et al. Optical Micro/Nano Fibers Enabled Smart Textiles for Human–Machine Interface. *Adv. Fiber Mater.* **2022**, *4*, 1108–1117. [CrossRef]
73. Dang, H.; Chen, M.; Li, J.; Liu, W. Sensing performance improvement of resonating sensors based on knotting micro/nanofibers: A review. *Measurement* **2021**, *170*, 108706. [CrossRef]
74. Zhang, L.; Tang, Y.; Tong, L. Micro-/Nanofiber Optics: Merging Photonics and Material Science on Nanoscale for Advanced Sensing Technology. *iScience* **2020**, *23*, 100810. [CrossRef]
75. Shi, S.; Si, Y.F.; Han, Y.; Wu, T.; Iqbal, M.I.; Fei, B.; Li, R.K.Y.; Hu, J.; Qu, J. Recent Progress in Protective Membranes Fabricated via Electrospinning: Advanced Materials, Biomimetic Structures, and Functional Applications. *Adv. Mater.* **2022**, *34*, 2107938–2107968. [CrossRef]
76. Cai, D.; Tong, T.; Zhang, Z.; Pan, J.; Zhang, L.; Tong, L. Functional Film Coated Optical Micro/Nanofibers for High-Performance Gas Sensing. *IEEE Sens. J.* **2019**, *19*, 9229–9234. [CrossRef]
77. Li, X.; Chen, N.; Zhou, X.; Gong, P.; Wang, S.; Zhang, Y.; Zhao, Y. A review of specialty fiber biosensors based on interferometer configuration. *J. Biophotonics* **2021**, *14*, e202100068. [CrossRef]
78. Ma, S.; Wang, Z.; Zhu, Y.; Tang, Y.; Fan, G.; Ma, B.; Ye, T.; Wei, L. Micro/nanofiber fabrication technologies for wearable sensors: A review. *J. Micromech. Microeng.* **2022**, *32*, 064002–064022. [CrossRef]
79. Li, T.L.; Su, Y.F.; Chen, F.; Zheng, H.; Meng, W.; Liu, Z.; Ai, Q.; Liu, Q.; Tan, Y.; Zhou, Z. Bioinspired Stretchable Fiber-Based Sensor toward Intelligent Human-Machine Interactions. *ACS Appl. Mater. Interfaces* **2022**, *14*, 22666–22677. [CrossRef] [PubMed]
80. Chen, M.H.; He, Y.C.; Liang, H.; Zhou, H.; Wang, X.; Heng, X.; Zhang, Z.; Gan, J.; Yang, Z. Stretchable and Strain-Decoupled Fluorescent Optical Fiber Sensor for Body Temperature and Movement Monitoring. *ACS Photonics* **2022**, *9*, 1415–1424. [CrossRef]
81. Wang, S.P.; Wang, X.Y.; Wang, S.; Yu, W.; Yu, L.; Hou, L.; Tang, Y.; Zhang, Z.; Ni, Y.; Cao, C.; et al. Optical-Nanofiber-Enabled Gesture-Recognition Wristband for Human-Machine Interaction with the Assistance of Machine Learning. *Adv. Intell. Syst.* **2023**, *5*, 2200412–2200419. [CrossRef]
82. Rao, H.C.; Luo, B.B.; Wu, D.; Yi, P.; Chen, F.; Shi, S.; Zou, X.; Chen, Y.; Zhao, M. Study on the Design and Performance of a Glove Based on the FBG Array for Hand Posture Sensing. *Sensors* **2023**, *23*, 8495. [CrossRef] [PubMed]
83. Liang, H.H.; He, Y.C.; Chen, M.; Jiang, L.; Zhang, Z.; Heng, X.; Yang, L.; Hao, Y.; Wei, X.; Gan, J.; et al. Self-Powered Stretchable Mechanoluminescent Optical Fiber Strain Sensor. *Adv. Intell. Syst.* **2021**, *3*, 2100035–2100043. [CrossRef]
84. Yang, W.F.; Gong, W.; Gu, W.; Liu, Z.; Hou, C.; Li, Y.; Zhang, Q.; Wang, H. Self-Powered Interactive Fiber Electronics with Visual-Digital Synergies. *Adv. Mater.* **2021**, *33*, 2104681–2104690. [CrossRef]
85. Ge, Q.-Q.; Zhou, T.; Gong, T.-Y.; Liang, Y.; Ngiejungbwen, L.A.; Chen, M.-Y. Highly sensitive measurement of finger joint angle based on a double-U tapered POF embedded in PDMS film. *Opt. Fiber Technol.* **2023**, *76*, 103236. [CrossRef]
86. Socorro-Leranoz, A.B.; Diaz, S.; Castillo, S.; Dreyer, U.J.; Martelli, C.; da Silva, J.C.C.; Uzqueda, I.; Gomez, M.; Zamarreno, C.R. Optical System Based on Multiplexed FBGs to Monitor Hand Movements. *IEEE Sens. J.* **2021**, *21*, 14081–14089. [CrossRef]

87. Li, Z.; Tang, X.; Zhao, T.; Chen, K.; Zhang, T. Highly sensitive skin-like wearable optical sensor for human physiological signals monitoring. *Opt. Fiber Technol.* **2024**, *82*, 103652. [CrossRef]
88. Guo, J.J.; Zhou, B.Q.; Zong, R.; Pan, L.; Li, X.; Yu, X.; Yang, C.; Kong, L.; Dai, Q. Stretchable and Highly Sensitive Optical Strain Sensors for Human-Activity Monitoring and Healthcare. *ACS Appl. Mater. Interfaces* **2019**, *11*, 33589–33598. [CrossRef]
89. Li, Y.P.; Tan, S.J.; Yang, L.; Li, L.; Fang, F.; Sun, Q. Optical Microfiber Neuron for Finger Motion Perception. *Adv. Fiber Mater.* **2022**, *4*, 226–234. [CrossRef]
90. Qu, X.Y.; Wu, Y.C.; Ji, P.; Wang, B.; Liang, Q.; Han, Z.; Li, J.; Wu, Z.; Chen, S.; Zhang, G.; et al. Crack-Based Core-Sheath Fiber Strain Sensors with an Ultralow Detection Limit and an Ultrawide Working Range. *ACS Appl. Mater. Interfaces* **2022**, *14*, 29167–29175. [CrossRef] [PubMed]
91. Ballaji, H.K.; Correia, R.; Korposh, S.; Hayes-Gill, B.R.; Hernandez, F.U.; Salisbury, B.; Morgan, S.P. A Textile Sleeve for Monitoring Oxygen Saturation Using Multichannel Optical Fibre Photoplethysmography. *Sensors* **2020**, *20*, 6568. [CrossRef]
92. Pant, S.; Umesh, S.; Asokan, S. A Novel Approach to Acquire the Arterial Pulse by Finger Plethysmography Using Fiber Bragg Grating Sensor. *IEEE Sens. J.* **2020**, *20*, 5921–5928. [CrossRef]
93. Song, X.D.; Wang, Q.; Liu, Q.; Yu, L.; Wang, S.; Yao, N.; Tong, L.; Zhang, L. Twisted Optical Micro/Nanofibers Enabled Detection of Subtle Temperature Variation. *ACS Appl. Mater. Interfaces* **2023**, *15*, 47177–47183. [CrossRef] [PubMed]
94. Li, L.; He, R.; Soares, M.S.; Savovic, S.; Hu, X.; Marques, C.; Min, R.; Li, X. Embedded FBG-Based Sensor for Joint Movement Monitoring. *IEEE Sens. J.* **2021**, *21*, 26793–26798. [CrossRef]
95. Liu, W.; Yin, Z.; Song, D.; Yan, X.; Zhang, X.; Wang, F.; Suzuki, T.; Ohishi, Y.; Cheng, T. A ratiometric fluorescent fiber sensor based on integrated silica and tellurite glass for real-time human thermal monitoring. *Ceram. Int.* **2023**, *49*, 26049–26056. [CrossRef]
96. Massari, L.; Fransvea, G.; D'abbraccio, J.; Filosa, M.; Terruso, G.; Aliperta, A.; D'alesio, G.; Zaltieri, M.; Schena, E.; Palermo, E.; et al. Functional mimicry of Ruffini receptors with fibre Bragg gratings and deep neural networks enables a bio-inspired large-area tactile-sensitive skin. *Nat. Mach. Intell.* **2022**, *4*, 425–435. [CrossRef]
97. Sun, S.; He, S.; Han, Y.; Long, Y.; He, Z. Soft Bionic Skin with Touch-Slip and Thermal Perception Using Double-Layer Distributed FBG Sensing Array. *IEEE Sens. J.* **2022**, *22*, 4096–4105. [CrossRef]
98. Yue, X.; Lu, R.; Yang, Q.; Song, E.; Jiang, H.; Ran, Y.; Guan, B.-O. Flexible Wearable Optical Sensor Based on Optical Microfiber Bragg Grating. *J. Light. Technol.* **2023**, *41*, 1858–1864. [CrossRef]
99. Yao, N.; Wang, X.; Ma, S.; Song, X.; Wang, S.; Shi, Z.; Pan, J.; Wang, S.; Xiao, J.; Liu, H.; et al. Single optical microfiber enabled tactile sensor for simultaneous temperature and pressure measurement. *Photonics Res.* **2022**, *10*, 2040–2046. [CrossRef]
100. Shang, C.; Fu, B.; Tuo, J.; Guo, X.; Li, Z.; Wang, Z.; Xu, L.; Guo, J. Soft Biomimetic Fiber-Optic Tactile Sensors Capable of Discriminating Temperature and Pressure. *ACS Appl. Mater. Interfaces* **2023**, *15*, 53264–53272. [CrossRef] [PubMed]
101. Li, T.; Su, Y.; Zheng, H.; Chen, F.; Li, X.; Tan, Y.; Zhou, Z. An Artificial Intelligence-Motivated Skin-Like Optical Fiber Tactile Sensor. *Adv. Intell. Syst.* **2023**, *5*, 2200460–2200470. [CrossRef]
102. Leal, A.; Avellar, L.; Jaimes, J.; Díaz, C.; dos Santos, W.; Siqueira, A.A.G.; Pontes, M.J.; Marques, C.; Frizera, A. Polymer Optical Fiber-Based Integrated Instrumentation in a Robot-Assisted Rehabilitation Smart Environment: A Proof of Concept. *Sensors* **2020**, *20*, 3199. [CrossRef] [PubMed]
103. To, C.; Hellebrekers, T.; Jung, J.; Yoon, S.J.; Park, Y.-L. A Soft Optical Waveguide Coupled with Fiber Optics for Dynamic Pressure and Strain Sensing. *IEEE Rob. Autom. Lett.* **2018**, *3*, 3821–3827. [CrossRef]
104. Leal, A.G.; Frizera, A.; Pontes, M.J. Dynamic Compensation Technique for POF Curvature Sensors. *J. Light. Technol.* **2018**, *36*, 1112–1117. [CrossRef]
105. Dhanabalan, S.S.; Sriram, S.; Walia, S.; Avaninathan, S.R.; Carrasco, M.F.; Bhaskaran, M. Wearable Label-Free Optical Biodetectors: Progress and Perspectives. *Adv. Photonics Res.* **2020**, *2*, 2000076–2000093. [CrossRef]
106. Zhang, C.; Ye, W.B. Bioinspired Artificial Sensory Nerve Based on Nafion Memristor. *Adv. Funct. Mater.* **2019**, *29*, 1808783–1808792. [CrossRef]
107. Jiang, Y.; Ji, S.; Sun, J.; Huang, J.; Li, Y.; Zou, G.; Salim, T.; Wang, C.; Li, W.; Jin, H.; et al. A universal interface for plug-and-play assembly of stretchable devices. *Nature* **2023**, *614*, 456–462. [CrossRef] [PubMed]
108. Lu, S.; Tan, Z.; Li, G.; Yang, J. A Sensitized Plastic Fiber Sensor for Multi-Point Bending Measurement Based on Deep Learning. *IEEE Photonics J.* **2021**, *13*, 1–7. [CrossRef]
109. Hou, B.; Yi, L.; Li, C.; Zhao, H.; Zhang, R.; Zhou, B.; Liu, X. An interactive mouthguard based on mechanoluminescence-powered optical fibre sensors for bite-controlled device operation. *Nat. Electron.* **2022**, *5*, 682–693. [CrossRef]
110. Wang, K.; Mizuno, Y.; Dong, X.; Kurz, W.; Koehler, M.; Kienle, P.; Lee, H.; Jakobi, M.; Koch, A.W. Multimode optical fiber sensors: From conventional to machine learning-assisted. *Meas. Sci. Technol.* **2023**, *35*, 022002–022027. [CrossRef]
111. Martínez-Manuel, R.; Esquivel-Hernández, J.; Valentín-Coronado, L.M.; LaRochelle, S. Fiber-Optic Sensor for Directional Bending Monitoring Based on a Pattern Recognition Algorithm. *IEEE Sens. J.* **2023**, *23*, 4819–4824. [CrossRef]
112. Zhong, T.; Cheng, M.; Lu, S.; Dong, X.; Li, Y. RCEN: A Deep-Learning-Based Background Noise Suppression Method for DAS-VSP Records. *IEEE Geosci. Remote Sens. Lett.* **2022**, *19*, 1–5. [CrossRef]
113. Kot, P.; Muradov, M.; Gkantou, M.; Kamaris, G.S.; Hashim, K.; Yeboah, D. Recent Advancements in Non-Destructive Testing Techniques for Structural Health Monitoring. *Appl. Sci.* **2021**, *11*, 2750. [CrossRef]
114. Ma, Z.; Song, Z.; Huang, X.; Guo, T.; Yuan, W.; Chen, H.; Zhang, T.; Wang, W. A Zero-Cross Detection Algorithm for Cavity-Length Interrogation of Fiber-Optic Fabry–Perot Sensors. *Sensors* **2019**, *19*, 38681. [CrossRef] [PubMed]

115. Lin, M.-T. Quadri-directional optical bending sensor using side-emitting POF integrated with wearable device technology in smart glove. *Int. J. Cloth. Sci. Tech.* **2022**, *34*, 502–515. [CrossRef]
116. Lu, M.; Wang, C.; Fan, R.; Lin, M.; Guang, J.; Peng, W. Review of Fiber-Optic Localized Surface Plasmon Resonance Sensors: Geometries, Fabrication Technologies, and Bio-Applications. *Photonic Sens.* **2024**, *14*, 240202. [CrossRef]
117. Tian, X.; Qiu, X.; Li, H.; Lu, J.; Yang, C. Dynamically light-switched polarization-sensitive absorber based on semiconductor-incorporated metamaterial structure. *Opt. Mater.* **2024**, *149*, 115139. [CrossRef]
118. Zhou, H.; Wang, X.; He, Y.; Liang, H.; Chen, M.; Liu, H.; Qasem, A.; Xiong, P.; Peng, D.; Gan, J.; et al. Distributed Strain Sensor Based on Self-Powered, Stretchable Mechanoluminescent Optical Fiber. *Adv. Intell. Syst.* **2023**, *5*, 2300482–2300495. [CrossRef]
119. Yuan, C.; Fu, J.; Qu, F.; Zhou, Q. A combined MEMS thermal vacuum sensor with a wide pressure range. *J. Micromech. Microeng.* **2023**, *34*, 015006–015014. [CrossRef]
120. Feng, Z.; Tang, Y.; Cheng, Y.; Sima, C.; Yuan, L. High-Spatial-Resolution Quasi-Distributed Fiber Sensing Technique Based on Cascaded FP Cavities and DSP Algorithm. *J. Light. Technol.* **2024**, *42*, 2186–2192. [CrossRef]

sensors

MDPI

Review

Flexible Triboelectric Mechanical Energy Harvesters for Wearable and Self-Powered Sensing Applications: A Review

Manchi Punnarao [1] and Hong-Joon Yoon [1,2,*]

[1] Department of Semiconductor Engineering, Gachon University, Seongnam 13120, Republic of Korea
[2] Department of Electronic Engineering, Gachon University, Seongnam 13120, Republic of Korea
* Correspondence: yoonhj1222@gachon.ac.kr

Abstract

Triboelectric nanogenerators (TENGs) have been gaining significant attention owing to their excellent energy conversion efficiency and their integration towards a large number of practical applications in energy harvesting, wearables, and self-powered sensing. In recent advancements, the utilization of flexible triboelectric composite films can help to enhance the TENG's electrical output performance, as they possess excellent mechanical and dielectric properties and tunable surface characteristics. Moreover, by combining flexible active layers with triboelectric nanogenerators, the advantages of each component result in sensor devices which offer superior characteristics, including high sensitivity, biocompatibility, less weight, and mechanical flexibility. This review mainly focuses on the applications of TENGs in mechanical energy harvesting, self-powered wearable sensor systems, as well as the latest research progress in the TENG field. The working principles of TENG will be first explained in detail, including four basic operational modes of TENG, simulation results, and the working mechanism of the contact–separation mode TENGs. The fabrication techniques of triboelectric flexible films, along with TENG construction, will then be introduced. Common applications of TENGs are based on mechanical energy harvesting and powering portable electronic devices, which will subsequently be classified and summarized. Additionally, the applications of various wearable and self-powered sensor applications are elucidated. Finally, the current limitations and future directions of the TENG will be explained in detail and proposed. By exploring these innovations, the review underscores the importance of triboelectric flexible film-based TENGs in driving the future of energy harvesting and sensor technologies.

Keywords: polymer/composite films; mechanical energy harvesting; wearable electronics; self-powered sensors; TENGs

1. Introduction

With the rapid expansion of smart, flexible, and wearable electronic devices and the growing energy crisis caused by the reduction in fossil fuels, mechanical energy harvesting from the living environment to power wearable and low-powered portable electronics has attracted increasing attention [1,2]. In parallel, the widespread evolution of portable and wearable electronics has raised the demand for continuous, dependable energy sources. Although modern batteries offer improved energy density and longer operating times, their fundamental limitations, such as storage capacity, weight, and size, and the continuous need for recharging, still pose significant challenges [3,4]. These limitations are particularly restrictive for mobile and remote applications that support the expansion of the Internet of

 https://doi.org/10.3390/s26041166

Things [5]. The development of green energy harvesting technologies currently plays a key role in addressing worldwide energy demands and environmental interests. Different types of energy harvesting technologies exist to convert different external energy sources, such as mechanical, wind, tidal, sound, and thermal energy, into electricity [6–8]. Examples of energy harvesting mechanisms in such technologies include the utilization of piezoelectric, triboelectric, thermoelectric, and electromagnetic effects.

Among various harvesting technologies, triboelectric nanogenerators (TENGs) are one of the promising techniques that convert mechanical energy into electricity, which is vital for self-powered systems. In the year 2012, Z. L. Wang and his co-workers proposed a TENG based on the contact electrification and electrostatic induction phenomena [9]. TENGs have attracted more interest due to their advantages, such as simple design, high energy conversion efficiency, lower cost, long-lasting nature, and a wide range of possible materials used in TENG fabrication [10–12]. For example, H. Luo et al. proposed a multi-phase rotating disk TENG (MPRD-TENG) with excellent DC electrical output for operating fire warning alarm systems and speed-monitoring applications [13]. Moreover, H. Luo et al. designed a highly stable rotary solid–liquid TENG (RSL-TENG) for wastewater detection applications, which are abundantly available in temperature, humidity, and light-sensitive monitoring modules [14]. Since the concept was invented, TENGs have been extensively investigated in various application areas, such as energy harvesting, healthcare monitoring, robotics, and sensor devices [15–18]. Also, several methodologies have been improving the TENG output power, such as structural design, charge injection, and material optimization by incorporating fillers, etc. [19–22]. So far, several organic polymers and inorganic materials, dielectrics, carbon materials, filler materials, and loaded polymers (composites) have been selected for designing flexible and high-output TENGs, such as $BaTiO_3$, $KNaNbO_3$, Mxene, CNTs, PVA, PVDF, PDMS, etc. [23–27]. Triboelectric charges are built on surfaces where dissimilar materials with opposite electron affinities contact by external friction. The polymers and composite films improve the surface charge density, mechanical strength, and electrical properties of materials, resulting in high electrical output from the corresponding TENG. Also, their essential beneficial factors include easy incorporation, biocompatibility, and high energy-conversion efficiency, which make TENGs highly promising for self-powered sensing and wearable applications [28–30]. Such devices can employ various forms of mechanical energies, like pressure, stretching, vibration, and rotations, and convert them into electrical energy. The compatibility of polymer and composite polymer film-based materials allows for high flexibility, lower cost, high mechanical strength, and easy fabrication, making them ideal for next-generation applications.

This review highlights recent progress in the area of polymer/composite films employed in flexible TENGs and their smart applications. The discussion includes operation modes, working principles of TENGs, and the mechanism of the contact–separation mode of TENGs was explained in detail. Also, the COMSOL multi-physics simulation results are explained. Further, various classes of polymer/composite films and TENG fabrication techniques are systematically analyzed to explain how their design structures contribute to charge-generation efficiency, mechanical properties, and interfacial stability. The capability of these engineered film-based TENGs delivers high output performance while maintaining flexibility and long-term operational stability, making them promising candidates for next-generation smart electronics. Therefore, the practical and real-time applications of the TENG under mechanical energy harvesting, wearable, and smart self-powered sensors were systematically investigated. However, significant challenges endure, such as optimizing filler materials in polymer matrix interfaces, confirming stable electric output under constant mechanical loading, and developing eco-friendly fabrication methods.

https://doi.org/10.3390/s26041166

Overcoming these limitations is vital for the development of flexible TENGs constructed from organic, inorganic, and hybrid composite films, thereby supporting their reliable use in advanced wearable sensing devices, soft robotic systems, and health-monitoring applications, as shown in Figure 1, refs [31–40].

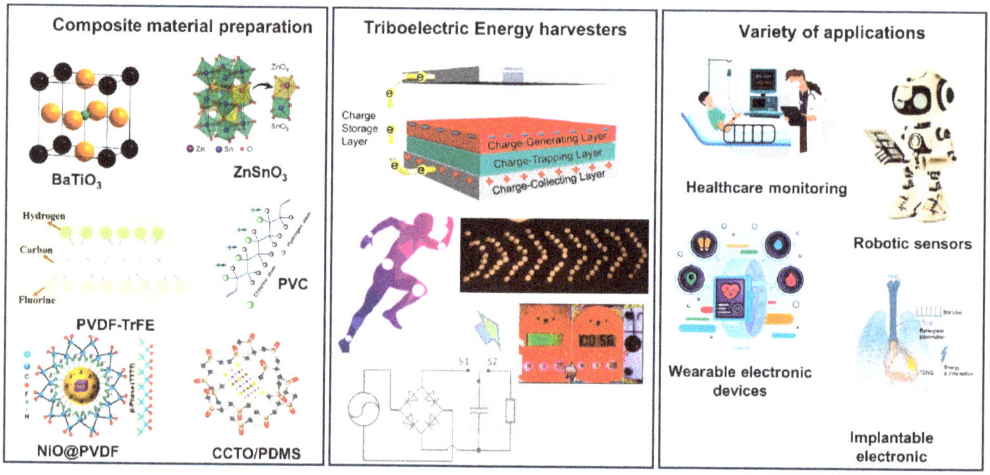

Figure 1. Overview of flexible TENG. Fabrication of organic, inorganic, and composite-based TENGs for energy harvesting and a variety of smart sensing applications [31–40]. All essential copyrights and permissions received.

2. TENG Working Principle

Self-powered electronic applications utilizing the triboelectric effect show significant capabilities in converting mechanical energies into electrical power, eliminating reliance on external power sources. These devices boast increased consistency, scalability, and compactness, offering unparalleled versatility in a variety of applications. In this section, starting from the triboelectric working principle, the operational working modes of various TENGs will be described, with a detailed emphasis on the working mechanism of the contact–separation-mode TENG. The triboelectric effect occurs abundantly in living environments and generates electrical charges through mechanical actions, such as contact and separation between a pair of materials. Prof. Z. L. Wang and his group devised the TENG in 2012 by combining the effects of contact electrification and electrostatic induction, thus providing a new direction through the development of a novel energy source for the era of green energy [9].

The principles of charge flow between the two triboelectric material surfaces have been extensively reviewed and can generally be split into three types: electron, ion, and material transfers [41–43]. Electron transfer is widely considered the main charge transfer process in solid–solid triboelectrification, particularly for insulating and semiconducting materials. When two surfaces come into close contact, their surface electronic states interact, causing electrons to move from the material with a higher Fermi level to the material with a higher work function [41]. Ion transfer becomes important in triboelectric processes that involve liquids, soft materials, or humid conditions [42]. Material transfer happens when tiny molecular parts, polymer chains, or nanoscale particles are physically shifted from one surface to another during contact. This effect is most often seen in soft polymer materials, elastomers, and rough surfaces [43]. Together, these mechanisms determine the surface charge and electric potential that enable energy generation in triboelectric nanogenerators. For example, G. Xu et al. proposed that electron transfer is the main

https://doi.org/10.3390/s26041166

process of contact electrification and put forward an electron cloud model to explain the relevant mechanism [44]. Figure 2a describes, on an atomic level, the electronic behavior when two originally separated layers, each containing a triboelectrically active material, come into contact. When separated, the electrons in each layer are strongly bound within their respective potential wells and specific orbitals as they are within the attractive region of their respective atomic potential. Their corresponding electron clouds, therefore, remain separated without overlapping, and charge escape and transfer are barred. When the atoms of the separated layers approach and come into contact, their respective electron clouds overlap, and charge "screening" lowers the energy barrier between the two materials, allowing the originally distinct potential wells of each layer to merge and transform into an asymmetric double-well potential. The chemical potentials of the surfaces of the layers also equilibrate during contact, facilitating electron transfer. When the layers are separated, if the rate of mechanical separation exceeds the rate of electron re-equilibration, then the transferred electrons would remain on the material surface as static charges, causing both triboelectric materials to exhibit positive and negative polarities, respectively.

3. Operational Modes of TENG

Depending on the structural design and the shape of the triboelectric materials in a TENG device, four types of fundamental operational modes have been defined, such as contact–separation TENG (CS-TENG), single-electrode (SE-TENG), lateral-sliding (LS-TENG), and freestanding (FS-TENG) modes, as shown in Figure 2b, refs [45–47]. According to different electrode configurations, it can be split into two types, i.e., single and double electrode types.

The most common and simple fundamental form of operation mode is the CS-mode TENG, as shown in Figure 2b(i), which is applied when the relative motion is perpendicular to the interface [48–50]. The two dissimilar dielectric materials are oppositely placed, and the electrodes are attached to each dielectric material. When the two dielectric materials come into contact by applying friction, a charge is generated on the active materials. When both dielectric layers are separated by removing friction, the surfaces of both dielectric layers preserve their electric charges, which can lead to a potential difference between the two dielectric materials. As a result, an electrostatic field is generated, and a charge transport starts from one electrode to another. When the dielectric materials come into contact again, the electrostatic field vanishes, and electron flows return. The result of the alternating current signal will be generated by this continuous contact–separation TENG process. CS-TENG provides a fundamental and effective route for converting mechanical energy, characterized by simple design, stable electrical output performance, and wide adaptability to mechanical stimuli. Advances in materials, structural design, and device integration continue to enhance their performance and support their application in self-powered and sustainable electronic systems.

In the lateral-sliding-mode TENG (LS-TENG), the two different dielectric materials remain in direct contact, and a frictional force induces sliding or rotational motion, thus leading the effective contact area to change constantly, as shown in Figure 2b(ii) [51–53]. This interaction generates triboelectric charges on the dielectric material surfaces, leading to a changing potential difference between the electrodes and generating an alternating output current. Compared with the CS mode, the LS mode configuration facilitates more effective charge generation, supplies higher electrical output, and does not require an air gap for surface separation. Nevertheless, LS mode is limited by the intense friction that occurs during operation, which can accelerate material degradation and reduce the overall lifespan of the TENG. LS-TENGs offer an efficient route for harvesting mechanical energy, supported by continuing developments in device architecture, surface modifi-

https://doi.org/10.3390/s26041166

cation, and power regulation. Also, continued interdisciplinary efforts are expected to improve their reliability and facilitate their integration into self-powered and sustainable electronic systems.

In the single-electrode-mode TENG (SE-TENG) (Figure 2b(iii)), a single grounded electrode, typically composed of a material with less affinity to attract electrons, functions as both the reference electrode and the triboelectric dielectric layer [54,55]. This electrode interacts with a freely moving dielectric layer that has a higher electron affinity. When the dielectric material approaches the grounded electrode, electrostatic induction drives electrons from the electrode toward the ground. As the dielectric material goes away, the induced potential reverses, causing electrons to go back from the ground to the electrode. Although it has a simple and versatile nature, the SE mode frequently suffers from limited electron-transfer efficiency, which reduces the overall power output. SE-TENGs offer a simple yet adaptable solution for mechanical energy harvesting and self-powered sensing. Continuing advances in triboelectric materials, device fabrication, and system integration are expected to enhance their output power and accelerate their adoption in wearable electronics and delivered sensing systems.

In the freestanding-mode TENG (FS-TENG), two symmetrical electrodes connected through an external circuit are positioned under a movable dielectric triboelectric material, as shown in Figure 2b(iv) [56,57]. When the dielectric material shifts relative to the electrodes, the spatial distribution of electrostatic potential is altered, driving electrons to move back and forth between the electrodes and generating an electrical output. Because charge transport occurs through electrostatic induction rather than direct mechanical rubbing between solid surfaces, this mode significantly reduces surface wear and improves the long-term durability of the TENG. FS-TENG efficiently harvests mechanical energy by coupling the moving triboelectric layer to the fixed electrodes, which minimizes mechanical wear and supports varied motion modes. This design enhances electrostatic charge induction, while continued advances in triboelectric layers, device structures, and system integration are expected to elevate performance further and expand applications in self-powered electronics, smart sensing, and sustainable energy harvesting.

In addition, the working mechanism of the CS-mode TENG was further investigated using COMSOL simulation results to show the obtained electric potential during the contact and separation operation. Based on the observed surface charge density and dielectric constant of the triboelectric materials, the generated electric potential from the TENG was observed in different stages by performing an analysis using a finite element analyzer, solver, and COMSOL simulation software. The detailed device simulation results are provided in Figure 2c(i–iv), ref. [58]. The highest electric potential was produced when the TENG was in approaching and separating stages, and the minimum potential was produced when the device was in idle, contact, and separated stages.

4. Working Mechanism of CS-Mode TENG

The working mechanism of the vertical contact–separation-mode TENG was explained in detail, and it depends on the coupling effects of triboelectrification and electrostatic induction, as shown in Figure 2d, refs [48–50,59]. This CS mode was mainly divided into four stages. First, both the dielectric layers fully contact by the friction applied on the TENG (Figure 2d(i)), and the positive and negative charges are produced on the surface of the dielectric materials. The produced charges do not transfer and are quickly neutralized owing to the insulating nature of the dielectric layers. When the dielectric materials start to separate due to the removal of friction on TENG (Figure 2d(ii)), the electric potential difference is created in both electrodes, leading to electron flow transfer from one electrode to another. As a result, the triboelectric positive electric output is generated from the TENG.

 https://doi.org/10.3390/s26041166

When the dielectric material is completely separated or reaches an equilibrium state by removing friction on TENG (Figure 2d(iii)), the potential difference is completely offset by the electron flow between the electrodes, and no electric potential is produced in the external circuit. Furthermore, both dielectric materials make contact again by applying friction on TENG (Figure 2d(iv)), creating a reverse electrical output signal via an external circuit because of the backflow of the electrons. Consequently, continuous external friction applied and removed on the TENG could produce a continuous alternative output current signal.

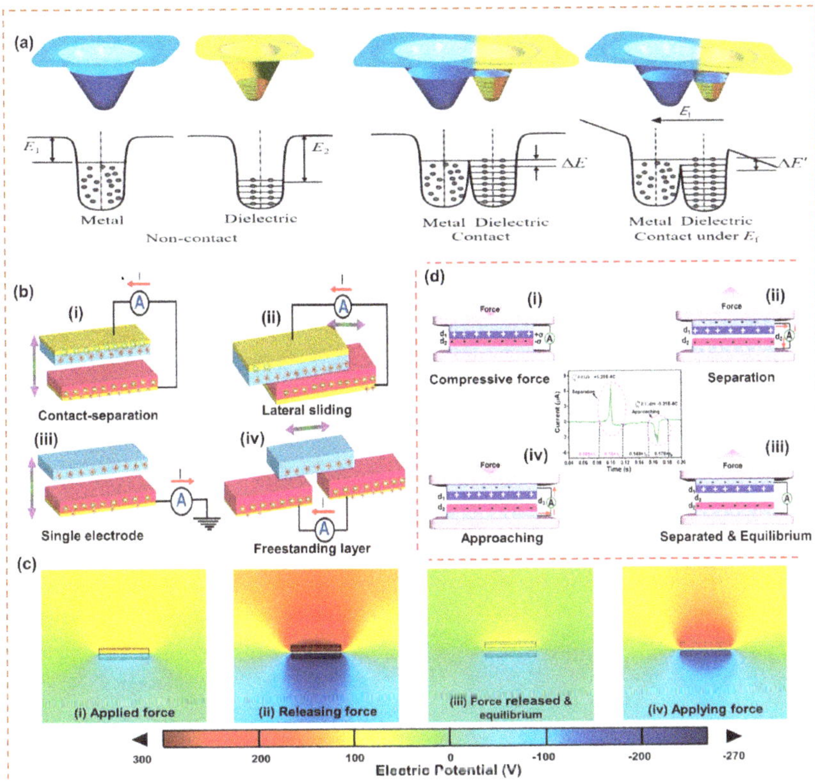

Figure 2. (**a**) Working principle of TENG [44]. (**b**) Different types of operational modes of TENG: (**i**) contact–separation mode, (**ii**) freestanding mode, (**iii**) sliding mode, and (**iv**) single-electrode mode [47]. (**c**) (**i–iv**) COMSOL Multiphysics simulation results of TENG [58]. (**d**) (**i–iv**) Working mechanism of the CS-mode TENG [59]. All essential copyrights and permissions received.

5. Fabrication Techniques of Triboelectric Flexible Films Along with TENG Design

Various effective materials synthesis techniques have been implemented to design high-performance TENG devices, along with flexible triboelectric films, addressing diverse energy harvesting and various self-powered sensing applications, as shown in Figure 3. These techniques include hydrothermal, vacuum filtration, plasma treating, tape casting, spin and dip coating, doctor blade, 3D printing, spray coating, electrospinning, etc., to prepare flexible triboelectric films and design a TENG. In this case, to prepare flexible triboelectric films and high-performance TENGs, various materials were selected, such as polymer materials, dielectric/ferroelectric materials, conductive fillers, and combined filler materials in polymers for forming composite films. For example, A. Gupta. et al.

 https://doi.org/10.3390/s26041166

proposed a novel metal-coupled $NaYF_4$ dielectric polymer matrix, which is prepared by an in situ hydrothermal process and doped with lanthanide ions (NaYF4: Yb, Er@Ag nanocomposites, NCs) mixed in Ecoflex and polyester cloth (PC) [60]. Further, the TENG is designed with and without a polymer spacer, and both triboelectric materials act as triboelectric positive and negative layers. The fabricated TENG exhibits excellent electrical output performance and harvests mechanical energy from human activities, as well as further utilizes it for wearable applications. W. Wang et al. prepared a Pd@ZnO/$MoSe_2$ nanoflower via a solvothermal synthesis reaction, and further, the Pd@ZnO/$MoSe_2$ solution was decorated on a Cu-Ni electrode using a spray-coating technique for the gas sensing and TENG applications [61]. E. Cho. et al. developed a hierarchical-wrinkled-architecture TENG (HWA-TENG) based on the plasma–polymer–fluorocarbon (PPFC) thin film [62]. The fabricated thin film is employed as a high-performance triboelectrification layer due to the high surface charge potential, water repellency, transparency, and ultrathin thickness. This proposed PPFC-based HWA-TENG demonstrated excellent electrical output and was successfully employed as a raindrop energy harvester. C.K. Chung et al. developed a micro-needle structure created on a PDMS/conductive polyester fabric-based MN-TENG device [63]. In this study, microstructures were designed using laser engraving machines. The fabricated MN-TENG exhibits a stable, high electrical output and is used to power various low-powered electronic devices. Also, C. Luo. et al. established an eco-friendly and recyclable bacterial cellulose-based flexible triboelectric nanogenerator (BC-TENG), as shown in Figure 3a, ref. [64]. Here, the surface potential of BC films was controlled and designed using a dip-coating hydroxyethyl cellulose solution to improve the electrical output performance of the BC-TENG. In addition, the BC-TENG exhibits outstanding electrical output stability and is effectively employed as a self-powered sensor to detect the mechanical movements in daily human life and power portable electronic gadgets. As shown in Figure 3b, F. Wang. et al. developed a TENG self-powered filter using triboelectric material ZIF-8 decorated on the surface of cellulose/graphene oxide aerogel (CN/GO aerogel) and grafting it with trimethoxy(1*H*,1*H*,2*H*,2*H*-heptadecafluorodecyl) silane using negative corona treatment [65]. The fabricated TENG generated excellent output performance via electrical energy converted from the mechanical energy through the breathing-driven mechanism. From these results, the TENG is utilized for different applications such as wearable and self-powered healthcare monitoring. Moreover, A. Kurakula et al. prepared a strontium bismuth titanate nanoparticle (SBT NPs) incorporated polyglucosamine (PGA) (SBT/PGA) composite film-based flexible TENG, as shown in Figure 3c, ref. [59]. The SBT NPs were synthesized via a solid-state synthesis technique. The prepared composite films show excellent dielectric properties, which can lead to improvement in the electrical performance of the corresponding TENG. S. He et al. established a TENG device based on the porous/patterned surface PE/PDMS and PA/PEG triboelectric films, which are prepared via a traditional 3D printing technique [66]. The fabricated TENG shows excellent flexibility, a short fabrication cycle, lower cost, and the possibility of large-area modeling properties, along with high electrical output. These results can stimulate the industrialization of the proposed TENG and make it more accessible for practical applications. Furthermore, B. Amrutha et al. prepared a copper oxide nanoparticle (CuO NPs) loaded with PVDF composite nanofibers—a TENG device for energy harvesting applications [67]. The CuO/PVDF composite nanofibers were fabricated via the conventional electrospinning technique. The proposed TENG exhibits excellent electrical output and harvests energy from daily human activities. In addition, J. Yin et al. have developed a flexible TENG based on hybrid-structured nanofiber membranes (HNMs) using TPU/CB beaded nanofibers and PVDF-TrFE nanofibers by a modified free-surface electrospinning technique (MFSE), as shown in Figure 3d, ref. [68]. The fabricated HNM-based TENG exhibited superior electrical output and was further employed as a mechanical-energy harvesting device to convert biomechanical energy from daily human activities. M. Kim et al. developed a high-performance TENG using a composite film of PEDOT:PSS-loaded PVDF-TrFE and $MAPbI_3$ particles [69]. The composite film was fabricated using a spin-coating technique. The fabricated perovskite thin film-based TENG exhibited an excellent electrical output, and the TENG was employed to harvest mechanical energy for powering various electronic devices. Moreover, J. Mao et al., H. Zhu et al., H. Patnam et al., P.S. Das et al., Y. Lui et al., and T. Huang et al. also designed a flexible polymer/composite film via different fabrication techniques, such as blade coating, casting/rolling, solution casting, vacuum filtration, spin coating, and electrospinning [70–75]. The prepared flexible films were utilized to fabricate TENG devices for energy harvesting and sensor applications. From these results, flexible TENGs are fabricated based on the flexible polymer/composite films via different techniques, which are easy, scalable, lightweight, budget-friendly, and environmentally friendly. Therefore, the flexible triboelectric polymer or composite films based on TENG have unique advantages in the various energy harvesting, wearable electronic, and self-powered sensor fields. A summary of a comprehensive list of different fabrication processes of polymer/composite film-based TENGs for various TENG applications is shown in Table 1. As material technologies continue to progress, constructing highly flexible triboelectric polymer and composite films will gradually become more important for enabling more efficient and flexible TENG constructions.

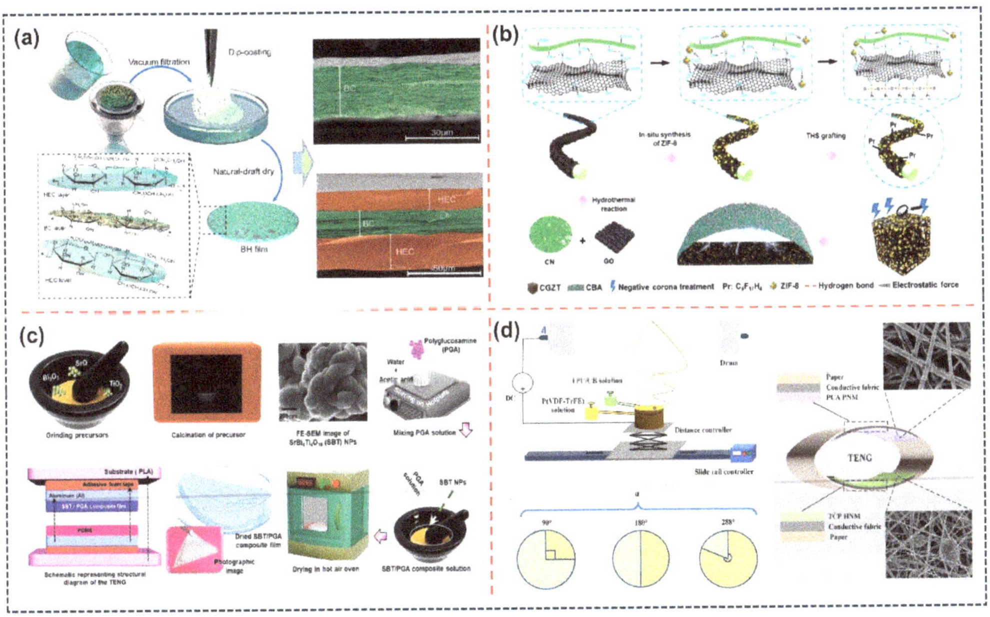

Figure 3. Fabrication techniques of polymer/composite films based on TENG. (**a**) Preparation process of bacterial cellulose fiber-based TENG device, which is fabricated using vacuum-filtration and dip-coating techniques [64]. (**b**) Schematic representation of ZIF-8 decorated carboxycellulose nanofiber/graphene oxide aerogels (CG aerogels) via a hydrothermal reaction for the construction of flexible TENG [65]. (**c**) Schematic diagrams and fabrication of SBT NPs loaded with PGS composite films via the solid-state synthesis technique for TENG design [59]. (**d**) MFSE preparation process of TCP HNMs with P(VDF-TrFE), two adjustable regions, and assembled TENG [68]. All essential copyrights and permissions received.

 https://doi.org/10.3390/s26041166

Table 1. Comprehensive list of different fabrication processes of polymer/composite films based on TENGs for various TENG applications.

Polymer/ Composite Film	Fabrication Method	Electrical Output			Application	Ref.
		Voltage	Current	Power		
$NaYF_4$: Yb,Er@ Au (Ln@Au) nanocomposite	Hydrothermal	424 V	8.5 μA	3.7 W/m^2	Mechanical energy harvesting	[60]
Pd@ZnO/$MoSe_2$@Cu-Ni electrode	Spray coating	200 V	--	17.5 μW	Mechanical energy harvesting and gas sensing	[61]
HWA-PPFC	Plasma treatment	200 V	30 μA	11 W/m^2	Raindrop energy harvesting	[62]
MN-PDMS	Laser-treated micropattern	73.6 V	36 μA	0.264 W/m^2	Human motion sensor	[63]
BC/HEC	Vacuum filtration/Dip-coating	76.61 V	8.68 μA	290.7 μW	Mechanical energy harvesting	[64]
CN/GO/ZIF-8/THS	Hydrothermal	184.5 V	22.8 μA	41.32 μW/cm^2	Wearable and self-powered sensor	[65]
SBT/PGA	Solid-state and drop-casting	~239 V	~7.5 μA	2 W/m^2	Energy harvesting and self-powered sensor	[59]
PE/PDMS and PA/PEG	3D printing	306 V	6.14 mA	236.67 W/m^3	Energy harvesting	[66]
PVDF-CuO	Electrospinning	7.5 V	--	--	Biomechanical energy harvesting	[67]
PCA PNMs-TCP HNMs	Electrospinning	49.29 V	--	0.054 W/m^2	Energy harvesting and self-powered sensor	[68]
PVDF-TrFE/PEDOT:PSS/$MAPbI_3$	Spin coating	--	--	12.5 mW/cm^2	Mechanical energy harvesting	[69]
PDMS/Silica	Blade coating	254 V	20.4 μA	4.37 W/m^2	Energy harvesting	[70]
PDMS/Ag NW Composite	Casting and Rolling	33.4 V	4.5 mA	0.162 W/m^2	Mechanical energy harvesting	[71]
PVDF/PVA	Solution casting	230 V	6 μA	3.1 W/m^2	Self-powered electronic	[72]
MWCNT/PEDOT:PSS@Nylon	Vacuum filtration	109.6 V	--	--	Wearable electronics	[73]
PDMS/Mxene	Spin coating	453 V	131 μA	--	Acoustic and wind energy harvesting	[74]
PVDF/GO@PHBV	Electrospinning	340 V	78 μA	0.141 mW	Energy harvesting	[75]

6. Applications of Flexible Polymer/Composite Films Based on TENG

The combination of TENGs designed from flexible polymer/composite films into specific applications offers significant potential to enhance their electrical output and large-scale practical applications. The films enable efficient energy conversion and open a wide range of applications owing to their adaptable structural properties. This combination plays a key role in the advancement of self-powered sensing devices and holds great promise for human biomechanical energy harvesting, wearable, and self-powered smart sensing applications.

7. Mechanical Energy Harvesting

Within the last couple of decades, several studies have been conducted on the flexible polymer/composite film-based TENGs for biomechanical energy harvesting, wearable, and self-powered sensing applications. P. Manchi et al. constructed a high-performance flexible TENG device based on the dielectric and ferroelectric bismuth tungstate (Bi_2WO_6) marigold flower-like structures by hydrothermal synthesis and further embedded them into a poly (vinylidene fluoride-co-hexafluoropropylene) (PVDF-HFP) polymer to form a flexible PVDF-HFP/Bi_2WO_6 composite polymer film (CPF), as shown in Figure 4a,

https://doi.org/10.3390/s26041166

ref. [76]. The fabricated flexible PVDF-HFP/Bi_2WO_6 CPF exhibited a high polar crystalline β-phase, a high piezoelectric coefficient, and ferroelectric properties, which can improve the TENG's electrical output. The proposed TENG shows a highly stable and durable electrical output over the long term, harvesting mechanical energy from everyday human body activities and powering various portable electronic devices. Also, Y.L. Yashaswini et al. designed high-performance TENGs using functionalized reduced graphene oxide (FrGO) in the polyvinyl alcohol (PVA) nanocomposite [77]. The fabricated flexible FrGO nanocomposite-based TENG exhibited an electrical output 17 times higher, as compared to pristine PVA-TENG. Further, the proposed TENG demonstrated a practical applicability, demonstrating its ability to power LEDs and digital timers via charging commercial capacitors using bridge rectifiers. Q. Ju et al. proposed a reduced graphene oxide (rGO) and polyaniline (PANI) decorated on a carbon fiber (CF) core electrode-based TENG, as shown in Figure 4b, ref. [78]. The fabricated TENG exhibits excellent electrical output and was also utilized for energy storage applications using a hybrid self-powered supercapacitor. The proposed model utilizes mechanical energy from daily human activities for harvesting and employs it to power electronic devices. Also, R. Umapathi et al. proposed lead-free strontium-doped bismuth ferrite (BSFO) nanoparticles with multiferroic and dielectric behavior via a solid-state synthesis technique and further incorporated them into PDMS to fabricate the BSFO/PDMS composite film-based TENG [79]. The fabricated BSFO-TENG delivered excellent power density and long-term operational stability. Furthermore, it illuminated various commercial LEDs, powered a digital clock, harvested electrical energy from daily human body motions, and was utilized for security alert applications. M. Mao et al. developed a biodegradable, recyclable, and biocompatible composite film-based TENG using a polyethyleneimine (PEI)/iron chloride ($FeCl_3$)/microcrystalline cellulose (MCC) composite film [80]. The fabricated MCC/PEI/$FeCl_3$ composite films based on TENG demonstrated exceptional pressure sensitivity and high-humidity sensing, fatigue resistance, and highly stable and long-term durable electrical output. Furthermore, the device exhibits high sensitivity to human body activities, such as walking, running, jumping, coughing, and breathing, etc. From these results, the proposed TENG is employed to harvest mechanical energy from the daily human actions and pressure-sensing applications. In addition, V. Mohan et al. fabricated a TENG using sonochemically prepared graphene/polydimethylsiloxane (SGP) nanocomposite and PEO layers [81]. The proposed TENG shows excellent electrical output, and the device can be successfully combined into practical applications such as human motion monitoring in daily activities, gaming interfaces, and power-point control applications. In a more fundamental study, C.M. Veerabhadraswamy et al. proposed a natural fiber-based TENG comprising PVA, and the agro-waste composites were prepared using a solution casting method [82]. The fabricated TENG exhibited high electrical output and powered various low-powered portable electronics, including LEDs and LCD timers. Also, it functioned biomechanical sensor for harvesting mechanical energy from daily human motions. M. Qu et al. established a flexible superhydrophobic MOF-based textile TENG for mechanical energy harvesting and wearable electronic applications, as shown in Figure 4c [83]. The fabricated TENG shows high electrical output and high durability even after long-term study over several weeks. The TENG demonstrated mechanical harvesting from human body activities and powering electronics, as well as self-powered sensors in human motion detection and Morse code information transmission. Furthermore, the conversion of mechanical energy into electrical energy from human mechanical movements by operating a TENG can be thought of as a sensing capability to be used in wearable applications. This functionality can be observed as high responsivity to voltage and current variations within a sensible time. In this case, flexible TENGs can be a perfect wearable sensor, with a quick response and sensitivity

https://doi.org/10.3390/s26041166

to mechanical stimuli. F. Jiang et al., in Figure 4d, developed a breathable, stretchable, and stable nanofiber composite (LPPS-NFC) via the electrospinning technique based on the $Cs_3Bi_2Br_9$ perovskites decorated in lead-free perovskite/poly (vinylidene fluoride-co-hexafluoropropylene) (PVDF-HFP) and styrene-ethylene-butylene-styrene (SEBS) [84]. The LPPS-NFC-based TENG demonstrated excellent electrical output, enabling it to show a robust wearable device that converts mechanical energy from the different mechanical motions into electricity to operate various portable electronics. These results demonstrate that the proposed LPPS-NFC-based TENG can endure extreme deformation while maintaining excellent electrical output, highlighting its strong potential for use in smart textiles and wearable power sources. Moreover, J. He et al. presented a novel superhydrophobic and conductive composite fabric (HPC) using polydopamine (PDA), carbon nanotubes (CNTs), and polypyrrole (PPy), which were loaded on the cotton fabric surface to create the hydrophobic nature of cetyltrimethoxysilane (HDTMS), as shown in Figure 4e, ref. [85]. The prepared HPC exhibits excellent wear resistance, breathability, and self-cleaning properties, as well as high and stable electrical output from the HPC-TENG. The proposed TENG demonstrated mechanical energy harvesting from human activities and powering various electronics, as well as utilizing human–machine interface sensors for gaming blankets.

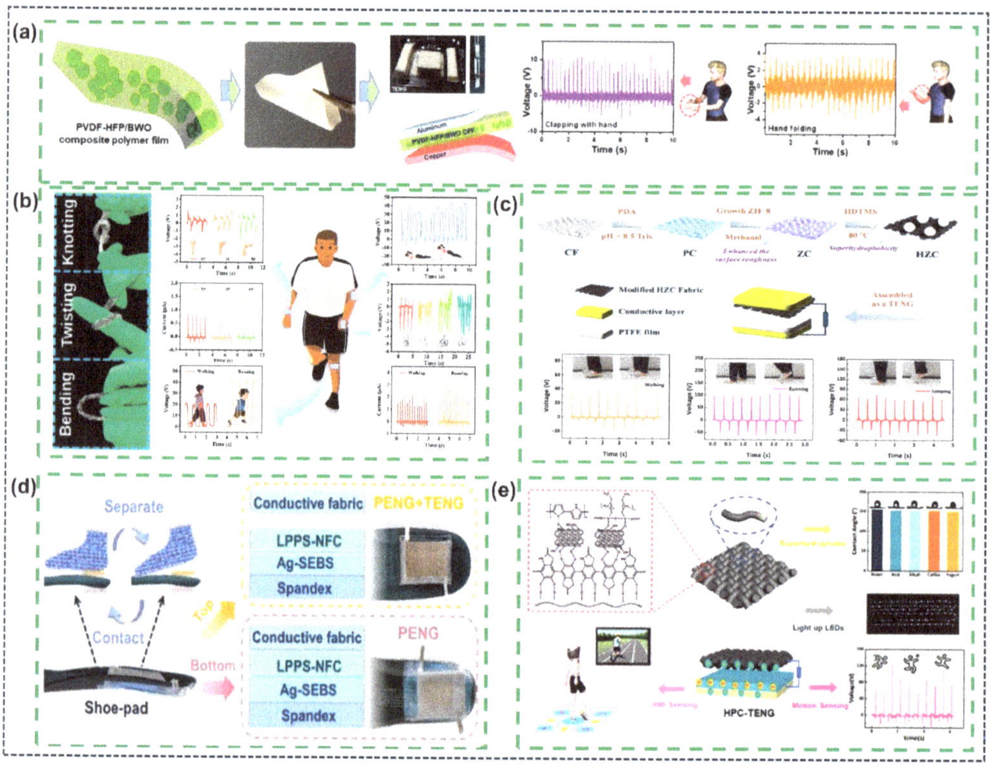

Figure 4. Practical applications of the TENG for mechanical energy harvesting. (**a**) Schematic illustration of the PVDF-HFP/BWO composite polymer film-based TENG for mechanical energy harvesting [76]. (**b**) Photographic images of the coaxial optical fiber self-powered system and energy harvesting from human body activities [78]. (**c**) Construction process of HZC fabric-based TENG device and its real-time mechanical energy harvesting applications [83]. (**d**) Schematic illustration of the LPPS-NFC-based flexible and wearable TENG device placed on the shoe pad actuated by the foot

https://doi.org/10.3390/s26041166

stamping for energy harvesting and sensing applications [84]. (e) Schematic diagram representation and output results of HPC fabric-based TENG for mechanical energy harvesting, wearable, and smart sensing applications [85]. All essential copyrights and permissions received.

8. Self-Powered Sensors

In addition to detecting mechanical stimuli, physiological signals, and pressure-sensing parameters, increasing research efforts are being directed toward monitoring a wide range of environmentally relevant factors, reflecting the rising attention to sustainable and eco-friendly sensing technologies. For example, Q. Tao et al. proposed a triboelectric fabric pressure mapping sensor with high sensitivity and free crosstalk using polytetrafluoroethylene/stainless steel (PTFE-SS) braided yarns and the polyamide 66/stainless steel (PA66-SS) braided yarns in the weave fabric, as shown in Figure 5a, ref. [86]. The fabricated fabric-based pressure sensor TENG demonstrated excellent sensitivity of 2.942 $V{\cdot}kPa^{-1}$ and a fast response time of 123 ms. This proposed device was also utilized as a sensor for human motion sensing while doing daily activities, as well as a touch sensor. T. Yang et al. proposed sheath-core PVDF/graphene/carbon fiber (PVDF/G-CF) yarns fabricated using a conjugate electrospinning technique, covering a commercial CF core and an electrospun graphene-doped PVDF sheath [87], which can enhance the fatigue resistance of electrospun nanofibers under prolonged friction and keep a high degree of freedom. The resulting electronic textile, woven with large-area electrospun PVDF/G-CF yarns, delivers a high-power density of 25.5 mW/m^2. The uniform distribution of PVDF/G nanofibers on the textile surface provides good softness, washability, and long-term durability. Further, the proposed electrospun PVDF/G-CF textile shows strong potential for pressure sensing, motion monitoring, and self-powered operation, making it a strong potential application for next-generation wearable electronic systems. X. Wang et al. proposed a soft and stretchable TENG skin for energy harvesting to operate portable electronics from biomechanical energy [88]. The fabricated TENG exhibits excellent electrical output and has desirable features like biocompatibility, simple fabrication, lightweight, and environmentally friendly nature. The TENG is integrated with a power management unit to drive electronic devices by harvesting energy from hand motion. Also, the proposed TENG provides a new insight for clean power sources of skin-mounted electronic gadgets and promotes the development of sustainable energy sources for wearable and portable electronics. J. Huang et al. established a novel self-powered pressure sensor based on the bacterial cellulose/chitosan (BC/CS) composites and copper nanoparticles-doped polydimethylsiloxane (PDMS/CU) films [89]. The fabricated device demonstrated excellent mechanical stability and high-pressure sensitivity of 0.24 V/kPa^{-1}. The device exhibits high sensitivity when it is attached to human body parts, and it detects joint motions. In addition, S. Wang et al. developed 3D microporous MXene/polyurethane (MXene/PU) composite gel-based triboelectric nanogenerators, as shown in Figure 5b, ref. [90]. The fabricated Mxene/PU exhibits exceptional performance characteristics such as a large compressive range, excellent cycling durability, long-term stability, and high sensitivity of 0.96144 kPa^{-1}. The proposed TENG generates excellent electrical output performance as well as remarkable effectiveness in applications of healthcare monitoring and rehabilitation. M. Qu et al. fabricated a superhydrophobic, humidity-resistant flexible TENG based on the PDMS/PTFE and AgNWs/PVA hydrogel electrodes, as shown in Figure 5c, ref. [91]. The fabricated PP/AgH-TENG generated a high electrical power density of 3.07 W/m^2 and can easily power several portable electronic devices. Additionally, the bracelet-type TENG was developed to harvest mechanical energy from human movements, as well as demonstrated its potential in the field of wearable sensing applications. Also, Y. Liu et al. have developed a rapid-response, highly sensitive, and self-powered artificial sensory

https://doi.org/10.3390/s26041166

memory actuated by a triboelectric sensory receptor (TESR) for real-time neuromorphic computing applications [92]. The flexible TESR is designed based on the DPP-DTT and ion-gel layers. The device exhibits high electrical output and pressure-sensing properties with a sensitivity of 0.192 kPa^{-1}. Further, several TESR devices were arranged in a matrix shape to demonstrate handwritten image recognition and neuromorphic computing results. These results from TESR sensors have strong potential for real-time broad applications in neuromorphic systems, human–machine interaction, and large-scale neural networks. Also, M.V. Paranjape et al. have proposed a 3D-printed box-type hybrid nanogenerator (BT-HNG) based on the acetone-dissolved microarchitectured strontium (Sr)-doped silver niobate (ASNb)/ecoflex composite film for efficient mechanical energy harvesting and sensor applications (Figure 5d), ref. [93]. The highly effective and stable electrical output exhibited by BT-HNG is further utilized for powering small-scale electronic devices. Moreover, the proposed BT-HNG successfully demonstrated its use as a real-time sensor that can be utilized for powering the stairs, corridors, streetlamps, etc., in human presence, as well as harvesting vibrational/linear movements into electrical energy.

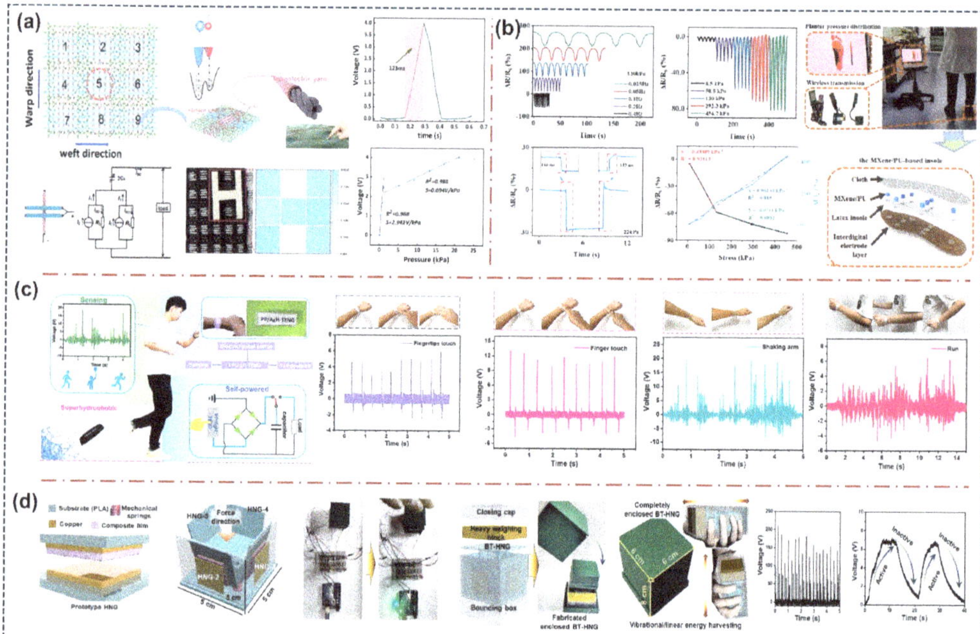

Figure 5. (**a**) Schematic diagram illustration and fabrication of triboelectric fabric sensor array (TFSA) for pressure sensor applications [86]. (**b**) 3D microporous MXene/polyurethane (MXene/PU) composite gels based on ultra-high stable pressure sensors and flexible triboelectric nanogenerators for gait monitoring applications [90]. (**c**) Core-shell superhydrophobic and flexible triboelectric nanogenerator based on the PDMS film surface modified with PTFE particles as the triboelectric layer and AgNWs/PVA hydrogel electrodes for biomechanical energy harvesting and wearable self-powered sensing applications [91]. (**d**) Schematic representing the structural composition of the fabricated HNG, along with the heavy-weighting block enclosed in a 3D-printed box for vibrational/linear mechanical energy harvesting applications [93]. All essential copyrights and permissions received.

The flexible polymer/composite film-based TENG devices have demonstrated substantial potential for the enhancement of electrical output and various self-powered sensing applications, including acoustic, tactile, etc. For example, G.M. Rani et al. successfully fabricated a nature-driven CF-CNT TENG device based on the carbon nanotubes (CNT)

https://doi.org/10.3390/s26041166

decorated on the cigarette filter (CFs) [94]. The assembled CF-CNT-based TENG demonstrated high and stable electrical output, and it employed various commercial portable electronics. Furthermore, the proposed device can be used as a sound-driven TENG that works efficiently across different bandwidth frequencies. These results demonstrate that the CF-CNT TENG has immense potential in various sensing, acoustic, and energy harvesting applications. Also, S.K. Ghosh et al. have developed a high-performance stretchable TENG based on a sandwich structure between the Ag NWs electrodes and hierarchically engineered spongy TPU composite polymer with ferroelectric barium titanate (BTO) and 2D MXene ($Ti_3C_2T_x$) nanosheets, as shown in Figure 6a, ref. [95]. The Mxene sheets increase the dielectric properties by coupling with BTO particles, resulting in high and stable electrical output from the TENG. The fabricated TENG exhibits a high electrical output power density of 6.65 W/m^2, high pressure sensitivity, and efficiency. The TENG enables a broad range of real-time applications, including mechanical energy harvesting, vital-sign monitoring, acoustic sensing, and gesture-sensing functionality of a robotic hand. S. Sardana et al. have designed an electrospun triboelectric nanogenerator (TENG) based on the MXenes loaded cellulose acetate NFs, as shown in Figure 6b, ref. [96]. The fabricated TENG generated a high output power density of 1.361 W/m^2. Further, they assembled an MXene/TiO_2/C-NFs heterojunction-based sensory component using cellulose nanofibers for the detection of NH_3. The sensor exhibits excellent reproducibility, high sensitivity, and fast response/recovery time of 76 s/62 s at room temperature. The proposed sensor demonstrated its possibilities as a gas-sensing network driven by human motion energy and external battery dependence for environment monitoring applications. L. Yang et al. have proposed a novel liquid metal (LM)-incorporated thermoplastic polyurethane (TPU) nanofiber-based TENG for mechanical energy harvesting and self-powered sensing properties [97]. Dual optimization of interfacial polarization and charge trapping capabilities was achieved by incorporating the LM nanoparticles into the nanofiber network. The proposed SM-TENG demonstrated the highest electrical output and ultralong-term stability results. This SM-TENG exhibits a high-pressure-sensitive triboelectric response of 6.11 V/kPa, and it enables multimodal sensing properties, including gait analysis, information transmission, physiological signal monitoring, and encrypted human–machine communication. These results are promising for next-generation smart healthcare systems and human–machine interfaces. On the other hand, M. Anithkumar et al. fabricated a novel hybrid nanogenerator (HNG) with multifunctional composite film (MFC), made up of Cu_2O-doped $0.3Ba_{0.7}Ca_{0.3}TiO_3$-$0.7BaSn_{0.12}Ti_{0.88}O_3$ nanoparticles (Cu_2O-BCST NPs) [98]. The HNG demonstrated an outstanding electrical output due to the synergistic properties of the MFC. Further, DL-HNG was assembled with a matrix structure using 3D-printed masks; its demonstrations are systematically operated. The DL-HNG can also be employed as a flexible self-powered trackpad (or touch) sensor to control the mouse pointer in computer or touch-screen display devices. Q. Yi et al. have established a novel 3D printer and integrated a self-powered wearable sensing system for real-time vital sign monitoring applications [99]. The TENG fabricated using Mxene ink is coupled with a styrene–ethylene–butylene–styrene (SEBS) substrate with a positive triboelectric property and high stretchability. The device exhibits excellent power density, high sensitivity of 6.03 kPa^{-1}, and a fast response time of 80 ms. The proposed TENG employed as a wearable system for continuous and real-time physiological biosignal monitoring under human motions and signaling exhibits potential in sensing applications. Also, in Figure 6c, Q. Liu et al. have developed a surface-patterned PVDF-HFP/PVC/TiO_2 composite film-based flexible triboelectric nanogenerator for tactile sensing applications [100]. The prepared composite films show excellent hydrophobic nature with a rough surface, which enables the device to overcome the effect of water molecules on charge transfer in the humid environment, thus

 https://doi.org/10.3390/s26041166

resulting in high and stable electrical output from the corresponding TENG device. The proposed TENG demonstrated the mechanical energy harvesting from human movements as well as a self-powered sensor in the tactile sensing applications. In addition, J. Ji et al. have designed a flexible tactile sensor based on the templated laser-induced graphene (TLIG) 3D-printed on Ag electrodes, as shown in Figure 6d, ref. [101]. The designed tactile sensor obtains a high sensitivity of 52,260.2 kPa^{-1}, a broad detection range up to 1000 kPa, a quick response and recovery time of 12 and 16 ms, and high stability over 10,000 cycles. The TLIG device acts as a sensory receptor for the soft robotic gripper, and its ability for texture recognition is further demonstrated through the implementation of a soft tactile sensing array. Moreover, M. Lou et al., C. Ning et al., F. Chen et al., W. Xu et al., Y. Yang et al., and Y. Li et al. also designed a flexible polymer/composite film-based TENG device for energy harvesting and real-time acoustic and tactile sensing applications [102–107]. From these results, the flexible polymer/composite film-based TENG have been widely used in different practical and real-time scenarios, including energy harvesting and various sensor applications. Tables 2 and 3 represent the summary of various TENG sensors with their electrical output results, sensing parameters, and applications. Furthermore, the comparison table of the TENG device with various important parameters, including materials, electrical output, stability/durability, cost of fabrication, and their application areas, was summarized in Table 4.

Table 2. A summary of TENGs and various sensor systems, highlighting their sensing parameters, and application fields.

Materials (Polymer/ Composite Film)	Electrical Output (Voltage, Current, and Power)			Sensing Parameters		Application	Ref.
	V	I	PD	Sensitivity	Response Time		
PVDF-HFP/BWO CPF	140	5 μA	43.3 mW/cm^2	0.88 $V\cdot cm^{-1}$	5 ms	Energy harvesting and sensor	[76]
PVA-NH_2 FrGO	322.46 V	78.22 μA	192 mW	--	--	Energy harvesting and self-powered sensor	[77]
rGo@PANI@CF	160 V	9 μA	0.576 W/m^2	--	--	Signal sensing and energy harvesting	[78]
PDMS-BSFO	152 V	10.6 μA	4.71 W/m^2	--	--	Energy harvesting and sensing	[79]
MCC/PEI/ $FeCl_3$	248.3 V	5.3 μA	0.575 W/m^2	3.136 V/N	60 ms	Energy harvesting and self-powered multi sensor	[80]
GO/PDMS	415 V	5.06 μA	552 μW	0.385 V/N	--	Energy harvesting	[81]
PVA-Natural fibers (NF)	154.64 V	6.29 μA	28.65 mW/m^2	--	--	Bio-mechanical sensor	[82]
Modified HZC fabric	110 V	18 μA	0.625 W/m^2		2 ms	Mechanical energy harvesting and multi-sensing applications	[83]
PVDF-HFP/SEBS@ $Cs_3Bi_2Br_9$	400 V	1.63 $\mu A/cm^2$	2.34 W/m^2	--	--	Wearable and self-powered sensor	[84]
DA/CNT/Py@ CC	300 V	50 μA	2.6 W/m^2	--	--	Wearable applications and self-powered sensor	[85]
PTFE-SS @PA66-SS	4 V	26.112 nA	--	2.942 $V\cdot kPa^{-1}$	123 ms	Wearable applications and a self-powered pressure sensor	[86]
PVDF/G-CF	71.5 V	0.36 mA/cm^2	25.5 mW/m^2	3.48286 $V\cdot kPa^{-1}$	0.16 s	Wearable sensor, pressure sensor, and human motion monitoring	[87]
Physiological saline/silicone rubber	57.0 V	1.7 mA/m^2	11.6 W/m^2	--	--	Wearable applications and a self-powered energy harvester	[88]

https://doi.org/10.3390/s26041166

Table 2. *Cont.*

Materials (Polymer/Composite Film)	Electrical Output (Voltage, Current, and Power)			Sensing Parameters		Application	Ref.
	V	I	PD	Sensitivity	Response Time		
Bacterial cellulose/chitosan (BC/CS)	23 V	500 nA	3.25 mW/m^2	0.24 V·kPa^{-1}	--	Mechanical energy harvesting and pressure sensor	[89]
MXene/PU	7.5 V	17 nA	--	2.64819 V·kPa^{-1}	140 ms	Pressure sensor	[90]
PP@AgH/PVA	--	--	3.07 W/m^2	22.86 V/N	--	Wearable self-powered pressure sensing and energy harvesting	[91]
n-Cu/m-PDMS	~13 V	--	~33 μW/cm^2	0.192 kPa^{-1}	8 ms	Self-powered pressure sensing and health monitoring	[92]
ASNb/ecoflex CF	270 V	9 μA	4 W/m^2	--	--	Mechanical energy harvesting and smart sensing	[93]
CF-CNT	60 V	1.805 μA	110.6 mW/m^2	--	--	Acoustic energy harvesting	[94]
TPU/BTO/MXene NCF	260 V	160 mA/m^2	6.65 W/m^2	4.6 V·kPa^{-1}		Energy harvesting, acoustic, and gesture sensing	[95]
MXene/TiO_2@CNF	140 V	92 μA	~1.36 W/m^2	--	76 s	NH_3 gas sensor	[96]
TPU@LM	162 V	3.3 μA	0.176 W/m^2	6.11 V·kPa^{-1}	4.9 s	Pressure sensor and health monitoring	[97]
Cu_2O-BCST@ PDMS	176.41 V	--	0.168 W/m^2	--	--	Self-powered touch sensor	[98]
Mxene/SEBS	--	--	~0.81 W/m^2	~ 6.03 kPa^{-1}	~80 ms	Wearable sensor for wireless healthcare monitoring	[99]
PVDF–HFP/PVC/ TiO_2 CF	235 V	35 μA	1.4 W/m^2	--	--	Energy harvesting and self-powered tactile sensing	[100]
Templated laser-induced graphene (TLIG)	--	--	--	52,260.2 kPa^{-1}	12 ms	Tactile sensor, health monitoring, and texture recognition	[101]
Nylon/PTFE	2.17 V	~22 nA	9.9 μW/m^2	1.33 V·kPa^{-1}	--	Pressure sensor, human motion, and pulse monitoring	[102]
CNT/AgNWs@PDMS	22 V	0.6 μA	21.5 μW/m^2	5.2 mV·Pa^{-1}	--	Pressure and tactile sensing applications	[103]
PVDF/CF	400 V	175 μA	7 W/m^2	--	--	Self powered sensor and acoustic energy harvesting	[104]
PVDF/Nylon	124 V	23.5 μA	1.28 W/m^2	53.6 V·Pa^{-1}	--	Pressure sensor and acoustic energy harvester	[105]
PDMS/ITO	−1000 V	8 mA/m^2	0.5 W/m^2	~0.29 V·kPa^{-1}	--	Pressure and tactile sensor system	[106]
AgNWs/SA	53 V	0.37 μA	4 μW	0.237 V/kPa	100 ms	Pressure and tactile sensor system	[107]

Table 3. A summary of TENGs and various sensor systems, highlighting their electrical output performance, energy conversion efficiency, sensitivity, and temperature fields.

Materials (Polymer/Composite Film)	Electrical Output (Voltage, Current, and Power)			Efficiency	Sensitivity	Temperature Dependency	Ref.
	V	I	PD				
PVC/TiO_2 NPs	121 V	11.1 μA	141 μW/cm^2	20%	2.03 V·kPa^{-1}	Stable output at 25–55 °C	[108]
P-ZnO@ PVDF/TBAHP TLNM	98 V	18.7 μA	362.5 mW/m^2	42.8%	--	Stable output at 10–45 °C	[109]

https://doi.org/10.3390/s26041166

Table 3. *Cont.*

Materials (Polymer/ Composite Film)	Electrical Output (Voltage, Current, and Power)			Efficiency	Sensitivity	Temperature Dependency	Ref.
	V	I	PD				
PVDF/MWCNT/$BaTiO_3$	48.46 V	1.22 $mA \cdot m^{-2}$	29.27 mW/m^2	31.62%	--	--	[110]
Liquid metal	679 V	9 μA	6.7 W/m^2	70.6%	--	--	[111]
PTFE	76 V	0.74 μA	28 μW	29.7%	--	--	[112]
Phosphorus-doped g-C_3N_4@PVDF	177 V	15 μA	118 $\mu W/cm^2$	--	1.48 $V \cdot kPa^{-1}$	--	[113]
PVDF/3wt.%PMMA	600 V	1.2 μA	0.750 W/m^2	--	0.45 $V \cdot kPa^{-1}$	--	[114]
CS/GNPs CFs	166.25 V	13.56 μA	44 mW/m^2	--	0.96 $V \cdot kPa^{-1}$	40 ± 5 °C	[115]
PVA/rGO NFs	728 V	22 μA	7.2 mW	73%	--	--	[116]
SBT/PGA	239	7.5 μA	2 W/m^2	48%	--	Almost stable output at 30–60 °C	[59]

Table 4. A summary of TENGs and various materials, highlighting their electrical output performance, durability, cost of fabrication, and their applications.

Materials (Polymer/ Composite Film)	Electrical Output (Voltage, Current, and Power)			Durability/Stability	Cost of Fabrication	Applications	Ref.
	V	I	PD				
P(VDF-TrFE)/$BaTiO_3$	315 V	6.7 $\mu A/cm^2$	141 $\mu W/cm^2$	10,000 cycles	Moderate cost (electro-spinning)	Wearable electronics and a pressure sensor	[116]
PVA/rGO NF@PTFE/PI	728 V	22 μA	~2.25 W/m^2	Not reported	Moderate cost (electro-spinning)	Energy harvesting for self-charging supercapacitor	[117]
Corn husk-based TENG	630 V	0.79 mA	32.75 mW/cm^2	Stable output ~7200 cycles	Very low (bio-materials)	Powering electronic devices	[118]
PA-66 fabric/ PP fabric	210 V	28.3 μA	0.901 W/m^2	Stable output ~6000 cycles	Low cost	Self-powered pedestrian volume and pugilism training monitor	[119]
Gelatin/polyimide@ PANI/$NiCo_2O_4$	400 V	49 μA	0.246 W/m^2	10,000 cycles	Low cost	Mechanical energy harvesting and gas sensor	[120]
MoS_2/CNT@ Nylon	300 V	11.5 μA	0.134 W/m^2	Stable 3000 cycles (6 months)	Moderate cost (electro-spinning)	Mechanical energy harvesting	[121]
TOCN/CCTO composite aerogel	152 V	33.8 μA	0.483 W/m^2	Stable ~50,000 cycles	Low-cost	Energy harvesting for electronic gadgets	[122]
$SrTiO_3$@PDMS Composite sponge	338 V	9.06 $\mu A/cm^2$	6.47 W/m^2	Stable ~15,000 cycles	Low-cost (blade coating)	Energy harvesting for electronic gadgets	[123]
PCL/CNTs	808 V	23 μA	54 W/m^2	Stable ~10,000 cycles (21 days)	Moderate cost (electro-spinning)	Mechanical energy harvesting and sensing	[124]
PTFE balls/Al	--	20.91 μA	34.65 W/m^3	--	High cost (complex design)	Harvesting wave energy, sensors, and robotics	[125]

https://doi.org/10.3390/s26041166

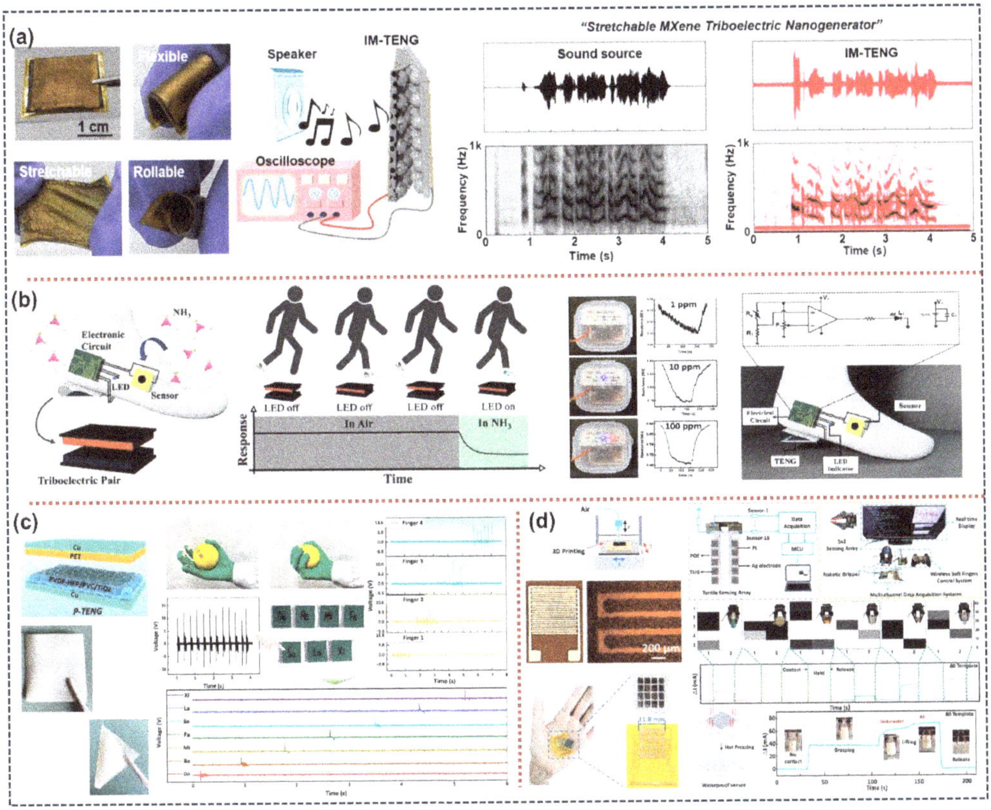

Figure 6. (**a**) Photographic images of the IM-TENG determining its flexible, stretchable, and rollable nature, further the experimental setup and output results of the acoustic wave sensing [95]. (**b**) Schematic representation of the MXene/TiO_2/cellulose nanofiber-based TENG integrated shoe for energy harvesting and self-powered gas sensing applications [96]. (**c**) Schematic diagram and photographic images of flexible PVDF–HFP/PVC/TiO_2 composite film-based TENG for environmental energy harvesting and self-powered tactile sensing applications [100]. (**d**) Templated laser-induced graphene (TLIG)-based 3D-printed Ag electrodes for tactile sensing applications, which are used in the tactile sensing array system on a wireless soft robotic gripper [101]. All essential copyrights and permissions received.

9. Conclusions and Feature Perspectives

In conclusion, this review highlights recent progress in the development of flexible triboelectric polymer/composite films for TENGs designed for energy harvesting, wearable technologies, and self-powered sensing applications. Advances in polymer engineering and composite-film preparation show that the addition of selected filler materials can substantially improve dielectric properties, charge-generation ability, mechanical strength, and robust behavior under various environmental conditions. The combined behavior of polymers and engineered nanofillers has enabled flexible TENGs that deliver enhanced electrical output while also providing consistent sensing functions for motion tracking and physiological monitoring. These developments highlight the role of material innovation in affecting the performance constraints of traditional triboelectric films and in expanding the operational versatility of TENG devices. Finally, practical and real-time applications enabled by TENG devices are also introduced, such as wearable, portable, health monitoring, acoustic, tactile, and self-powered pressure sensor systems. These findings show that

https://doi.org/10.3390/s26041166

continuity and multifunctional features can significantly improve and broaden an active device's capabilities in practical and real-time application scenarios.

With these advances, ongoing innovation has generated a wide range of studies offering practical strategies for the realization of continuous and multimodal active devices, thereby highlighting their strong potential for deployment in diverse real-world applications. Over the past decades, flexible triboelectric polymer/composite films have played a vital role in advancing TENG devices, mainly due to their ability to deliver higher electrical output, mechanical strength, and support for distinct functional parameters. These strengths position polymer or composite film-based TENGs as robust candidates for wearable energy harvesters and self-powered sensor devices, where devices must remain lightweight, flexible, deformable, and reliable during continuous use. However, applying these materials in practical or real-world application scenarios introduces some challenges. Wearable and textile systems require films that can endure repeated bending, stretching, and surface friction without losing electrical performance. Self-powered sensing applications must also ensure that the materials operate stably under different humidity/temperature ranges and long-term cyclic tests.

Moving forward, several research areas merit focused research to accelerate the practical use of these flexible polymers and composite materials. Enhancing electrical output performance remains a priority and can be achieved by designing these composite films by incorporating high dielectric permittivity fillers, strengthening interactions at polymer and filler interfaces, and producing multiscale structures that promote efficient charge generation and movement. Improving mechanical stability and long-term durability is equally important, as composite films can degrade when exposed to moisture, heat, or sustained mechanical stress. Integrating protective surface layers, reinforcing polymer networks, or using materials capable of employing a self-healing nature may help address these concerns. Another promising direction is the development of flexible polymer composite films with built-in sensing functions. By tailoring the composition or integrating responsive nanomaterials, a TENG device could simultaneously harvest energy and monitor mechanical or environmental changes, which would be valuable for advanced wearables, motion tracking, tactile pressure sensing, and health-monitoring technologies. Also, the integration of TENGs with energy storage, wireless communication modules, and multifunctional sensing components will be essential for realizing fully self-powered systems. Overall, continued innovations in polymer/composite material fabrication, structural design, and device integration are expected to accelerate the development of flexible composite film-based TENGs, paving the way for their potential applications in energy harvesting, wearable electronics, self-powered smart sensing networks, and health-care monitoring.

Author Contributions: Conceptualization, Methodology, Software, Formal analysis, Writing—original draft preparation, M.P.; supervision, visualization, Funding acquisition, project administration, writing—review and editing, H.-J.Y. All authors have read and agreed to the published version of the manuscript.

Funding: This research was funded by the National Research Foundation of Korea (NRF); grant funded by the Korean government (MSIT) (No. RS-2025-00521476 and No. RS-2024-00445819).

Data Availability Statement: Data will be made available on request.

Conflicts of Interest: The authors declare no conflict of interest.

https://doi.org/10.3390/s26041166

References

1. Mao, Y.; Li, Y.; Xie, J.; Liu, H.; Guo, C.; Hu, W. Triboelectric nanogenerator/supercapacitor in-one self-powered textile based on PTFE yarn wrapped PDMS/MnO_2NW hybrid elastomer. *Nano Energy* **2021**, *84*, 105918. [CrossRef]
2. Li, C.; Luo, R.; Bai, Y.; Shao, J.; Ji, J.; Wang, E.; Li, Z.; Meng, H.; Li, Z. Molecular doped biodegradable triboelectric nanogenerator with optimal output performance. *Adv. Funct. Mater.* **2024**, *34*, 2400277. [CrossRef]
3. Sasikumar, M.; Sreenivasan, S.; Vijayakumar, P.; Manikandan, S. Multi-Objective Energy Management in Microgrids with Hybrid Renewable Energy Sources and Battery Energy Storage Systems Using Hybrid Optimization Algorithm. *Trans. Electr. Electron. Mater.* **2025**, 1–18. [CrossRef]
4. Polat, S.; Mashrah, M.; Maksur, A. Evaluation of weight, area, and volumetric specific capacitance performance of high graphene content $ZnFe_2O_4$ electrode for supercapacitors. *Trans. Electr. Electron. Mater.* **2024**, *25*, 801–810. [CrossRef]
5. Singh, G.; Singh, A.P.; Ali, Z.; Sharma, N.K.; Singh, V.K. An Innovative DGS Based Textile Antenna with Semi-Circular Slot for Future IoT's and AI Applications. *Trans. Electr. Electron. Mater.* **2025**, *26*, 78–103. [CrossRef]
6. Khalifa, M.; Kumar, M.; Subramanian, G.; Anandhan, S. A Facile Strategy to Achieve High Piezoelectric Performance in Electrospun Poly (Vinylidene Fluoride) Non-woven Nanofabrics. *Trans. Electr. Electron. Mater.* **2024**, *25*, 201–209. [CrossRef]
7. Ghemari, Z.; Belkhiri, S. Exploring Novel Piezoelectric Materials with Enhanced Thermal Stability for High-Performance Sensors. *Trans. Electr. Electron. Mater.* **2025**, *26*, 289–304. [CrossRef]
8. Shwetha, M.; Lakshmi, S. Harmless Sustainable Energy Resource using TEG with MPPT Based DC-DC Converter for Power Gain in Implantable Medical Devices. *Trans. Electr. Electron. Mater.* **2024**, *25*, 608–635. [CrossRef]
9. Niu, S.; Wang, Z.L. Theoretical systems of triboelectric nanogenerators. *Nano Energy* **2015**, *14*, 161–192. [CrossRef]
10. Fu, H.; Liu, L.; Liu, W.; Long, Z.; Huang, P.; Fu, K.; Ou, H.; Hu, L.; Liao, M.; Gong, J. Polydopamine-Ag engineered $BaTiO_3$ core-double-shells nanohybrids in PDMS composites for synergistically enhanced triboelectric output and antibiosis in self-sustaining wearables. *Chem. Eng. J.* **2025**, *525*, 170081. [CrossRef]
11. Ge, X.; Hu, N.; Yan, F.; Wang, Y. Development and applications of electrospun nanofiber-based triboelectric nanogenerators. *Nano Energy* **2023**, *112*, 108444. [CrossRef]
12. Guo, W.T.; Lei, Y.; Zhao, X.H.; Li, R.; Lai, Q.T.; Liu, S.Z.; Chen, H.; Fan, J.C.; Xu, Y.; Tang, X.G.; et al. Printed-scalable microstructure $BaTiO_3$/ecoflex nanocomposite for high-performance triboelectric nanogenerators and self-powered human-machine interaction. *Nano Energy* **2024**, *131*, 110324. [CrossRef]
13. Luo, H.; Ni, X.; Zhang, C.; Cui, Y.; Yang, T.; Shao, J.; Jing, X. Multi-Phase Rotating Disk Triboelectric Nanogenerator with DC Output for Speed Monitoring. *Small* **2024**, *20*, 2406091. [CrossRef] [PubMed]
14. Luo, H.; Ni, X.; Cui, Y.; Huang, C.; Yuan, P.; Yang, T.; Shao, J.; Huang, X. High stability rotary solid-liquid triboelectric nanogenerator for ionic liquid detection. *Nano Energy* **2025**, *138*, 110870. [CrossRef]
15. Liu, W.; Wang, X.; Yang, L.; Wang, Y.; Xu, H.; Sun, Y.; Nan, Y.; Sun, C.; Zhou, H.; Huang, Y. Swing Origami-Structure-Based Triboelectric Nanogenerator for Harvesting Blue Energy toward Marine Environmental Applications. *Adv. Sci.* **2024**, *11*, 2401578. [CrossRef] [PubMed]
16. Guo, M.; Xia, Y.; Liu, J.; Zhang, Y.; Li, M.; Wang, X. Wearable Pressure Sensor Based on Triboelectric Nanogenerator for Information Encoding, Gesture Recognition, and Wireless Real-Time Robot Control. *Adv. Funct. Mater.* **2025**, *35*, 2419209. [CrossRef]
17. Wang, Y.; Chu, L.; Meng, S.; Yang, M.; Yu, Y.; Deng, X.; Qi, C.; Kong, T.; Liu, Z. Scalable and ultra-sensitive nanofibers coaxial yarn-woven triboelectric nanogenerator textile sensors for real-time gait analysis. *Adv. Sci.* **2024**, *11*, 2401436. [CrossRef]
18. Ma, C.; Matin Nazar, A.; Moradi, A.H.; Goharian, H.; Mao, G.; Yari, M.; Ji, X.; Dong, S. Advanced Triboelectric Nanogenerator Sensing Technologies for High-Efficiency Cardiovascular Monitoring. *Energy Technol.* **2025**, *13*, 2401863. [CrossRef]
19. Sengupta, C.; Mondal, S.K.; Pati, B.; Mondal, T. Anisotropic functionalized filler driven interfacial charge orchestration in elastomer composite based TENG for wireless communication. *Chem. Eng. J.* **2025**, *523*, 168504. [CrossRef]
20. Tan, L.; Zeng, Q.; Xu, F.; Zhao, Q.; Chen, A.; Wang, T.; Tao, X.; Yang, Y.; Wang, X. Controllable Manipulation of Large-Volume Droplet on Non-Slippery Surfaces Based on Triboelectric Contactless Charge Injection. *Adv. Mater.* **2024**, *36*, 2313878. [CrossRef]
21. Wang, J.; Xu, S.; Li, G.; Zhang, X.; Zhao, Q.; Wu, H.; Li, K.; Fu, S.; Shan, C.; He, W.; et al. Effectively inhibiting charge injection and dielectric loss of high permittivity inorganic materials by rationally coating organic polymer for achieving high output charge density. *Adv. Energy Mater.* **2024**, *14*, 2400429. [CrossRef]
22. Manchi, P.; Paranjape, M.V.; Kurakula, A.; Graham, S.A.; Kavarthapu, V.S.; Yu, J.S. PDA-Ag/TiO_2 Nanoparticles-Loaded Electrospun Nylon Composite Nanofibrous Film-Based Triboelectric Nanogenerators for Wearable Biomechanical Energy Harvesting and Multifunctional Sensors. *Adv. Funct. Mater.* **2025**, *35*, 2416018. [CrossRef]
23. Sahu, P.; Choudhary, R.N.P.; Samal, S.K. Investigation of Electrical, Dielectric, and Ferroelectric Properties of Complex Compounds $BaTiO_3$-$Bi_{0.5}K_{0.5}TiO_3$ for some Device Applications. *Trans. Electr. Electron. Mater.* **2024**, *25*, 577–588. [CrossRef]
24. Agnihotri, P.; Hawaldar, R.; Rai, R. Study the structural, electrical and ferroelectric behaviour of lead-free sodium potassium niobate (KNN) ceramics in view of energy storage capacity. *Trans. Electr. Electron. Mater.* **2024**, *25*, 745–759. [CrossRef]

 https://doi.org/10.3390/s26041166

25. Ghemari, Z.; Bendaikha, A.; Belkhiri, S.; Saad, S. Fabrication of Flexible Resistive Pressure Sensors Using Graphene/Polydimethylsiloxane Composites. *Trans. Electr. Electron. Mater.* **2025**, *26*, 323–344. [CrossRef]
26. Nabeel, N.; Jain, A.; Juglan, K.C.; Naeem, S. Synthesis and Optimization of Polypyrrole/Reduced Graphene Oxide/Nickel Cobalt Ferrite (PRNC) Ternary Composites for the Application in High-Performance Supercapacitors. *Trans. Electr. Electron. Mater.* **2025**, *26*, 890–905. [CrossRef]
27. Manchi, P.; Yoon, H.J. Electrospun Nanofibers Based Flexible Triboelectric Nanogenerators for Mechanical Energy Harvesting and Smart Sensing Applications: A Review. *Trans. Electr. Electron. Mater.* **2025**, *27*, 90–99. [CrossRef]
28. Lai, Y.C.; Deng, J.; Zhang, S.L.; Niu, S.; Guo, H.; Wang, Z.L. Single-thread-based wearable and highly stretchable triboelectric nanogenerators and their applications in cloth-based self-powered human-interactive and biomedical sensing. *Adv. Funct. Mater.* **2017**, *27*, 1604462. [CrossRef]
29. Guan, X.; Xu, B.; Wu, M.; Jing, T.; Yang, Y.; Gao, Y. Breathable, washable and wearable woven-structured triboelectric nanogenerators utilizing electrospun nanofibers for biomechanical energy harvesting and self-powered sensing. *Nano Energy* **2021**, *80*, 105549. [CrossRef]
30. Wu, F.; Li, C.; Yin, Y.; Cao, R.; Li, H.; Zhang, X.; Zhao, S.; Wang, J.; Wang, B.; Xing, Y.; et al. A Flexible, Lightweight, and Wearable Triboelectric Nanogenerator for Energy Harvesting and Self-Powered Sensing. *Adv. Mater. Technol.* **2019**, *4*, 1800216. [CrossRef]
31. Smith, M.B.; Page, K.; Siegrist, T.; Redmond, P.L.; Walter, E.C.; Seshadri, R.; Brus, L.E.; Steigerwald, M.L. Crystal structure and the paraelectric-to-ferroelectric phase transition of nanoscale $BaTiO_3$. *J. Am. Chem. Soc.* **2008**, *130*, 6955–6963. [CrossRef]
32. Biswas, A.; Saha, S.; Jana, N.R. $ZnSnO_3$ nanoparticle-based piezocatalysts for ultrasound-assisted degradation of organic pollutants. *ACS Appl. Nano Mater.* **2019**, *2*, 1120–1128. [CrossRef]
33. Zhang, J.W.; Meng, X.; Fu, G.; Han, T.; Xu, F.; Putson, C.; Liu, X. Enhanced polarization effect of flexible magnetoelectric PVDF-TrFE/Fe_3O_4 smart nanocomposites for nonvolatile memory application. *ACS Appl. Electron. Mater.* **2023**, *5*, 1844–1852. [CrossRef]
34. He, Y.; Zhu, B.; Tan, S.; Liu, Z.; Zhang, X.; Zhang, Z. Improving the energy storage performance of a chlorinated poly (vinyl chloride) film at elevated electric field and temperature via a simple annealing process. *ACS Appl. Energy Mater.* **2023**, *6*, 5407–5415. [CrossRef]
35. Venkatesan, H.M.; Arun, A.P. Nickel-oxide-doped polyvinylidene fluoride nanofiber-based flexible triboelectric nanogenerator for energy harvesting and healthcare monitoring applications. *ACS Appl. Electron. Mater.* **2024**, *6*, 1161–1173. [CrossRef]
36. Manchi, P.; Graham, S.A.; Paranjape, M.V.; Kurakula, A.; Kavarthapu, V.S.; Yu, J.S. Calcium copper titanate incorporated polydimethylsiloxane flexible composite film-based triboelectric nanogenerator for energy harvesting and self-powered sensing applications. *J. Mater. Sci. Technol.* **2024**, *190*, 56–66. [CrossRef]
37. Kim, D.W.; Lee, J.H.; Kim, J.K.; Jeong, U. Material aspects of triboelectric energy generation and sensors. *NPG Asia Mater.* **2020**, *12*, 6. [CrossRef]
38. Lai, Y.C.; Wu, H.M.; Lin, H.C.; Chang, C.L.; Chou, H.H.; Hsiao, Y.C.; Wu, Y.C. Entirely, intrinsically, and autonomously self-healable, highly transparent, and superstretchable triboelectric nanogenerator for personal power sources and self-powered electronic skins. *Adv. Funct. Mater.* **2019**, *29*, 1904626. [CrossRef]
39. Paranjape, M.V.; Graham, S.A.; Manchi, P.; Kurakula, A.; Yu, J.S. Lead-free silver niobate microparticles-loaded PDMS composite films for high-performance clip-like hybrid mechanical energy harvesters. *J. Mater. Sci. Technol.* **2024**, *176*, 145–154. [CrossRef]
40. Ouyang, H.; Liu, Z.; Li, N.; Shi, B.; Zou, Y.; Xie, F.; Ma, Y.; Li, Z.; Li, H.; Zheng, Q.; et al. Symbiotic cardiac pacemaker. *Nat. Commun.* **2019**, *10*, 1821. [CrossRef]
41. Xu, C.; Zi, Y.; Wang, A.C.; Zou, H.; Dai, Y.; He, X.; Wang, P.; Wang, Y.C.; Feng, P.; Li, D.; et al. On the electron-transfer mechanism in the contact-electrification effect. *Adv. Mater.* **2018**, *30*, 1706790. [CrossRef]
42. Yan, D.; Ye, J.; Zhou, Y.; Lei, X.; Deng, B.; Xu, W. Research progress of fabrics with different geometric structures for triboelectric nanogenerators in flexible and wearable electronics. *Adv. Fiber Mater.* **2023**, *5*, 1852–1878. [CrossRef]
43. Zhang, R.; Hummelgård, M.; Örtegren, J.; Olsen, M.; Andersson, H.; Yang, Y.; Zheng, H.; Olin, H. The triboelectricity of the human body. *Nano Energy* **2021**, *86*, 106041. [CrossRef]
44. Xu, G.; Guan, D.; Fu, J.; Li, X.; Li, A.; Ding, W.; Zi, Y. Density of surface states: Another key contributing factor in triboelectric charge generation. *ACS Appl. Mater. Interfaces* **2022**, *14*, 5355–5362. [CrossRef] [PubMed]
45. Kang, X.; Li, P.; Yurchenko, D.; Dai, S.; Wang, J. Fundamental theory and cutting-edge applications of TENGs. *Mater. Futures* **2025**, *4*, 042101. [CrossRef]
46. Tao, L.; Chen, T.; Wu, J.; Zhang, T.; Shao, L.; Zhang, H.; Liu, L.; Wu, H.; Chen, T.; Ji, J. Principles of Operation and Application Extensions of Triboelectric Nanogenerators: Structure and Material Optimization. *Micromachines* **2025**, *16*, 1127. [CrossRef] [PubMed]
47. Babu, A.; Mandal, D. Roadmap to human–machine interaction through triboelectric nanogenerator and machine learning convergence. *ACS Appl. Energy Mater.* **2024**, *7*, 822–833. [CrossRef]

48. Hasan, S.; Kouzani, A.Z.; Adams, S.; Long, J.; Mahmud, M.P. Comparative study on the contact-separation mode triboelectric nanogenerator. *J. Electrost.* **2022**, *116*, 103685. [CrossRef]
49. Zhang, H.; Quan, L.; Chen, J.; Xu, C.; Zhang, C.; Dong, S.; Lü, C.; Luo, J. A general optimization approach for contact-separation triboelectric nanogenerator. *Nano Energy* **2019**, *56*, 700–707. [CrossRef]
50. Zhao, H.; Wang, H.; Yu, H.; Xu, Q.; Li, X.; Guo, J.; Shao, J.; Wang, Z.L.; Xu, M.; Ding, W. Theoretical modeling of contact-separation mode triboelectric nanogenerators from initial charge distribution. *Energy Environ. Sci.* **2024**, *17*, 2228–2247. [CrossRef]
51. Niu, S.; Liu, Y.; Wang, S.; Lin, L.; Zhou, Y.S.; Hu, Y.; Wang, Z.L. Theory of sliding-mode triboelectric nanogenerators. *Adv. Mater.* **2013**, *25*, 6184–6193. [CrossRef]
52. Zhang, H.; Zhang, C.; Zhang, J.; Quan, L.; Huang, H.; Jiang, J.; Dong, S.; Luo, J. A theoretical approach for optimizing sliding-mode triboelectric nanogenerator based on multi-parameter analysis. *Nano Energy* **2019**, *61*, 442–453. [CrossRef]
53. Shao, J.; Jiang, T.; Tang, W.; Xu, L.; Kim, T.W.; Wu, C.; Chen, X.; Chen, B.; Xiao, T.; Bai, Y.; et al. Studying about applied force and the output performance of sliding-mode triboelectric nanogenerators. *Nano Energy* **2018**, *48*, 292–300. [CrossRef]
54. Niu, S.; Liu, Y.; Wang, S.; Lin, L.; Zhou, Y.S.; Hu, Y.; Wang, Z.L. Theoretical investigation and structural optimization of single-electrode triboelectric nanogenerators. *Adv. Funct. Mater.* **2014**, *24*, 3332–3340. [CrossRef]
55. Shao, J.; Willatzen, M.; Shi, Y.; Wang, Z.L. 3D mathematical model of contact-separation and single-electrode mode triboelectric nanogenerators. *Nano Energy* **2019**, *60*, 630–640. [CrossRef]
56. Jiang, T.; Chen, X.; Han, C.B.; Tang, W.; Wang, Z.L. Theoretical study of rotary freestanding triboelectric nanogenerators. *Adv. Funct. Mater.* **2015**, *25*, 2928–2938. [CrossRef]
57. Ren, X.; Fan, H.; Wang, C.; Ma, J.; Zhao, N. Coaxial rotatory-freestanding triboelectric nanogenerator for effective energy scavenging from wind. *Smart Mater. Struct.* **2018**, *27*, 065016. [CrossRef]
58. Kavarthapu, V.S.; Paranjape, M.V.; Manchi, P.; Kurakula, A.; Lee, J.K.; Graham, S.A.; Yu, J.S. Wireless Alerts and Data Monitoring from BNNO-MWCNTs/PDMS Composite Film-Based TENG Integrated Inhaler for Smart Healthcare Application. *Small* **2024**, *20*, 2403218. [CrossRef] [PubMed]
59. Kurakula, A.; Paranjape, M.V.; Manchi, P.; Arbaz, S.J.; Gujjala, L.K.S.; Graham, S.A.; Kavarthapu, V.S.; Yu, J.S. Aurivillius-type $SrBi_4Ti_4O_{15}$/PGA composite film-based flexible triboelectric nanogenerators for energy harvesting/storage and multipurpose tap-indication transducer applications. *ACS Appl. Mater. Interfaces* **2024**, *16*, 48246–48256. [CrossRef] [PubMed]
60. Gupta, A.; Gajula, P.; Yoon, J.U.; Lee, S.H.; Kim, H.; Adusumalli, V.N.; Bae, J.W.; Park, Y.I. Revolutionizing energy harvesting: Eco-friendly and adaptable triboelectric sensors with recycled cloth and metallo-dielectric loaded Ecoflex hybrid films. *Nano Energy* **2024**, *122*, 109346. [CrossRef]
61. Wang, W.; Wang, D.; Zhang, X.; Yang, C.; Zhang, D. Self-powered nitrogen dioxide sensor based on Pd-decorated ZnO/$MoSe_2$ nanocomposite driven by triboelectric nanogenerator. *Nanomaterials* **2022**, *12*, 4274. [CrossRef]
62. Cho, E.; Kim, K.N.; Yong, H.; Choi, W.J.; Park, J.S.; Lee, S.J. Highly transparent and water-repellent hierarchical-wrinkled-architecture triboelectric nanogenerator with ultrathin plasma-polymer-fluorocarbon film for artificial triboelectric skin. *Nano Energy* **2022**, *103*, 107785. [CrossRef]
63. Chung, C.K.; Huang, Y.J.; Wang, T.K.; Lo, Y.L. Fiber-based triboelectric nanogenerator for mechanical energy harvesting and its application to a human–machine interface. *Sensors* **2022**, *22*, 9632. [CrossRef]
64. Luo, C.; Shao, Y.; Yu, H.; Ma, H.; Zhang, Y.; Yin, B.; Yang, M.B. Improving the output performance of bacterial cellulose-based triboelectric nanogenerators by modulating the surface potential in a simple method. *ACS Sustain. Chem. Eng.* **2022**, *10*, 13050–13058. [CrossRef]
65. Wang, F.; Wang, S.; Liu, Y.; Ouyang, S.; Sun, D.; Yang, X.; Li, J.; Wu, Z.; Qian, J.; Zhao, Z.; et al. Cellulose nanofiber-based triboelectric nanogenerators for efficient air filtration in harsh environments. *Nano Lett.* **2024**, *24*, 2861–2869. [CrossRef]
66. He, S.; Yu, Z.; Zhou, H.; Huang, Z.; Zhang, Y.; Li, Y.; Li, J.; Wang, Y.; Li, D. Polymer tubes as carrier boats of thermosetting and powder materials based on 3D printing for triboelectric nanogenerator with microstructure. *Nano Energy* **2018**, *52*, 134–141. [CrossRef]
67. Amrutha, B.; Prasad, G.; Sathiyanathan, P.; Reza, M.S.; Kim, H.; Pathak, M.; Prabu, A.A. Fabrication of CuO-NP-doped PVDF composites based electrospun triboelectric nanogenerators for wearable and biomedical applications. *Polymers* **2023**, *15*, 2442. [CrossRef] [PubMed]
68. Yin, J.; Li, J.; Ramakrishna, S.; Xu, L. Hybrid-structured electrospun nanofiber membranes as triboelectric nanogenerators for self-powered wearable electronics. *ACS Sustain. Chem. Eng.* **2023**, *11*, 14020–14030. [CrossRef]
69. Kim, M.; Lee, S.; Cao, V.A.; Kim, M.C.; Nah, J. Performance enhancement of triboelectric nanogenerators via photo-generated carriers using a polymer-perovskite composite. *Nano Energy* **2023**, *112*, 108474. [CrossRef]
70. Mao, J.; Wen, Y.; Feng, X.; Xu, W.; Seo, S. Multi-Sized porous silica/PDMS composite layer for enhanced performance in triboelectric nanogenerators. *Appl. Surf. Sci.* **2025**, *687*, 162291. [CrossRef]
71. Zhu, H.; Liang, J.; Long, W.; Zeng, F.; Zhang, X.; Chen, Z. Enhancing PDMS-based triboelectric nanogenerator output by optimizing the microstructure and dielectric constant. *J. Mater. Chem. C* **2024**, *12*, 1782–1791. [CrossRef]

72. Patnam, H.; Dudem, B.; Graham, S.A.; Yu, J.S. High-performance and robust triboelectric nanogenerators based on optimal microstructured poly (vinyl alcohol) and poly (vinylidene fluoride) polymers for self-powered electronic applications. *Energy* **2021**, *223*, 120031. [CrossRef]
73. Das, P.S.; Park, J.Y.; Kim, D.H. Vacuum filtered conductive nylon membrane-based flexible TENG for wearable electronics. *Micro Nano Lett.* **2017**, *12*, 697–700. [CrossRef]
74. Liu, Y.; Li, E.; Yan, Y.; Lin, Z.; Chen, Q.; Wang, X.; Shan, L.; Chen, H.; Guo, T. A one-structure-layer PDMS/Mxenes based stretchable triboelectric nanogenerator for simultaneously harvesting mechanical and light energy. *Nano Energy* **2021**, *86*, 106118. [CrossRef]
75. Huang, T.; Lu, M.; Yu, H.; Zhang, Q.; Wang, H.; Zhu, M. Enhanced power output of a triboelectric nanogenerator composed of electrospun nanofiber mats doped with graphene oxide. *Sci. Rep.* **2015**, *5*, 13942. [CrossRef]
76. Manchi, P.; Graham, S.A.; Patnam, H.; Paranjape, M.V.; Yu, J.S. High-Efficiency Poly (Vinylidene Fluoride-Co-Hexafluoropropylene) Loaded 3D Marigold Flower-Like Bismuth Tungstate Triboelectric Films for Mechanical Energy Harvesting and Sensing Applications. *Small* **2022**, *18*, 2200822. [CrossRef] [PubMed]
77. Yashaswini, V.L.; Farheen, S.R.; Mahadevaswamy, B.P.; Madhukar, B.S.; Sangamesha, M.A.; Krishnaveni, S. Synergistic effects of rGO functionalization in nanocomposite-based triboelectric nanogenerators for enhanced energy harvesting. *Sens. Actuators A Phys.* **2024**, *370*, 115200. [CrossRef]
78. Ju, Q.; Zu, G.; Wu, H.; Yang, X. Flexible Fiber Shaped Self-Powered System Based on Conductive PANI for Signal Sensing and Energy Harvesting. *ACS Appl. Mater. Interfaces* **2024**, *16*, 54412–54422. [CrossRef]
79. Umapathi, R.; Pammi, S.V.N.; Han, S.; Haribabu, J.; Safarkhani, M.; Rani, G.M.; Huh, Y.S. Designing smart anti-theft alarm system via lead-free BSFO-PDMS composite based triboelectric nanogenerator. *Chem. Eng. J.* **2025**, *511*, 161799. [CrossRef]
80. Mao, M.; Zhao, Q.; Kang, J.; Mu, Q.; Tao, J.; Zhao, X.; He, X.; Li, J. Humidity-adaptive microcrystalline cellulose triboelectric composite film with engineered microstructures for sensing and energy harvesting. *Chem. Eng. J.* **2025**, *515*, 163921. [CrossRef]
81. Mohan, V.; Reji, R.P.; Krishnamoorthy, K.; Sivalingam, Y.; Jayaraman, S.V.; Kim, S.J. Non-Metallic Triboelectric Patch as a Haptic Sensor for Diversified Applications. *Carbon Neutraliz.* **2025**, *4*, 70038. [CrossRef]
82. Veerabhadraswamy, C.M.; Rashmi, S.N.; Tazleem, S.M.; Puneeth, S.; Farheen, S.R.; Sangamesha, M.A.; Krishnaveni, S. Novel approach to bio-inspired triboelectric nanogenerators employing recycled natural fibres for sustainable energy harvesting. *Sens. Actuators A Phys.* **2024**, *377*, 115678. [CrossRef]
83. Qu, M.; Liu, Z.; Liu, H.; Li, J.; Liu, Q.; Xue, K.; Zhang, R.; Wei, X.; Huang, Z.; He, J. A Flexible Superhydrophobic MOF-Based Textile Triboelectric Nanogenerator with Multifunctions: Mechanical Energy Harvesting, Self-Powered Human Motion Sensing, and Human–Computer Interaction. *ACS Appl. Mater. Interfaces* **2025**, *17*, 64591–64605. [CrossRef] [PubMed]
84. Jiang, F.; Zhou, X.; Lv, J.; Chen, J.; Chen, J.; Kongcharoen, H.; Zhang, Y.; Lee, P.S. Stretchable, breathable, and stable lead-free perovskite/polymer nanofiber composite for hybrid Triboelectric and piezoelectric energy harvesting. *Adv. Mater.* **2022**, *34*, 2200042. [CrossRef]
85. He, J.; Xue, Y.; Liu, H.; Li, J.; Liu, Q.; Zhao, Y.; Mu, L.; Sun, C.L.; Qu, M. Humidity-resistant, conductive fabric-based triboelectric nanogenerator for efficient energy harvesting and human–machine interaction sensing. *ACS Appl. Mater. Interfaces* **2023**, *15*, 43963–43975. [CrossRef]
86. Tao, Q.; Zheng, C.; Wang, R.; Hu, J.; Jiang, J. Scalable-Manufacturing Triboelectric Fabric Sensor for Pressure Distribution Mapping during Human Body Activities. *ACS Appl. Mater. Interfaces* **2025**, *17*, 55107–55120. [CrossRef]
87. Yang, T.; Wan, C.; Zhang, X.; Liu, T.; Niu, L.; Fang, J.; Liu, Y. High-efficiency preparation of multifunctional conjugated electrospun graphene doped PVDF/CF yarns for energy harvesting and human movement monitoring in TENG textile. *Nano Res.* **2024**, *17*, 4478–4488. [CrossRef]
88. Wang, X.; Yin, Y.; Yi, F.; Dai, K.; Niu, S.; Han, Y.; Zhang, Y.; You, Z. Bioinspired stretchable triboelectric nanogenerator as energy-harvesting skin for self-powered electronics. *Nano Energy* **2017**, *39*, 429–436. [CrossRef]
89. Huang, J.; Hao, Y.; Zhao, M.; Qiao, H.; Huang, F.; Li, D.; Wei, Q. Biomass-based wearable and Self-powered pressure sensor for human motion detection. *Compos.-A Appl. Sci. Manuf.* **2021**, *146*, 106412. [CrossRef]
90. Wang, S.; Chen, Z.; Zhou, H.; Fu, X.; Chen, B.; Wang, F.; Liu, A.; Guo, W.; Ye, M. Microporous MXene/polyurethane gels derived from iron foam templates for ultra-high stable pressure sensors and triboelectric nanogenerators. *ACS Appl. Electron. Mater.* **2024**, *6*, 3491–3500. [CrossRef]
91. Qu, M.; Shen, L.; Wang, J.; Zhang, N.; Pang, Y.; Wu, Y.; Ge, J.; Peng, L.; Yang, J.; He, J. Superhydrophobic, humidity-resistant, and flexible triboelectric nanogenerators for biomechanical energy harvesting and wearable self-powered sensing. *ACS Appl. Nano Mater.* **2022**, *5*, 9840–9851. [CrossRef]
92. Liu, Y.; Yang, W.; Yan, Y.; Wu, X.; Wang, X.; Zhou, Y.; Hu, Y.; Chen, H.; Guo, T. Self-powered high-sensitivity sensory memory actuated by triboelectric sensory receptor for real-time neuromorphic computing. *Nano Energy* **2020**, *75*, 104930. [CrossRef]

 https://doi.org/10.3390/s26041166

93. Paranjape, M.V.; Graham, S.A.; Manchi, P.; Kurakula, A.; Lee, J.K.; Kavarthapu, V.S.; Yu, J.S. Microarchitectured strontium doped silver niobate embedded ecoflex composite films for highly efficient box-type mechanical energy harvesters. *Nano Energy* **2024**, *119*, 109005. [CrossRef]
94. Rani, G.M.; Wu, C.M.; Motora, K.G.; Umapathi, R.; Jose, C.R.M. Acoustic-electric conversion and triboelectric properties of nature-driven CF-CNT based triboelectric nanogenerator for mechanical and sound energy harvesting. *Nano Energy* **2023**, *108*, 108211. [CrossRef]
95. Ghosh, S.K.; Kim, J.; Kim, M.P.; Na, S.; Cho, J.; Kim, J.J.; Ko, H. Ferroelectricity-coupled 2D-MXene-based hierarchically designed high-performance stretchable triboelectric nanogenerator. *ACS Nano* **2022**, *16*, 11415–11427. [CrossRef]
96. Sardana, S.; Kaur, H.; Arora, B.; Aswal, D.K.; Mahajan, A. Self-powered monitoring of ammonia using an MXene/TiO_2/cellulose nanofiber heterojunction-based sensor driven by an electrospun triboelectric nanogenerator. *ACS Sens.* **2022**, *7*, 312–321. [CrossRef] [PubMed]
97. Yang, L.; Hu, W.; Cao, R.; Zhou, Q.; Hu, C.; Dong, B.; Song, H.; Xu, L. Liquid metal modified all-nanofiber triboelectric nanogenerator for energy harvesting and multi-functional self-powered sensing. *Chem. Eng. J.* **2025**, *517*, 164372. [CrossRef]
98. Anithkumar, M.; Prasanna, A.P.S.; Alluri, N.R.; Kim, S.J. Self-powered smart touchpad using novel intrinsic Piezo–Tribo hybrid nanogenerator. *Adv. Funct. Mater.* **2023**, *33*, 2213907. [CrossRef]
99. Yi, Q.; Pei, X.; Das, P.; Qin, H.; Lee, S.W.; Esfandyarpour, R. A self-powered triboelectric MXene-based 3D-printed wearable physiological biosignal sensing system for on-demand, wireless, and real-time health monitoring. *Nano Energy* **2022**, *101*, 107511. [CrossRef]
100. Liu, Q.; Xue, Y.; He, J.; Li, J.; Mu, L.; Zhao, Y.; Liu, H.; Sun, C.L.; Qu, M. Highly moisture-resistant flexible thin-film-based triboelectric nanogenerator for environmental energy harvesting and self-powered tactile sensing. *ACS Appl. Mater. Interfaces* **2024**, *16*, 38269–38282. [CrossRef]
101. Ji, J.; Zhao, W.; Wang, Y.; Li, Q.; Wang, G. Templated laser-induced-graphene-based tactile sensors enable wearable health monitoring and texture recognition via deep neural network. *ACS Nano* **2023**, *17*, 20153–20166. [CrossRef] [PubMed]
102. Lou, M.; Abdalla, I.; Zhu, M.; Wei, X.; Yu, J.; Li, Z.; Ding, B. Highly wearable, breathable, and washable sensing textile for human motion and pulse monitoring. *ACS Appl. Mater. Interfaces* **2020**, *12*, 19965–19973. [CrossRef]
103. Ning, C.; Dong, K.; Cheng, R.; Yi, J.; Ye, C.; Peng, X.; Sheng, F.; Jiang, Y.; Wang, Z.L. Flexible and stretchable fiber-shaped triboelectric nanogenerators for biomechanical monitoring and human-interactive sensing. *Adv. Funct. Mater.* **2021**, *31*, 2006679. [CrossRef]
104. Chen, F.; Wu, Y.; Ding, Z.; Xia, X.; Li, S.; Zheng, H.; Diao, C.; Yue, G.; Zi, Y. A novel triboelectric nanogenerator based on electrospun polyvinylidene fluoride nanofibers for effective acoustic energy harvesting and self-powered multifunctional sensing. *Nano Energy* **2019**, *56*, 241–251. [CrossRef]
105. Xu, W.; Guo, J.; Wen, H.; Meng, X.; Hong, H.; Yuan, J.; Gao, J.; Liu, D.; Ran, Q.; Wang, Y.; et al. Laminated triboelectric acoustic energy harvester based on electrospun nanofiber towards real-time noise decibel monitoring. *Nano Energy* **2022**, *99*, 107348. [CrossRef]
106. Yang, Y.; Zhang, H.; Lin, Z.H.; Zhou, Y.S.; Jing, Q.; Su, Y.; Yang, J.; Chen, J.; Hu, C.; Wang, Z.L. Human skin based triboelectric nanogenerators for harvesting biomechanical energy and as self-powered active tactile sensor system. *ACS Nano* **2013**, *7*, 9213–9222. [CrossRef]
107. Li, Y.; Chen, S.; Yan, H.; Jiang, H.; Luo, J.; Zhang, C.; Pang, Y.; Tan, Y. Biodegradable, transparent, and antibacterial alginate-based triboelectric nanogenerator for energy harvesting and tactile sensing. *Chem. Eng. J.* **2023**, *468*, 143572. [CrossRef]
108. Park, H.; Ryu, Y.; Joo, H.; Gwak, S.; Gbadam, G.S.; Niu, S.; Lee, J.H. High-output, thermally resilient Nano-TiO_2 dielectric gel triboelectric nanogenerator for energy harvesting and reliable temperature-independent pressure sensing. *J. Mater. Chem. A* **2025**, *13*, 4197–4206. [CrossRef]
109. Li, Z.; Lu, Y.; Wang, X.; Li, T.; Xiao, Y.; Li, Y.; Kang, W. All-Weather Triboelectric Nanogenerator with Bidirectional Temperature Regulation and Hydrophobicity for Energy Harvesting and Motion Detection. *Chem. Eng. J.* **2025**, *522*, 5334035. [CrossRef]
110. Yang, X.; Daoud, W.A. Synergetic effects in composite-based flexible hybrid mechanical energy harvesting generator. *J. Mater. Chem. A* **2017**, *5*, 9113–9121. [CrossRef]
111. Tang, W.; Jiang, T.; Fan, F.R.; Yu, A.F.; Zhang, C.; Cao, X.; Wang, Z.L. Liquid-metal electrode for high-performance triboelectric nanogenerator at an instantaneous energy conversion efficiency of 70.6%. *Adv. Funct. Mater.* **2015**, *25*, 3718–3725. [CrossRef]
112. Lin, Z.; Zhang, B.; Xie, Y.; Wu, Z.; Yang, J.; Wang, Z.L. Elastic-connection and soft-contact triboelectric nanogenerator with superior durability and efficiency. *Adv. Funct. Mater.* **2021**, *31*, 2105237. [CrossRef]
113. Chiu, Y.S.; Rinawati, M.; Chang, Y.H.; Aulia, S.; Chang, C.C.; Chang, L.Y.; Hung, W.S.; Mizuguchi, H.; Haw, S.C.; Yeh, M.H. Enhancing self-induced polarization of PVDF-based triboelectric film by P-doped g-C_3N_4 for ultrasensitive triboelectric pressure sensors. *Nano Energy* **2024**, *131*, 110207. [CrossRef]
114. Mubarak, A.; Sarsembayev, B.; Serik, Y.; Onabek, A.; Kappassov, Z.; Bakenov, Z.; Tsuchiya, K.; Kalimuldina, G. Quenched PVDF/PMMA Porous Matrix for Triboelectric Energy Harvesting and Sensing. *Energy Environ. Mater.* **2025**, *8*, 12808. [CrossRef]

 https://doi.org/10.3390/s26041166

115. Maity, S.K.; Tyagi, U.; Sharma, A.K.; Bisht, P.; Sirohi, S.; Kumar, K.; Sheoran, N.; Singh, S.; Kumar, G. Enhancing the electrical performance of chitosan-based triboelectric nanogenerator using graphene nanoplatelets for real-time sports application. *Cellulose* **2025**, *32*, 1787–1804. [CrossRef]
116. Abid, Y.; Shuja, A.; Murtaza, I. Nanofiber-based triboelectric nanogenerator integrated with supercapacitor using power processing circuit for energy harvesting applications. *J. Mater. Sci. Mater. Electron.* **2025**, *36*, 1363. [CrossRef]
117. Min, G.; Pullanchiyodan, A.; Dahiya, A.S.; Hosseini, E.S.; Xu, Y.; Mulvihill, D.M.; Dahiya, R. Ferroelectric-assisted high-performance triboelectric nanogenerators based on electrospun P (VDF-TrFE) composite nanofibers with barium titanate nanofillers. *Nano Energy* **2021**, *90*, 106600. [CrossRef]
118. Kheirabadi, N.R.; Karimzadeh, F.; Enayati, M.H.; Kalali, E.N. Sustainable and photoresponse triboelectric nanogenerators based on 2D-gC_3N_4 and agricultural wastes. *J. Mater. Sci. Mater. Electron.* **2023**, *34*, 1571. [CrossRef]
119. Peng, F.; Liu, D.; Zhao, W.; Zheng, G.; Ji, Y.; Dai, K.; Mi, L.; Zhang, D.; Liu, C.; Shen, C. Facile fabrication of triboelectric nanogenerator based on low-cost thermoplastic polymeric fabrics for large-area energy harvesting and self-powered sensing. *Nano Energy* **2019**, *65*, 104068. [CrossRef]
120. Zhang, D.; Yang, Y.; Xu, Z.; Wang, D.; Du, C. An eco-friendly gelatin based triboelectric nanogenerator for a self-powered PANI nanorod/$NiCO_2O_4$ nanosphere ammonia gas sensor. *J. Mater. Chem. A* **2022**, *10*, 10935–10949. [CrossRef]
121. Sun, C.; Zu, G.; Wei, Y.; Song, X.; Yang, X. Flexible triboelectric nanogenerators based on electrospun poly (vinylidene fluoride) with MoS_2/carbon nanotube composite nanofibers. *Langmuir* **2022**, *38*, 1479–1487. [CrossRef]
122. Song, Y.; Liu, M.; Bao, J.; Hu, Y.; Xu, M.; Yang, Z.; Yang, Q.; Cai, H.; Xiong, C.; Shi, Z. TOCN/copper calcium titanate composite aerogel films as high-performance triboelectric materials for energy harvesting. *Carbohydr. Polym.* **2022**, *298*, 120111. [CrossRef] [PubMed]
123. Chen, J.; Guo, H.; He, X.; Liu, G.; Xi, Y.; Shi, H.; Hu, C. Enhancing performance of triboelectric nanogenerator by filling high dielectric nanoparticles into sponge PDMS film. *ACS Appl. Mater. Interfaces* **2016**, *8*, 736–744. [CrossRef] [PubMed]
124. Zhong, J.; Hou, X.; He, J.; Xue, F.; Yang, Y.; Chen, L.; Yu, J.; Mu, J.; Geng, W.; Chou, X. Asymmetric permittivity enhanced bilayer polycaprolactone nanofiber with superior inner interfacial polarization and charge retention for high-output and humidity-resistant triboelectric nanogenerators. *Nano Energy* **2022**, *98*, 107289. [CrossRef]
125. Wang, H.; Fan, Z.; Zhao, T.; Dong, J.; Wang, S.; Wang, Y.; Xiao, X.; Liu, C.; Pan, X.; Zhao, Y.; et al. Sandwich-like triboelectric nanogenerators integrated self-powered buoy for navigation safety. *Nano Energy* **2021**, *84*, 105920. [CrossRef]

 https://doi.org/10.3390/s26041166

MDPI AG
Grosspeteranlage 5
4052 Basel
Switzerland
Tel.: +41 61 683 77 34

Sensors Editorial Office
E-mail: sensors@mdpi.com
www.mdpi.com/journal/sensors

www.ingramcontent.com/pod-product-compliance
Lightning Source LLC
Chambersburg PA
CBHW040813250926
51850CB00007BA/1160
9783725877058